HYDROCRACKING SCIENCE AND TECHNOLOGY

CHEMICAL INDUSTRIES

A Series of Reference Books and Textbooks

Consulting Editor

HEINZ HEINEMANN

1. *Fluid Catalytic Cracking with Zeolite Catalysts,* Paul B. Venuto and E. Thomas Habib, Jr.
2. *Ethylene: Keystone to the Petrochemical Industry,* Ludwig Kniel, Olaf Winter, and Karl Stork
3. *The Chemistry and Technology of Petroleum,* James G. Speight
4. *The Desulfurization of Heavy Oils and Residua,* James G. Speight
5. *Catalysis of Organic Reactions,* edited by William R. Moser
6. *Acetylene-Based Chemicals from Coal and Other Natural Re-sources,* Robert J. Tedeschi
7. *Chemically Resistant Masonry,* Walter Lee Sheppard, Jr.
8. *Compressors and Expanders: Selection and Application for the Process Industry,* Heinz P. Bloch, Joseph A. Cameron, Frank M. Danowski, Jr., Ralph James, Jr., Judson S. Swearingen, and Marilyn E. Weightman
9. *Metering Pumps: Selection and Application,* James P. Poynton
10. *Hydrocarbons from Methanol,* Clarence D. Chang
11. *Form Flotation: Theory and Applications,* Ann N. Clarke and David J. Wilson
12. *The Chemistry and Technology of Coal,* James G. Speight
13. *Pneumatic and Hydraulic Conveying of Solids,* O. A. Williams
14. *Catalyst Manufacture: Laboratory and Commercial Preparations,* Alvin B. Stiles
15. *Characterization of Heterogeneous Catalysts,* edited by Francis Delannay
16. *BASIC Programs for Chemical Engineering Design,* James H. Weber
17. *Catalyst Poisoning,* L. Louis Hegedus and Robert W. McCabe
18. *Catalysis of Organic Reactions,* edited by John R. Kosak
19. *Adsorption Technology: A Step-by-Step Approach to Process Evaluation and Application,* edited by Frank L. Slejko

20. *Deactivation and Poisoning of Catalysts,* edited by Jacques Oudar and Henry Wise
21. *Catalysis and Surface Science: Developments in Chemicals from Methanol, Hydrotreating of Hydrocarbons, Catalyst Preparation, Monomers and Polymers, Photocatalysis and Photovoltaics,* edited by Heinz Heinemann and Gabor A. Somorjai
22. *Catalysis of Organic Reactions,* edited by Robert L. Augustine
23. *Modern Control Techniques for the Processing Industries,* T. H. Tsai, J. W. Lane, and C. S. Lin
24. *Temperature-Programmed Reduction for Solid Materials Characterization,* Alan Jones and Brian McNichol
25. *Catalytic Cracking: Catalysts, Chemistry, and Kinetics,* Bohdan W. Wojciechowski and Avelino Corma
26. *Chemical Reaction and Reactor Engineering,* edited by J. J. Carberry and A. Varma
27. *Filtration: Principles and Practices, Second Edition,* edited by Michael J. Matteson and Clyde Orr
28. *Corrosion Mechanisms,* edited by Florian Mansfeld
29. *Catalysis and Surface Properties of Liquid Metals and Alloys,* Yoshisada Ogino
30. *Catalyst Deactivation,* edited by Eugene E. Petersen and Alexis T. Bell
31. *Hydrogen Effects in Catalysis: Fundamentals and Practical Applications,* edited by Zoltán Paál and P. G. Menon
32. *Flow Management for Engineers and Scientists,* Nicholas P. Cheremisinoff and Paul N. Cheremisinoff
33. *Catalysis of Organic Reactions,* edited by Paul N. Rylander, Harold Greenfield, and Robert L. Augustine
34. *Powder and Bulk Solids Handling Processes: Instrumentation and Control,* Koichi Iinoya, Hiroaki Masuda, and Kinnosuke Watanabe
35. *Reverse Osmosis Technology: Applications for High-Purity-Water Production,* edited by Bipin S. Parekh
36. *Shape Selective Catalysis in Industrial Applications,* N. Y. Chen, William E. Garwood, and Frank G. Dwyer
37. *Alpha Olefins Applications Handbook,* edited by George R. Lappin and Joseph L. Sauer
38. *Process Modeling and Control in Chemical Industries,* edited by Kaddour Najim
39. *Clathrate Hydrates of Natural Gases,* E. Dendy Sloan, Jr.
40. *Catalysis of Organic Reactions,* edited by Dale W. Blackburn
41. *Fuel Science and Technology Handbook,* edited by James G. Speight
42. *Octane-Enhancing Zeolitic FCC Catalysts,* Julius Scherzer
43. *Oxygen in Catalysis,* Adam Bielański and Jerzy Haber
44. *The Chemistry and Technology of Petroleum: Second Edition, Revised and Expanded,* James G. Speight
45. *Industrial Drying Equipment: Selection and Application,* C. M. van't Land

46. *Novel Production Methods for Ethylene, Light Hydrocarbons, and Aromatics*, edited by Lyle F. Albright, Billy L. Crynes, and Siegfried Nowak
47. *Catalysis of Organic Reactions*, edited by William E. Pascoe
48. *Synthetic Lubricants and High-Performance Functional Fluids*, edited by Ronald L. Shubkin
49. *Acetic Acid and Its Derivatives*, edited by Victor H. Agreda and Joseph R. Zoeller
50. *Properties and Applications of Perovskite-Type Oxides*, edited by L. G. Tejuca and J. L. G. Fierro
51. *Computer-Aided Design of Catalysts*, edited by E. Robert Becker and Carmo J. Pereira
52. *Models for Thermodynamic and Phase Equilibria Calculations*, edited by Stanley I. Sandler
53. *Catalysis of Organic Reactions*, edited by John R. Kosak and Thomas A. Johnson
54. *Composition and Analysis of Heavy Petroleum Fractions*, Klaus H. Altgelt and Mieczyslaw M. Boduszynski
55. *NMR Techniques in Catalysis,* edited by Alexis T. Bell and Alexander Pines
56. *Upgrading Petroleum Residues and Heavy Oils*, Murray R. Gray
57. *Methanol Production and Use*, edited by Wu-Hsun Cheng and Harold H. Kung
58. *Catalytic Hydroprocessing of Petroleum and Distillates*, edited by Michael C. Oballah and Stuart S. Shih
59. *The Chemistry and Technology of Coal: Second Edition, Revised and Expanded,* James G. Speight
60. *Lubricant Base Oil and Wax Processing*, Avilino Sequeira, Jr.
61. *Catalytic Naphtha Reforming: Science and Technology*, edited by George J. Antos, Abdullah M. Aitani, and José M. Parera
62. *Catalysis of Organic Reactions*, edited by Mike G. Scaros and Michael L. Prunier
63. *Catalyst Manufacture,* Alvin B. Stiles and Theodore A. Koch
64. *Handbook of Grignard Reagents,* edited by Gary S. Silverman and Philip E. Rakita
65. *Shape Selective Catalysis in Industrial Applications: Second Edition, Revised and Expanded*, N. Y. Chen, William E. Garwood, and Francis G. Dwyer
66. *Hydrocracking Science and Technology*, Julius Scherzer and A. J. Gruia
67. *Hydrotreating Technology for Pollution Control: Catalysts, Catalysis, and Processes*, edited by Mario L. Occelli and Russell Chianelli
68. *Catalysis of Organic Reactions*, edited by Russell Malz

ADDITIONAL VOLUMES IN PREPARATION

Synthesis of Porous Materials: Zeolites, Clays, and Nanostructures, edited by Mario L. Occelli and Henri Kessler

Methane and Its Derivatives, Sunggyu Lee

HYDROCRACKING SCIENCE AND TECHNOLOGY

Julius Scherzer
Consultant
Anaheim, California

A. J. Gruia
UOP
Des Plaines, Illinois

Marcel Dekker, Inc. New York • Basel • Hong Kong

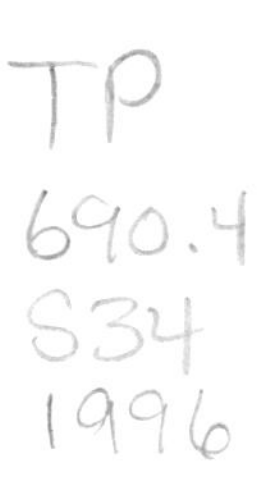

Library of Congress Cataloging-in-Publication Data

Scherzer, Julius
Hydrocracking science and technology / Julius Scherzer, A. J. Gruia.
p. cm. — (Chemical industries ; 66)
Includes bibliographical references and index.
ISBN 0-8247-9760-4 (alk. paper)
1. Cracking process. I. Gruia, A. J. II. Title. III. Series: Chemical industries ; v. 66.
TP690.4.S34 1996
665.5'33—dc20 96-26487
CIP

The publisher offers discounts on this book when ordered in bulk quantities. For more information, write to Special Sales/Professional Marketing at the address below.

This book is printed on acid-free paper.

MARCEL DEKKER, INC.
270 Madison Avenue, New York, New York 10016

Current printing (last digit):
10 9 8 7 6 5 4 3 2 1

PRINTED IN THE UNITED STATES OF AMERICA

In memory of my mother

—JS

To my sweetheart,
who made my participation
in this project possible

—AJG

Preface

Catalytic hydrocracking plays a key role in petroleum processing, providing high-value transportation fuels and upgrading low-quality feedstocks to make them suitable for further processing. New, reformulated fuels require extensive hydrocracking operations. Both catalyst performance and process design have seen significant improvements over the last decades. While particular aspects of this technology are described in detail in various articles and patents, a comprehensive and unified view of the field is not available. This book is an attempt to fill that void.

Over the last decade, much progress has been made in the development of new zeolitic materials, which have found applications in modern commercial hydrocracking catalysts. Superior catalyst characterization and testing methods have been developed, allowing a better understanding of the correlation between catalyst composition, structure, and catalytic performance. The study of model compound conversion over a variety of hydrocracking catalysts has led to the development of new reaction mechanisms and pathways.

Hydrocracking processes have been improved as a result of new designs and the use of more active and selective catalysts. New catalyst reclamation processes have been developed. The increasing use of residue feedstocks in petroleum refining has led to the development of new catalysts and processes able to upgrade these heavy feedstocks.

This book covers both the science and the technology of hydrocracking. After an overview of the petroleum-refining process and a historical introduction to hydrocracking, Chapters 3 through 9 discuss topics related to hydrocracking catalysts: composition, preparation methods, reaction mechanisms, deactivation and reactivation, correlations between composition and performance, and catalyst characterization methods. Chapters 10 through 14 review the different processes used in hydrocracking. Specific applications of hydrocracking technology in catalytic dewaxing, residue upgrading, and fluid catalytic cracking (FCC) feedstock improvement are also reviewed. Chapter 15 describes the technologies used for reclamation of hydroprocessing catalysts. The numerous references and an extensive glossary of terms commonly used in hydrocracking should be useful, especially to those who are new in this field.

Chapters 1–3, 5–9 and 15 have been written by J. Scherzer. Chapters 4 and 10–14 have been written jointly by A. J. Gruia and J. Scherzer. Many chapters start out with basic concepts and an overview of the subject before describing the specific applications to hydrocracking. Throughout we have attempted to explain catalysts and processes in terms of "how" and "why" to the extent currently known. Such a presentation of the material should make this volume a useful reference and textbook for those involved in hydrocracking research and development; for petroleum refinery chemists, engineers, and managers; for graduate students in chemistry and chemical engineering; and for those interested in catalysis and catalytic processes applied to petroleum refining in general.

We express our gratitude to UOP for supporting the publication of this book and gratefully acknowledge the contributions of D. Ackelson, G. Antos, J. Hammerman, E. Houde, T. Nguyen, J. Koepke, J. Sexton, M. Skripek, V. Thakkar, and E. Yuh, who provided helpful suggestions and discussions. Special thanks are due to Sharon Savord and Debby Shapiro for extensive secretarial assistance.

Although we have planned and edited this edition with great care, minor errors are inevitable, especially in the first printing. If these are discovered, we would very much appreciate hearing from you so that corrections can be effected promptly. Questions and comments are also welcome. Julius Scherzer may be reached at 4301 East Lamp Post Way, Anaheim, California 92807 or via fax (714) 283-1275, and Adrian Gruia may be reached via Internet ajgruia@uop.com.

Julius Scherzer
Adrian Gruia

Contents

HYDROCRACKING SCIENCE AND TECHNOLOGY

1

INTRODUCTION

Hydrocracking is a catalytic petroleum refining process that converts heavy, high-boiling feedstock molecules to smaller, lower boiling products through carbon–carbon bond breaking, accompanied by simultaneous or sequential hydrogenation. Hydrocracking is a process of considerable flexibility because it allows the conversion of a wide range of feedstocks to a variety of products.

The actual and projected worldwide hydrocracking capacity until the year 2000 is plotted in Figure 1.1 [1]. While the Asia–Pacific countries, Europe, and the Middle East are likely to continue to increase their hydrocracking capacity, the United States is not likely to show significant changes.

Hydrocracking plays a key role in petroleum refining. For a better understanding of that role, it is necessary first to describe the crude oils processed in the refinery and to provide an overview of the refining process.

Petroleum composition and processing has been described in numerous publications. For a general description of crude oils, petroleum processing, and petroleum products, see [2–4].

1.1. Composition and Characterization of Crude Oils

Petroleum or crude oil is a naturally occurring liquid mixture consisting primarily of hydrocarbons but also containing smaller amounts of organic compounds with heteroatoms such as sulfur, nitrogen, oxygen, and metals (mostly nickel and vanadium). It contains varying amounts of water, inorganic matter (e.g., elemental sulfur, salts), and gas. Nickel and vanadium are present mostly in the form of porphyrins and related structures. The distribution of type of compounds and of molecular weights varies greatly with location and the geological formation from which the crude oil was obtained.

The hydrocarbons in petroleum are primarily paraffins, aromatics, and naphthenes. No olefins or acetylenes are present. Crude oils can be broadly classified by composition into paraffinic (waxy), naphthenic, aromatic, asphaltic, and mixed-based crudes. Paraffinic crudes contain large quantities of linear and

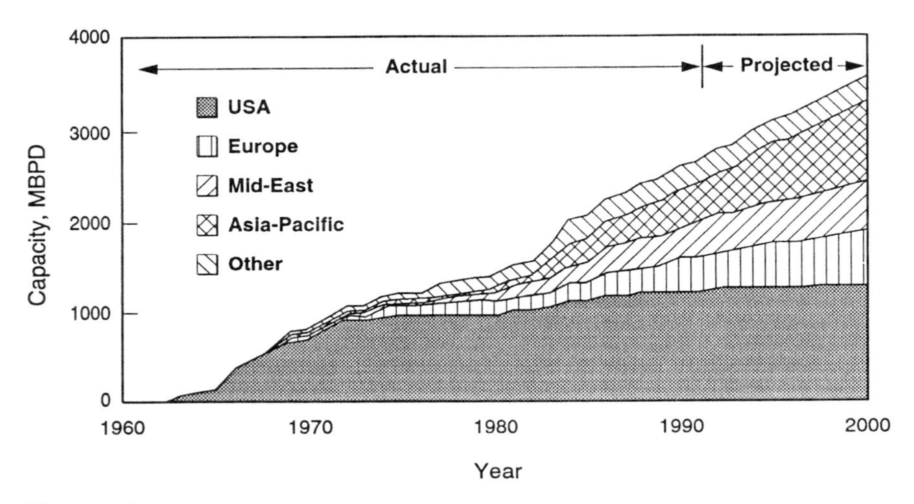

Figure 1.1 Actual and projected worldwide hydrocracking capacity [1]. (Reprinted with permission by Chevron Research and Technology Company, a division of Chevron U.S.A. Inc. Copyright Chevron Research Corporation 1993.)

branched paraffins (alkanes). These are more concentrated in the lower boiling fractions. The major portion of crude, however, consists of ring compounds, aromatic and/or naphthenic. Asphaltic crudes contain a larger quantity of asphaltenes. These are complex, colloidal substances consisting essentially of three-dimensional clusters of polycyclic aromatics, aliphatic chains, and naphthenic rings. Crude oils can also be classified by certain physical properties [2]. For example, based on specific gravity, crude oils can be divided into light and heavy oils.

Crude oils are submitted to distillation under atmospheric pressure and under vacuum. In this process, a series of distillate fractions or cuts are obtained, characterized by their boiling range, i.e., by an initial boiling point and an endpoint. The initial boiling point and endpoint of a fraction increases with the average molecular weight of the fraction. Gasoline and lighter fractions (C_1–C_4), i.e., fractions with a boiling range below about 175°C) (~350°F), are considered low-boiling fractions. Those with a boiling range higher than that of gasoline and up to about 370°C (~700°F) are considered middle distillates. The remaining fractions with boiling range beyond 345 or 370°C (650 or 700°F) are called residue. It should be noted that the boiling ranges listed for different fractions may show significant variations, depending on refinery process and specifications.

Many of the fractions obtained from crude oil distillation contain organic sulfur and nitrogen compounds. These compounds, although present in small amounts in crude oils, are undesirable components in distillate fractions derived from these oils. Organic sulfur compounds occur in all crudes, and their concentration can vary from 0.05 wt % to ~5 wt % sulfur. The sulfur content increases with the average molecular weight and the boiling range of a distillation fraction. A residual fraction will have a considerably higher sulfur content than the crude oil from which it is derived. In the low-boiling fractions, the major organic sulfur compounds are organic sulfides, disulfides, mercaptans, and thiophenes. In higher boiling fractions, sulfur is present largely in the form of thiophene derivatives, such as benzo- and dibenzothiophenes.

The organic nitrogen content of most crude oils is ~ 0.1 wt % or less. However, some heavy, asphaltic crude oils may contain up to 1 wt % nitrogen. Shale oil and synthetic crudes derived from coal or tar sands may have higher levels of nitrogen. The organic nitrogen compounds in crude oils are classified into basic compounds (e.g., pyridines, quinolines, and acridines) and nonbasic compounds (derivatives of indole, pyrrole, and carbazole).

Oxygenated compounds present in crude oils can be acidic (carboxylic acids, phenols, cresols) or nonacidic (esters, amides, ketones, benzofurans). Residual fractions usually have a higher concentration of oxygenated compounds.

The concentration of nickel and vanadium in crude oils can vary from ~3 ppm in light crude oils to ~600 ppm in heavy crude oils. These metals are found mostly in asphaltenes and therefore are concentrated mostly in heavy fractions and residue. The weight ratio of nickel to vanadium can vary from 1:3 to 1:10. Other heavy metals present in crude oils are iron and copper. The heavy metals, as well as most organic sulfur and nitrogen compounds, are removed from different fractions prior to catalytic processing because they have a deleterious effect on most of the catalysts used. A catalytic process, called hydrotreating, is used to remove these undesirable components from petroleum fractions (see below).

Crude oils are commonly characterized by a series of physical properties. In addition to knowing the crude oil composition, an inspection of certain physical properties is important in order to make a judicious choice of a crude oil to obtain a specific product or to establish the optimum process conditions for a given crude oil. Most methods used for inspection and evaluation of crude oils and their products have been standardized. In the United States, the methods developed by the American Society for Testing and Materials (ASTM) are being used for these evaluations.

Some of the more important physical properties used to characterize crude oils are density or specific gravity, viscosity, surface and interfacial tension, refractive index, cloud point, pour point, and flash point (volatility). Other properties frequently measured are carbon residue, aniline point, specific heat, latent heat of vaporization, and heat of combustion. Some of these properties are also measured on petroleum products and used for the evaluation of their quality.

1.2. Processing of Crude Oils

Crude oils are submitted to refining processes, designed to convert the oils to high-value products such as transportation fuels, chemicals, lubricants, and so forth. The simplified flow diagram of a refinery operation is shown in Figure 1.2. This is just one of many refinery configurations that can be used for oil processing. The crude oil is first fractionated by distillation (Table 1.1) and the different fractions are subsequently upgraded by a variety of processes.

Crude oil fractions frequently used as hydrocracker feedstocks are gas oils obtained from atmospheric distillation, from vacuum distillation, and from the coker unit. The most common feedstocks are vacuum gas oils (VGOs). Vacuum distillation yields a light vacuum gas oil (LVGO) (boiling range: ~345–430°C; ~650–800°F) and a heavy vacuum gas oil (HVGO) (boiling range: ~430–540°C; 800–1000°F). The boiling ranges may vary depending on refinery objectives. The content in aromatics and heterocompounds increases with boiling range.

Some key catalytic processes used in petroleum refining, besides hydrocracking, are cracking, hydrotreating, reforming, alkylation, and isomerization (see Glossary).

The petroleum products obtained from processing of crude oil vary considerably, depending on market demand and refinery objectives. Most refineries in the United States are centered on manufacturing transportation fuels, with the emphasis on gasoline. Other transportation fuels produced are diesel and jet fuel. Lubricating oils, heating oils, coke, and gaseous products are also endproducts of different refining processes. The carbon number and boiling point range of several key petroleum-refining products are shown in Figure 1.3 [5]. A brief description of the major transportation fuels follows.

Gasoline is a petroleum product used as fuel in internal combustion, spark ignition engines. It is one of the products obtained in the hydrocracking unit and is usually characterized by boiling range, octane rating, and volatility. Commercially available gasoline is obtained by blending different streams resulting from various refining processes (see Figure 1.2) and the addition of small amounts of additives (antioxidant, deicing, antiwear additives, etc.). The boiling range may vary but is generally in the range of 28–200°C (80–390°F). The major components of gasoline are paraffins, olefins, and aromatics, ranging from C_5 to C_{11}.

A key characteristic of gasoline and of its major components is the octane rating, which measures its knocking characteristics in an internal combustion engine. Higher octane numbers are indicative of superior gasoline quality. Another important characteristic is gasoline volatility, often expressed numerically as Reid vapor pressure (RVP). Reformulated gasolines contain small amounts of high-octane oxygenates (typically 2 wt % oxygen), such as MTBE (methyl tertiary-butyl ether) and TAME (tertiary-amylmethyl ether). These additives enhance gasoline octane and reduce environmentally harmful emissions.

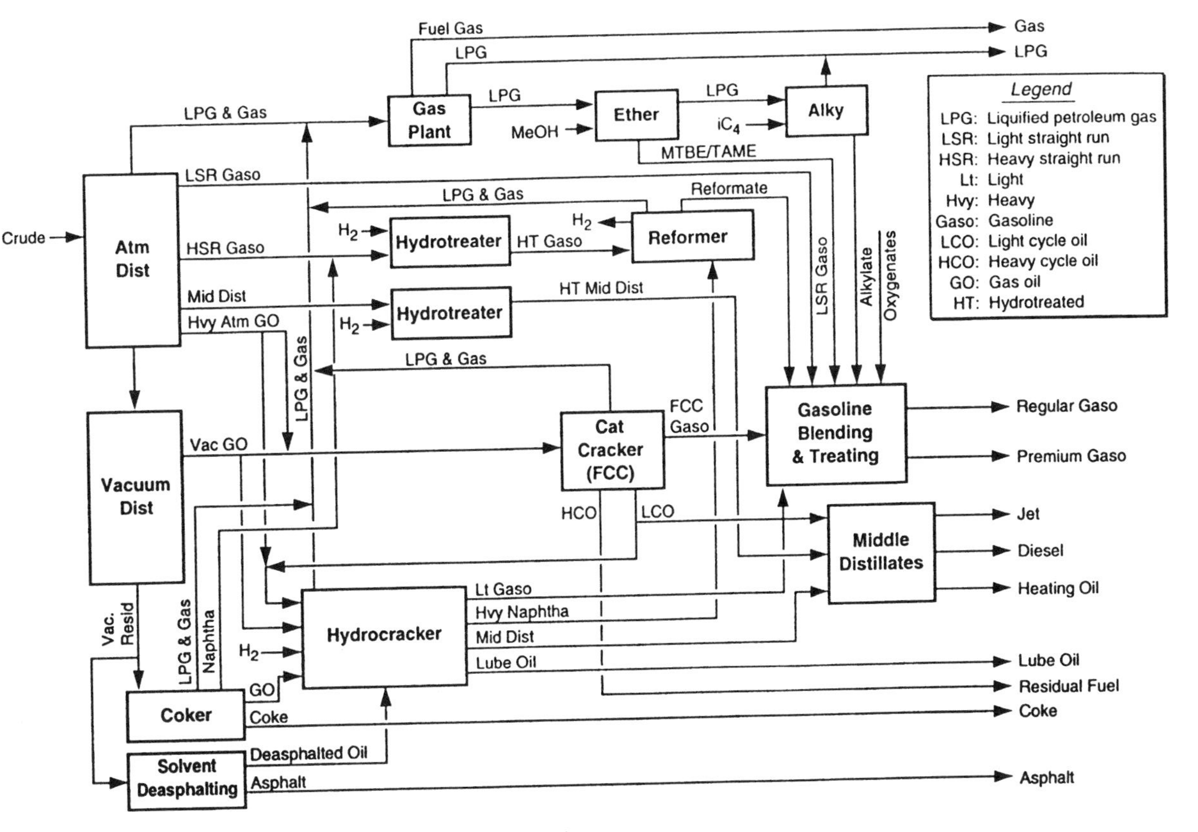

Figure 1.2 Simplified flow diagram of refinery operation.

Table 1.1 Typical Fractions from Distillation of Petroleum

Fraction	Boiling range, °C (°F)	Typical use
Gas (up to C_4)	Up to 10 (50)	Fuel, LPG (C_3 + C_4), olefins
Lt. naphtha	10–90 (50–195)	Gasoline blend
Hv. naphtha	90–200 (195–400)	Gasoline blend (after reforming), jet fuel, source of aromatics
Kerosene	200–260 (400–500)	Jet fuel, kerosene
Atm. gas oil	260–345 (500–650)	FCC feed, fuel oil, diesel fuel
Atm. residue	~345+ (650+)	Source of VGO (by vacuum distil.) coke, fuel oil; some hydroprocessed
Vacuum residue	~560+ (1040+)	Asphalt, coke, lube oil; some hydroprocessed

Diesel fuel is a middle distillate with a boiling range of ~175–370°C (350–700°F). Depending on desired product, diesel fuel may include most or all of the jet fuel fraction. Whereas gasoline engines rely on spark-induced ignition, diesel engines rely on compression-induced ignition. Thus, the properties of diesel fuels are opposite to the antiknock properties of gasoline. This is reflected in composition differences: diesel fuels require high levels of *n*-paraffins whereas aromatics must be minimized.

An important specification for diesel fuels is the cetane number. Higher numbers indicate better quality over the range of about 30–65. Typical diesel fuel has a cetane number in the range of 40–50. Different diesel fuel grades are available for different seasons of the year and for different applications (e.g., for trucks, buses, or railroad engines). Other specifications for diesel fuels refer to distillation range, flash point, pour point, aromatic and sulfur content. Environmental regulations have imposed limits on the aromatic and sulfur content of

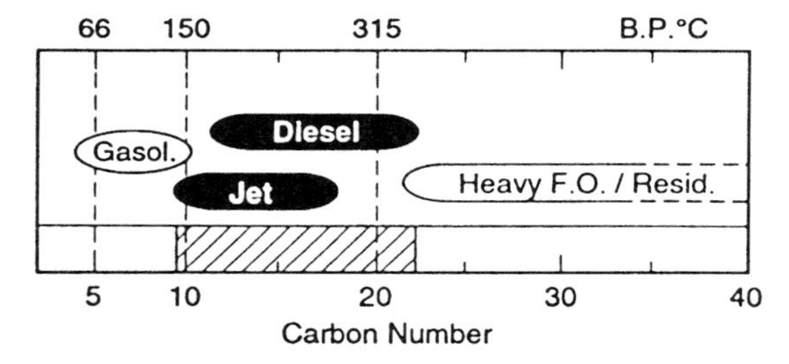

Figure 1.3 Carbon number and boiling range of several key petroleum refining products [5].

diesel fuels in order to reduce emissions of carcinogens and sulfur oxides into the atmosphere.

Jet fuel or aviation turbine fuel is another important transportation fuel produced in a refinery. A major source of jet fuel is the hydrocracking unit. Jet fuels are considered middle-distillate products, with a boiling range of 150–290°C (300–550°F). In some instances, lighter fractions are also included. They are obtained by blending different refinery streams (see Figure 1.2) and are characterized by flash point, freezing point, cloud point, smoke point, vapor pressure, viscosity, heat of combustion, aromatic and sulfur content.

1.3. Role of Hydrocracking in Refining

The location of the hydrocracking unit in the overall refining process is shown schematically in Figure 1.2. Hydrocracking is commonly applied to upgrade the heavier fractions obtained from the distillation of crude oils, including residue. It is also used to upgrade some of the products obtained from other processes, such as coker gas oil, deasphalted oil, fluid catalytic cracking (FCC) cycle oils, and tower bottoms. Hydrocracking is used to upgrade distillates and lube oil bases by removing long-chain paraffins (dewaxing). Furthermore, very heavy hydrocarbons, such as those extracted from tar sands or shale, can be upgraded by hydrocracking. Feedstocks used in hydrocracking as well as products obtained in the process are shown in Table 1.2.

The products obtained in the hydrocracking process are commonly of high quality. The light gasoline has a research octane number between 78 and 85, and can be used as blending stock for the gasoline pool. The heavy naphtha is a high-quality reformer feedstock and is used to produce aromatics for petrochemicals (BTX) or high-octane gasoline. Hydrocracking can yield jet fuel that is low in aromatics and has a high smoke point. Diesel fuel has a relatively high cetane

Table 1.2 Hydrocracking Feedstocks and Products

Feedstock	Products
• Straight run gas oils • Vacuum gas oils • FCC cycle oils and decant oils • Coker gas oils • Thermally cracked stock • Deasphalted oils • Straight-run and cracked naphthas	• LPG • Motor gasolines • Reformer feeds • Jet fuels • Diesel fuels • Heating oils • Olefin plant feedstocks • Lube oils • FCC feedstock

Table 1.3 Selected Hydrocracking Processes

Process	Company
Unicracking	UOP
Isocracking	Chevron
Ultracracking	Amoco
Shell	Shell Development Co.
BASF-IFP Hydrocracking	Badische Anilin und Soda Fabrik, and Institut Francais du Pétrole
MAK	Mobil-Akzo Chem.–Kellogg
Dewaxing:	
MDDW (for distillates)	Mobil
MLDW (for lube oil)	Mobil
Unicracking/DW	UOP
Residue processing:	
H-Oil	Texaco
Hycon	Shell
LC-Fining	Lummus/Amoco
CANMET	Lavalin, Canada

number and a very low sulfur content (a few ppm). High-quality lube oils and cracking feedstocks can also be obtained by this process. Some of the more recent developments in hydrocracking technology have been described in several reviews [6–9].

Selected hydrocracking processes are listed in Table 1.3. Unicracking is the dominant hydrocracking technology in North America. Many of the major oil companies have developed their own hydrocracking technology.

References

1. A. G. Bridge, D. R. Cash, and J. F. Mayer, NPRA Annual Mtg., San Antonio, TX, March 1993, AM-93-60.
2. J. G. Speight, *The Chemistry and Technology of Petroleum*, 2nd ed., Marcel Dekker, New York, 1991.
3. J. J. McKetta, (ed.), *Petroleum Processing Handbook*, Marcel Dekker, New York, 1992.
4. J. H. Gary and G. E. Handwerk, *Petroleum Refining: Technology and Economics*, 3d ed., Marcel Dekker, New York, 1994.
5. N. Y. Chen, W. E. Garwood, and F. G. Dwyer, *Shape Selective Catalysis in Industrial Applications*, Marcel Dekker, New York, 1988,p. 176.
6. H. Choudhary and D. N. Saraf, *Ind. Eng. Chem. Prod. Res. Dev.14*(2):74 (1975).
7. I. E. Maxwell, *Ingenieursblad 52*:53 (1983).
8. J. F. Kriz, M. Ternan, and J. M. Denis, *J. Can. Pet. Tech. 22*:29 (1983).
9. S. Mohanty, D. Kunzru, and D. N. Saraf, *Fuel 69*:1467 (1990).

2

HISTORICAL ASPECTS

Hydrocracking is one of the oldest hydrocarbon conversion processes. An extensive hydrocracking technology for coal conversion was developed in Germany between 1915 and 1945 in order to secure a supply of liquid fuels derived from domestic deposits of coal [1]. The first Bergius plant for hydrogenation of brown coal, put on stream in Leuna, Germany in 1927, applied what may be considered the first commercial hydrocracking process. Similar though less extensive efforts to convert coal to liquid fuels took place in Great Britain, France, and several other countries. Conversion of coal to liquid fuels was a catalytic process, operating at high pressures (200–700 atm) and high temperatures (375–525°C). Between 1925 and 1930, I. G. Farbenindustrie in Germany, in collaboration with Standard Oil of New Jersey, developed a hydrocracking technology designed to convert heavy gas oils to lighter fuels [2]. Attempts were also made in the United States to develop a hydrocracking technology capable of upgrading heavier petroleum fractions [3]. Such hydrocracking processes required high pressures (200–300 atm) and high temperatures (over 375°C). Among the earliest hydrocracking catalysts, pelleted tungsten sulfide was the most successful [4]. Other hydrocracking catalysts used before and during World War II were iron or nickel supported on fluorinated montmorillonite [5] and nickel supported on amorphous silica-alumina [6].

After World War II, hydrocracking became less important. The availability of Middle Eastern crude oil removed the incentive to convert coal to liquid fuels. Newly developed catalytic cracking processes proved more economical for converting heavy petroleum fractions to gasoline. Only in the early 1960s did hydrocracking of petroleum fractions become a commercial reality. Chevron Research Company, then known as California Research Co., announced a new hydrocracking process called "Isocracking" [7]. Unocal, then known as Union Oil Co., in collaboration with ESSO, introduced the hydrocracking process called "Unicracking-JHC" [8]. Universal Oil Products (UOP) announced the "Lomax" hydrocracking process [9]. Nickel or nickel-tungsten on silica-alumina was used as catalyst in these processes. Other major petroleum oil companies also had a significant research effort in hydrocracking. In the mid 1960s, seven different hydrocracking processes were offered for license [10].

Several factors contributed to the development of new hydrocracking processes in the early 1960s. The automobile industry started the manufacture of high-performance cars with high-compression ratio engines that required high-octane gasoline. Catalytic cracking expanded rapidly and generated, in addition to gasoline, large quantities of refractory cycle stock, i.e., effluent that was difficult to convert to gasoline and lighter products. Therefore there was a need to convert refractory stock to quality gasoline. Hydrocracking could fill that need. Furthermore, the switch of railroads from steam to diesel engines after World War II and the introduction of commercial jet aircraft in the 1950s increased the demand for diesel fuel and low-freeze-point jet fuel. The flexibility of the newly developed hydrocracking processes made possible the production of such fuels from heavier feedstocks.

The rapid growth of hydrocracking in the 1960s was accompanied by the development of new, zeolite-based hydrocracking catalysts [11,12]. R. Hansford at Unocal, then Union Oil Co., did the pioneering research work that lead to the development of zeolite-based catalysts. These catalysts were commercialized in 1964–1966. They showed a significant improvement in certain performance characteristics as compared to amorphous catalysts: zeolite-based catalysts had higher activity, had better ammonia tolerance, and offered higher gasoline selectivity.

The 1960s also saw the development of stable, large-pore catalysts used in the cracking of very heavy feedstocks [13]. For example, the H-oil process, commercialized by Hydrocarbon Research, Inc., was developed in the 1960s as a residue hydrocracking process [14]. In order to satisfy new market demands, new products were generated by hydrocracking. In addition to gasoline, jet fuel, and diesel fuel, products such as lubricating oil, low-sulfur fuel oil, liquefied petroleum gas (LPG), and feedstocks for chemicals were obtained by hydrocracking.

Both single-stage and two-stage commercial hydrocracking processes were developed in the 1960s. Single-stage processes were used to produce mainly middle distillates. The Unicracking process introduced by Union Oil Co. made it possible to use a two-catalyst system in series without intermediate removal of ammonia and hydrogen sulfide that resulted from the decomposition of heterocompounds [15].

The early 1970s saw a continuation of the rapid growth of hydrocracking in the United States. By the mid-1970s, hydrocracking had become a mature process. In the late 1970s the rate of growth became more moderate. This was due in part to the high cost of hydrogen, which made hydrocracking a more expensive process than catalytic cracking for gasoline production.

In the 1980s and early 1990s, hydrocracking continued to grow in the United States, but at a slow pace. At the same time, hydrocracking showed significant growth in other parts of the world, such as the Middle East, the Asia–Pacific region, and Europe (Figure 1.1). The increase was due primarily to increased demand for middle distillates and the processing of heavier feedstocks [16,17]. That time period also saw a continuing development of new catalysts, designed to improve catalyst activity and selectivity. Some "flexible" catalysts were developed that made it possible to maximize the yield of different products by using the same catalyst but changing operating conditions. A better understanding of the correlation between zeolite composition and catalytic performance also leads to the development of improved catalysts. Further details regarding the history of hydrocracking can be found in several earlier reviews [18–20].

Worldwide hydrocracking capacity is expected to grow from ~2.5 MM barrels per day in 1990 to 3.5 MM barrels per day in the year 2000 (Figure 1.1). Most of the new hydrocracking capacity is expected to be added outside North America, with continuing emphasis on the production of middle distillates.

References

1. Ministry of Fuel and Power, *Report on the Petroleum and Synthetic Oil Industry of Germany*, B.I.O.S. Overall Report No. 1, Section C, Hydrogenation Processes, 1947, pp. 46–73.
2. H. Heinemann, in *Catalysis: Science and Technology*, Vol. 1 (J. R. Anderson and M. Boudart, eds.), Springer-Verlag, New York, 1981, p. 1.
3. E. V. Murphree, C. L. Brown, and E. J. Gohr, *Ind. Eng. Chem. 32*(9):1203–1212 (1940).
4. M. Pier, *Z. Elektrochem. 53*:291 (1949).
5. A. Voorhies, Jr., and W. M. Smith, *Oil Gas J. 60*(13):184 (1962).
6. H. Clough, *Ind. Eng. Chem. 49*:673 (1957).
7. D. H. Stormont, *Oil Gas J. 57*(44):48–49 (1959).
8. *Oil Gas J. 58*(16):104–106 (1960).
9. M. J. Sterba and C. H. Watkins, *Oil Gas J. 58*(21):102–106 (1960).
10. J. W. Scott and N. J. Patterson, Proceedings of the 7th World Petroleum Congress, Mexico City, 1967, *4*, 97–111.
11. W. J. Baral and H. C. Huffman, Eighth World Petroleum Congress, Moscow, 1971, *4*, 119–127.
12. A. P. Bolton, *Zeolite Chemistry and Catalysis*, A.C.S. Monograph Series No. 117, Washington, D.C., 1976, p. 714.
13. J. W. Scott and A. G. Bridge, *Origin and Refining of Petroleum*, A.C.S. Adv. Chem. Series 103, Washington, D.C., 1971, pp. 113–129.
14. A. R. Johnson, J. E. Papso, R. Happel, and R. Wolk, NPRA Annual Mtg., San Francisco, March 21–23, 1971, AM-71-17.
15. J. H. Duir, *Hydrocarbon Proc. 46*(9):127–134 (1967).

16. H. Qabazard and R. Adarme, *Catalysts in Petroleum Refining 1989*, Elsevier, Amsterdam, 1990, p. 61.
17. A. Hook, T. Huizing, A. A. Esener, I. E. Maxwell, and W. Stork, *Oil Gas J. 89*(16):77 (1991).
18. J. W. Scott and A. G. Bridge, *Adv. Chem. Ser. 103*:113 (1971).
19. N. Choudhary and D. N. Saraf, *Ind. Eng. Chem. Prod. Res. Dev. 14*(2):74 (1975).
20. R. F. Sullivan and J. W. Scott, *Heterogeneous Catalysis: Selected American Histories*, A.C.S. Symposium Series 222, Washington, D.C., 1983, p. 293.

3

HYDROCRACKING CATALYSTS

Hydrocracking catalysts are dual-function catalysts. The concept of a dual-function catalyst having two distinctly different kinds of sites was introduced by Mills et al. in 1953 [1] and later expanded by Weisz et al. [2] and Sinfeld [3,4]. These studies showed that for certain reactions (e.g., hydrocracking, hydroisomerization, dehydrocyclization) to occur, both metallic sites and acidic sites must be present on the catalyst surface.

Hydrocracking catalysts have a cracking function and a hydrogenation-dehydrogenation function (Figure 3.1). The cracking function is provided by an acidic support, whereas the hydrogenation-dehydrogenation function is provided by metals. The acidic support consists of (a) amorphous oxides (e.g., silica-alumina), (b) a crystalline zeolite (mostly modified Y zeolite) plus binder (e.g., alumina), or (c) a mixture of crystalline zeolite and amorphous oxides. Cracking and isomerization reactions take place on the acidic support.

The metals providing the hydrogenation-dehydrogenation function can be noble metals (palladium, platinum), or nonnoble metal sulfides from group VIA (molybdenum, tungsten) and group VIIIA (cobalt, nickel). These metals catalyze the hydrogenation of the feedstock, making it more reactive for cracking and heteroatom removal, as well as reducing the coking rate. They also initiate the cracking by forming a reactive olefin intermediate via dehydrogenation.

The ratio between the catalyst's cracking function and hydrogenation function can be adjusted in order to optimize activity and selectivity. The relative strength of different hydrogenation components and cracking (acid) components in hydrocracking catalysts is shown in Table 3.1 [5,6].

For a hydrocracking catalyst to be effective, it is important that there be a rapid molecular transfer between the acid sites and hydrogenation sites in order to avoid undesirable secondary reactions. Rapid molecular transfer can be achieved by having the hydrogenation sites located in the proximity of the cracking (acid) sites [5].

Prior to the formulation of hydrocracking catalysts with zeolites, a variety of solid acids were used as catalyst supports [7,8]. The most common solid acids

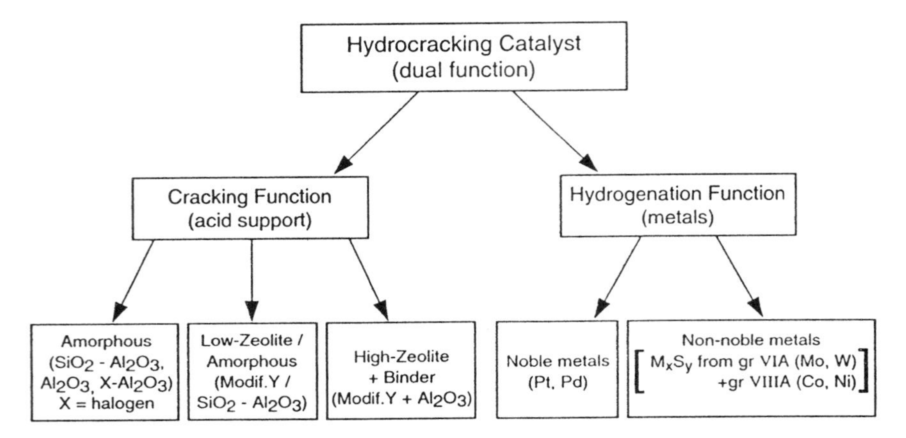

Figure 3.1 Composition of hydrocracking catalysts.

used were HF-treated montmorillonite, amorphous silica-alumina, and alumina. The acidity of solid acids was often enhanced by the addition of small amounts of acid halides, such as HF, NH_4F, BF_3, SiF_4, and the like.

Catalysts with amorphous supports are still in commercial use, primarily where maximizing the production of middle distillates or conversion to lube oil blending stock is the objective [9,10]. Amorphous hydrocracking catalysts contain primarily amorphous silica-alumina [11]. Other amorphous supports reported are silica-magnesia, silica-titania, silica-alumina-titania, titania-alumina, silica-titania-zirconia, silica-alumina dispersed in alumina, alumina-boria, and other acidic mixed oxides. Hydrocracking catalysts containing fluorinated inorganic oxides as supports have also been reported [12]. The use of molten metal chlorides as hydrocracking catalysts has been described in several patents [126,127].

The hydrocracking catalysts described so far are used primarily for hydrocracking gas oils and FCC cycle oils. For hydroprocessing residues, amorphous hydrocracking catalysts as well as specially designed hydrotreating catalysts and iron-containing catalysts are being used.

Table 3.1 Strength of Hydrogenation and Cracking Function in Bifunctional Catalysts [5,6]

Hydrogenation function:	Co/Mo < Ni/Mo < Ni/W < Pt(Pd) → increasing hydrogenation activity (in low-S environment)
Cracking function:	Al_2O_3 < Al_2O_3 - halogen < Si_2-Al_2O_3 < zeolite → increasing cracking activity (acidity)

Catalysts used for mild hydrocracking have a composition similar to that of hydrotreating catalysts. They consist of group VI and VIII nonnoble metals supported on γ-alumina. The metals commonly used are cobalt, nickel, molybdenum, and tungsten in sulfided form. Under mild process conditions, gas oil hydrocracking catalysts may also be used for mild hydrocracking.

Dewaxing catalysts usually consist of a hydrogenation metal (Pt, Pd, Ni) supported on a medium-pore zeolite (e.g., ZSM-5), combined with a binder, commonly alumina.

A detailed description of major commercial catalyst components follows. Methods for preparing commercial hydrocracking catalysts and catalyst supports are described in Chapter 4.

3.1. Zeolite Component

Zeolites are natural or synthetic microporous, crystalline aluminosilicates with ion exchange, sorption, and molecular sieving properties. Most zeolites are synthesized from a slurry containing a silica source (e.g., sodium silicate, silica sols), an alumina source (e.g., sodium aluminate), and caustic [13,14]. Other compounds, such as organic templates (polyamines, polyalcohols), may also be present during synthesis. The synthesized zeolite is modified by ionic exchange and thermal or chemical treatment in order to obtain an active catalyst.

In hydrocracking catalysts with high zeolite content, the zeolite is primarily responsible for the cracking activity of the catalyst. In catalysts with low zeolite content, both the zeolite and acidic amorphous component (e.g., silica-alumina) are responsible for the cracking activity of the catalyst. In amorphous catalysts, only the amorphous acidic component is responsible for cracking activity. Compared to amorphous catalysts, zeolite-based hydrocracking catalysts have the following advantages [15,16]:

1. Greater acidity, resulting in greater cracking activity
2. Better thermal/hydrothermal stability
3. Better naphtha selectivity
4. Better resistance to nitrogen and sulfur compounds
5. Low coke-forming tendency
6. Easy regenerability

Zeolites and nonzeolitic molecular sieves reported to be effective as hydrocracking catalysts are synthetic X, Y, mordenite [17], ZSM-5 [18], offretite, erionite [17], beta [19,20], omega [21], L [22], SSZ materials [23–25], mesoporous M41S materials (e.g., MCM-41) [26,27], silicoaluminaphosphates [28–30], metalaluminophosphates [31], different framework-substituted zeolites (e.g., with gallium substituted for aluminum) [32], and so forth. The use of zeolite mixtures has also been reported [33]. The high surface area and porosity of the zeolites facilitates the access of reactive molecules to a large number of acid sites. It also favors the dispersion of hydrogenation metals, when their precursors have been ion-exchanged into the zeolite. The strong acidity favors the cracking of heavier fractions of the feedstock (with 10–20 carbon atoms per molecule) to lighter fractions, such as naphtha and liquefied petroleum gas (LPG). The higher thermal and hydrothermal stability of low-sodium zeolites allows catalyst regeneration under severe conditions without deleterious effects on the zeolite structure.

High-boiling, high molecular weight hydrocarbon molecules (with over 25 carbon atoms per molecule), such as those present in resid feedstocks, do not enter the zeolite pores due to steric hindrance. These molecules are usually cracked on the acid, amorphous component of the catalyst, such as silica-alumina gel. The fragments obtained from amorphous cracking may undergo secondary cracking on the zeolite.

The structural characteristics of several types of zeolites are shown in Table 3.2. The zeolites most frequently used in commercial hydrocracking catalysts are Y-type zeolites (Figure 3.2), in hydrogen– or rare earth–exchanged form [34]. The zeolites are often used in a modified form, in which some of the aluminum has been removed from the zeolite framework (dealuminated or high-silica Y zeolites) [35,36]. Furthermore, the dealuminated zeolites are used in a low-sodium, rare earth–exchanged form [37] or in the hydrogen form. Such modifications have a significant impact on catalyst activity, selectivity, and stability. Zeolite crystal size also affects catalyst selectivity. A decrease in zeolite crystal size below 0.5 μm results in an increase in liquid product yield [38].

When zeolite mixtures are used as hydrocracking catalysts, the mixtures

Table 3.2 Structural Characteristics of Selected Zeolites

	Pore system		
Zeolite	Structure	Pore	Dimensions (nm)
FAU	3D	12 MR	0.74
(X,Y)		cage	1.3
MOR	1D	12 MR	0.65 × 0.70
(Mordenite)	1D	8 MR	0.26 × 0.57
OFF	1D	12 MR	0.67
(Offretite)	2D	8 MR	0.36 × 0.49
BETA	3D	12 MR	0.66 × 0.81
			0.56 × 0.65
MFI	3D	10 MR	0.53 × 0.56
(ZSM5)			0.56 × 0.65

Note: 1D, 2D, 3D—one, two, or three-dimensional; 12MR—12-membered ring.

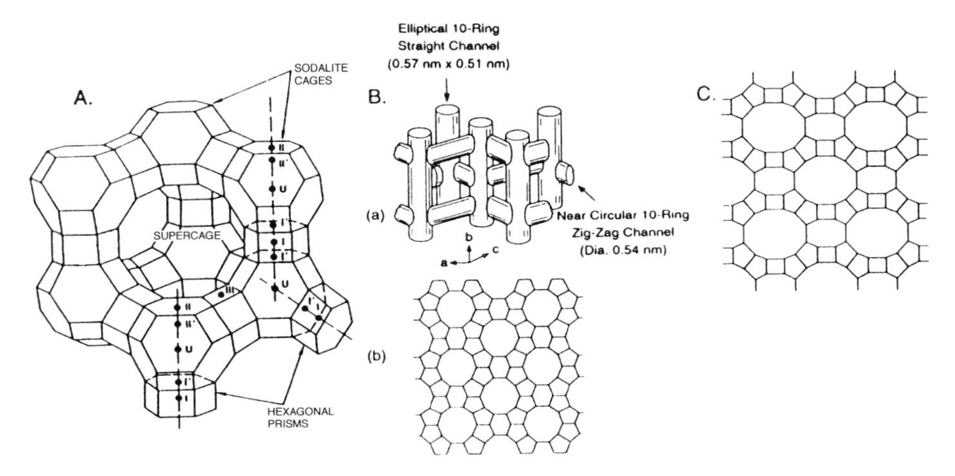

Figure 3.2 Structure of selected zeolites used in hydrocracking catalysts. A: Cubic faujasite-type structure showing nonframework locations (●) (II); B: ZSM-5 structure: (a) Channel system, (b) skeletal diagram of ZSM-5 layer [66]. C: Skeletal diagram of mordenite layer viewed along C axis.

may contain zeolites of the same type but of different composition (e.g., stabilized Y zeolites with different unit cell sizes or different cationic forms), or different types of zeolite (e.g., Y and β zeolite) (see Section 7.2).

The preparation, structure, and properties of conventional Y zeolites, dealuminated Y zeolites, and other zeolite types are described in detail in the literature (see, for example, [13,14,35,39]). Some zeolites, such as X, Y, or L, are prepared by hydrothermal crystallization of reactive alkali metal aluminosilicate gels at low temperatures ($\leqslant$100°C) and atmospheric pressure. A flow diagram for preparation of NaY zeolites is shown in Figure 3.3. The manufacturing of Y zeolites is described in Section 4.3.1. High-silica zeolites, such as ZSM-5, are usually synthesized at higher temperatures and pressures in the presence of templates (e.g., quaternary ammonium salts) (see Section 4.3.1).

Due to their extensive use in hydrocracking catalyst formulations, the major preparation methods and key properties of dealuminated Y zeolites are described in this chapter. ZSM-5 zeolite, a key component of some dewaxing catalysts, is also described.

3.1.1. Preparation of Dealuminated Y Zeolites

Starting with a conventional Y zeolite (usually in the ammonium-exchanged form), dealuminated Y zeolites can be prepared by one of the following methods [35]:

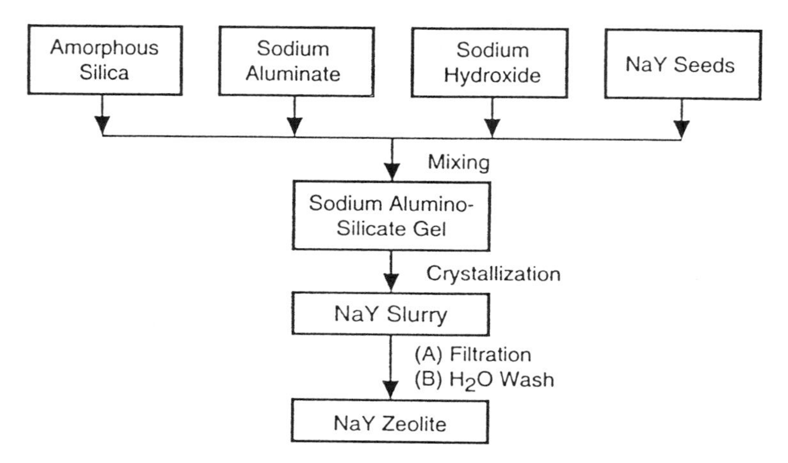

Figure 3.3 Flow diagram for preparation of NaY zeolites.

1. Thermal and hydrothermal dealumination
2. Chemical dealumination
3. Combination of hydrothermal and chemical dealumination

a. Thermal and hydrothermal dealumination. Hydrothermally dealuminated Y zeolites were first described in the mid-1960s by McDaniel and Maher [40]. Calcination of an ammonium-exchanged Y zeolite in the presence of steam results in the expulsion of tetrahedral aluminum from the framework into nonframework positions. This leads to an increase in the framework silica-to-alumina ratio, and a decrease in zeolite unit cell size and ion exchange capability. By varying the severity of the hydrothermal treatment, the degree of dealumination and the corresponding unit cell size decrease can be controlled. The severity of hydrothermal treatment is determined by treatment temperature, time, and partial pressure of water vapors.

A single- or multiple-calcination procedure may be used. For example, when using a double-calcination procedure, the partially ammonium exchanged zeolite is hydrothermally treated, further ammonium exchanged, and hydrothermally treated a second time [40,41]. Multiple hydrothermal treatments result in a significant reduction of unit cell size (less than 24.30 Å) and of sodium content (less than 0.2 wt %). Another dealumination procedure reported is the hydrothermal treatment of a zeolite after it has been mixed with an amorphous oxide component [42,43].

The unit cell size of a conventional Y zeolite may vary from 24.65 to 24.70 Å, whereas that of dealuminated Y zeolite may be as low as 24.20 Å. Low unit cell

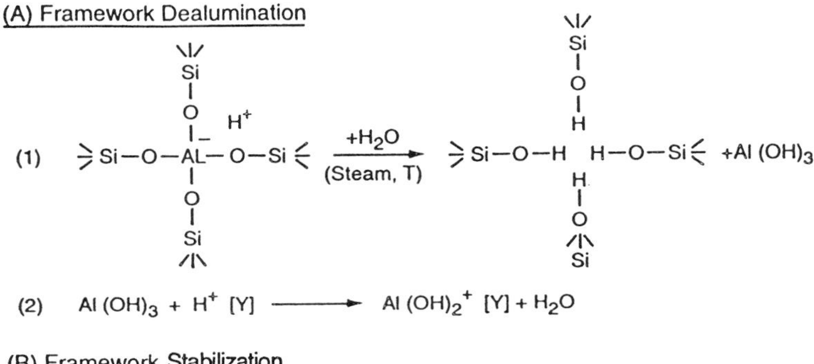

Figure 3.4 Reaction mechanism for hydrothermal dealumination (A) and stabilization (B) of Y zeolite [39].

size, dealuminated zeolites are used in middle-distillate hydrocracking catalysts [41,44].

The defect sites left by steam dealumination are filled to a large extent by silica, resulting in a very stable, highly siliceous framework (Figure 3.4) [39]. Due to their high thermal stability, they are also known as "ultrastable" Y zeolites (USY zeolites). Hydrothermally dealuminated zeolites are widely used in commercial hydrocracking catalysts [41].

b. Chemical dealumination. Dealuminated zeolites can be prepared by partial dealumination of conventional Y zeolites with different reagents. Dealumination may take place by reacting the zeolite with a suitable reagent in solution (e.g., EDTA [45], $(NH_4)_2\ SiF_6$ [46]) or by passing the reagent in vapor phase over the zeolite at high temperature (e.g., $SiCl_4$) [47]. Depending on the reagent used, dealumination can be carried out with silicon enrichment or without silicon enrichment of the zeolite. In the former case, silicon from the reagent is inserted into the vacancies left by dealumination; in the latter case, no such insertion takes place. An example of zeolite dealumination with silicon enrichment is the following reaction with a solution of ammonium fluorosilicate [46]:

$$\equiv Si-O-\overset{-}{Al}(NH_4^+)-O-Si\equiv \; + \; (NH_4)_2SiF_6 \longrightarrow$$

$$\equiv Si-O-Si-O-Si\equiv \; + \; (NH_4)_3AlF_6$$

Hydrocracking catalysts containing hydrogen– or rare earth–exchanged zeolites dealuminated with $(NH_4)_2SiF_6$ have been described in the patent literature [48–50].

c. Combination of hydrothermal and chemical treatment. This two-stage treatment consists of the initial dealumination of a conventional, ammonium-exchanged Y zeolite by hydrothermal treatment, followed by a chemical treatment designed to remove nonframework aluminum. Dealuminated zeolites prepared by a multiple calcination procedure can also have nonframework aluminum removed by chemical treatment [44,51]. The treatment can be carried out with acidic [51,52] or basic solutions [53], as well as with solutions of salts (e.g., KF) [54] or chelating agents (e.g., ethylenediaminetetraacetic acid [EDTA]) [55]. The severity of the treatment is controlled by varying the concentration of the reagent, treatment temperature, and reaction time. At high concentrations, these reagents will also react with the remaining framework aluminum. Mild acid leaching removes not only nonframework aluminum but most of the remaining sodium ions from the zeolite [56]. Some of the methods described have found commercial applications.

3.1.2. Properties of Dealuminated Y Zeolites

a. Common properties. The general characteristics of dealuminated Y zeolites are summarized in Table 3.3 [35]. A brief description of these characteristics follows. The properties of dealuminated Y zeolites are discussed in more detail in [35].

The expulsion of aluminum from the zeolite framework into the zeolite cages in the presence of steam or the extraction of aluminum by chemical reagents increases the SiO_2/Al_2O_3 ratio of the framework. This ratio is further increased by filling the resulting framework vacancies with silicon atoms originating from partially collapsed zeolite crystallites under steam (Figure 3.4), or from the chemical reagents used for dealumination, such as $(NH_4)_2SiF_6$ or $SiCl_4$. The increase in framework SiO_2/Al_2O_3 ratio affects a series of other zeolite properties. The unit cell size decreases as a result of replacing trivalent Al atoms with smaller tetravalent Si atoms (1.43 vs. 1.32 Å atomic radius). The ion exchange capacity

Table 3.3 General Characteristics of Dealuminated Y Zeolites (vs. Parent Y Zeolite) [54]

1. Increased SiO_2/Al_2O_3 ratio in framework
2. Decrease in unit cell size
3. Decrease in ion exchange capacity
4. Increase in thermal/hydrothermal stability
5. Predominant Si(OAl) groups
6. Composition gradient
7. Increase in θ value of XRD peaks
8. Increase in frequency of IR lattice vibrations
9. Decrease in intensity of acid OH bands in IR spectrum
10. Decrease in total acidity
11. Increase in strong acidity
12. Decrease in catalytic site density
13. Decrease in hydrophilic character

decreases because the removal of tetrahedral Al from the framework leaves fewer negative charges that require compensation by cationic species. A more siliceous framework conveys higher thermal and hydrothermal stability to the zeolite. The ^{29}Si-MASNMR (magic angle spinning nuclear magnetic resonance) spectra of dealuminated Y zeolites show a predominance of Si(OAl) groups because most Si atoms have no Al atoms in their immediate neighborhood (see Chapter 9). Most dealuminated zeolite crystals have an aluminum gradient, since the degree of dealumination close to the crystal surface is usually different from that in the interior of the crystal regardless of dealumination method used (see below). The higher silicon content of the framework also results in a shift of XRD peaks to higher 2θ values.

The infrared (IR) spectra of dealuminated Y zeolites show significant differences from those of corresponding conventional Y zeolites both in the OH stretching region and in the framework region. The original acid bands in the OH stretching region (at about 3640 and 3540 cm^{-1}) are reduced in intensity upon dealumination, whereas new absorption bands are generated around 3700 and 3600 cm^{-1}. Furthermore, the acid OH bands and some bands in the framework region of the IR spectra shift to different frequencies due to the decrease in aluminum content of the zeolite framework (see Chapter 9). The decrease in intensity of acid OH bands (assigned to ≡Si(OH)Al≡ framework structures) is indicative of the lower total acidity of dealuminated zeolites. However, acid strength measurements have shown an increase in acid strength of the remaining acid sites. Since acid sites make the zeolite catalytically active for carbenium ion reactions, the lower concentration of acid sites in dealuminated Y zeolites is indicative of a lower density of catalytic sites. This results, in general, in lower catalytic activity. Advanced dealumination also results in a decrease in the hydrophilic character of the zeolite due to the presence of fewer polar groups.

b. Properties affected by preparation method. Some key properties of dealuminated zeolites are affected by the preparation method used (Table 3.4) [57]. Differences in these properties can explain differences in catalytic behavior of dealuminated Y zeolites prepared by different methods. Several of the properties affected by the preparation method are briefly described. A more detailed description of these properties is given in the literature [35,36].

Framework and nonframework aluminum. Depending on the dealumination method used, the aluminum removed from the zeolite framework may remain in the zeolite pores or may be removed from the zeolite. For example, dealuminated Y zeolites prepared by hydrothermal dealumination contain nonframework aluminum in the zeolite pores, usually concentrated near the surface of the crystal [58]. However, chemical dealumination usually does not lead to the formation of nonframework aluminum species in the zeolite pores due to the solubility of the resulting aluminum compound [59,60].

The correlation between framework Si/Al ratio and the unit cell size of a steam-dealuminated Y zeolite is shown in Figure 3.5 [61]. The figure shows that an increase in the severity of framework dealumination results in a smaller unit cell size. Such structural changes in the zeolite have a significant impact on its catalytic properties (see Section 7.5).

Table 3.4 Effect of Preparation Method on Structural and Adsorption Properties of Dealuminated Y Zeolites [54]

	Preparation method				
Property	Hydrothermal	Hydrothermal + acid	$SiCl_4$	$(NH_4)_2SiF_6$	EDTA
Composition gradient	Al-rich surface	Al distribution near-uniform	Al-rich surface	Al-deficient surface	Al-deficient surface
Type of residual Al[a]	Al(T) + Al(E)	Al(T)	Al(T) + Al(E)	Al(T)	Al(T)
Pore system	Micro + secondary pores	Micro- + secondary pores	Micropores	Micropores	Micro- + secondary pores
Nitrogen adsorption	Type IV isotherm	Type IV isotherm	Type I isotherm	Type I isotherm	Type IV isotherm

[a]Al(T) = tetrahedral framework Al; Al(E) = extraframework Al.

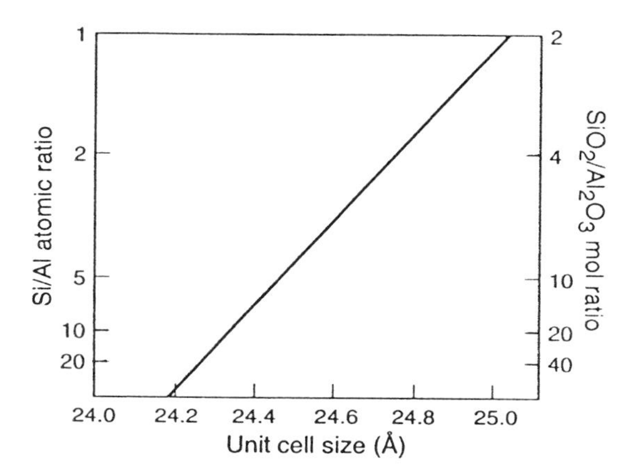

Figure 3.5 Correlation between zeolite unit cell and structural Si/Al ratio of dealuminated Y zeolite [60].

Leaching of nonframework aluminum from steam-dealuminated zeolites (e.g., with an acid) increases the zeolite surface area and results in a zeolite with a nearly uniform distribution of alumina [60].

The catalytic application of mildly dealuminated Y zeolites, with a framework silica-to-alumina ratio in the 6.5–12 range, is described in the patent literature [44]. While chemical analysis provides the bulk aluminum content of the zeolite, different physical methods (e.g., unit cell size measurements, ^{27}Al-MASNMR, etc.) are used to measure the amount of framework aluminum in the zeolite (see Chapter 9).

Pore system and diffusion properties. The diffusion properties of dealuminated zeolites depend on the preparation method used. In addition to micropores ($\phi = 7.4$ Å), dealuminated Y zeolites prepared by hydrothermal treatment contain mesopores ($\phi = 20$–500 Å). The mesopores result from the collapse of some portions of the zeolite crystal during steaming. The size and number of mesopores increases with the degree of framework dealumination (Figure 3.6) [62]. The presence of mesopores facilitates molecular transfer of reacting molecules and reaction products in the zeolitic catalyst due to higher diffusion rates in larger pores. The effect of preparation method on zeolite pore system is shown in Table 3.4.

When used as catalysts, zeolites are diffusion-limited catalysts, and diffusion is the rate-limiting step for many hydrocarbon reactions. The presence of mesopores in dealuminated zeolites plays an important role especially in the hydrocracking of heavy feedstocks [63] because it makes possible the diffusion of bulky hydrocarbon molecules into the zeolite crystals.

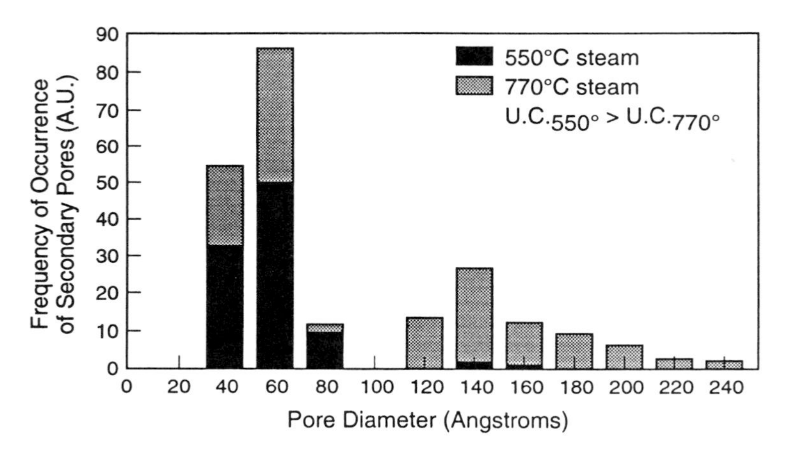

Figure 3.6 Effect of steaming severity on the development of mesopores in Y zeolites [62]. (Reprinted by permission of Elsevier Science Inc.)

Acidity. Both Broensted and Lewis acidity is present in dealuminated Y zeolites [64]. Broensted acidity is associated with framework aluminum atoms and is due primarily to acidic hydroxyl groups attached to the framework. Lewis acidity is associated with nonframework aluminum species as well as with coordinatively unsaturated tetrahedral aluminum in the framework.

Dealumination reduces the number of acid sites associated with framework aluminum. However, the acid strength increases as the number of acid sites decreases [65]. The number, strength, and distribution of acid sites in zeolites plays a key role in determining the catalytic properties of the zeolite. In general, a decrease in total acidity results in a decrease in catalyst activity and a change in selectivity (see Section 7.5). Methods used to measure zeolite acidity are described in Chapter 9.

3.1.3. ZSM-5 Zeolites

ZSM-5 zeolites are medium-pore zeolites used in some dewaxing catalysts. Their preparation is described in Section 4.3.1. ZSM-5 zeolites are members of the "pentasil" family of high-silica zeolites, whose structure is characterized by two types of intersecting channels with 10-membered ring openings (Figure 3.2) [66]. One channel system is sinusoidal and has near-circular (5.4 × 5.6 Å) openings. The other channels are straight and have elliptical (5.1 × 5.7 Å) openings [67]. The channel intersections have a diameter of about 9 Å and are probably the locus of strong acid sites and of catalytic activity [68]. These zeolites have been synthesized with silica-to-alumina ratios varying from 20 to 8000 [69].

In the hydrogen form, ZSM-5 zeolites have Broensted and Lewis acidity, with acidic OH groups located at channel intersections [68,70]. The presence of acidic sites explains the cracking activity of the zeolite. The concentration of strong acid sites is proportional to the concentration of framework aluminum. Steaming results in framework dealumination and a decrease in Broensted acidity.

ZSM-5 zeolites show unique shape-selective properties due to the presence of medium-size pores. These properties play a key role in their use as dewaxing catalysts. Zeolite pore geometry and diffusion limitations are often the controlling factors that determine catalytic activity and selectivity. Shape selectivity controls the size and shape of molecules diffusing into and out of the zeolite pore network. In catalytic dewaxing with ZSM-5-based catalysts, only normal and lightly branched paraffins (e.g., with one CH3 branch) diffuse into the zeolite pores, where they are cracked. Isoparaffins with longer or multiple branches remain practically unaffected. Such selective cracking results in a product with improved properties (see Chapter 14). Shape-selective hydrocracking of *n*-paraffins takes place also over catalysts containing other medium-pore size zeolites, such as offretite or erionite [17].

3.2. Nonzeolitic Components

In amorphous hydrocracking catalysts, the amorphous component serves as metal support and performs the cracking function. It usually consists of amorphous oxides with acidic properties. Most commercial amorphous catalysts contain amorphous silica-alumina.

In many hydrocracking catalysts, the support consists of zeolite–metal oxide mixtures. In these catalysts, the zeolite is embedded in a high-surface-area matrix. The matrix may consist of one or several oxides (mixed oxides), such as alumina, silica-alumina, or silica-alumina-titania. It may also consist of fluorinated alumina or fluorinated clay [12,71]. Mixtures of alumina and silica-alumina are also used [72]. During catalyst preparation, some of the amorphous oxides in the matrix are converted by calcination to crystalline forms. For example, amorphous alumina is converted to γ-alumina. The acidity of the single or mixed oxide contributes to the cracking function of the catalyst, whereas the high surface area facilitates the dispersion of the hydrogenation metals. The matrix should have adequate porosity to provide good molecular diffusion to and from the zeolite crystals. Furthermore, the matrix or some of its components act as binders and provide the catalyst with good crushing strength and abrasion resistance.

3.2.1. Alumina

Alumina is a major component of many zeolite-containing hydrocracking catalysts. It is also the support of catalysts used in mild hydrocracking and a component of many residue processing and dewaxing catalysts. Alumina is obtained commercially from bauxite ore by the Bayer process. Alumina from the ore is solubilized with caustic, the solid impurities (mostly iron oxide) are separated, and alumina trihydrate is crystallized from solution.

Aluminas are known to occur in a variety of forms, both amorphous and crystalline. Hydrated aluminas include amorphous hydroxide; crystalline trihydrates or trihydroxides $Al_2O_3 \cdot 3H_2O$ or $AlO(OH)_3$ (gibbsite, bayerite, and nordstrandite); and crystalline monohydrates or oxyhydroxides, $Al_2O_3 \cdot H_2O$ or AlO(OH) (boehmite and diaspor). Anhydrous aluminas are known in several low-temperature transition forms (ρ, χ, η, and γ), which can be converted to high-temperature transition forms (δ, κ, and θ). The "anhydrous" transition aluminas still contain minor amounts of water. The final stage in these transformations is α-Al_2O_3 (corundum), which is thermodynamically the most stable form. The transformation sequence of aluminas is shown in Figure 3.7 [73].

Of the different commercial aluminas available, pseudoboehmite is most frequently used in the formulation of hydrocracking catalysts. Its role is primarily that of a binder, although its acidity may also play a catalytic role. Pseudoboehmite can be prepared by the following methods:

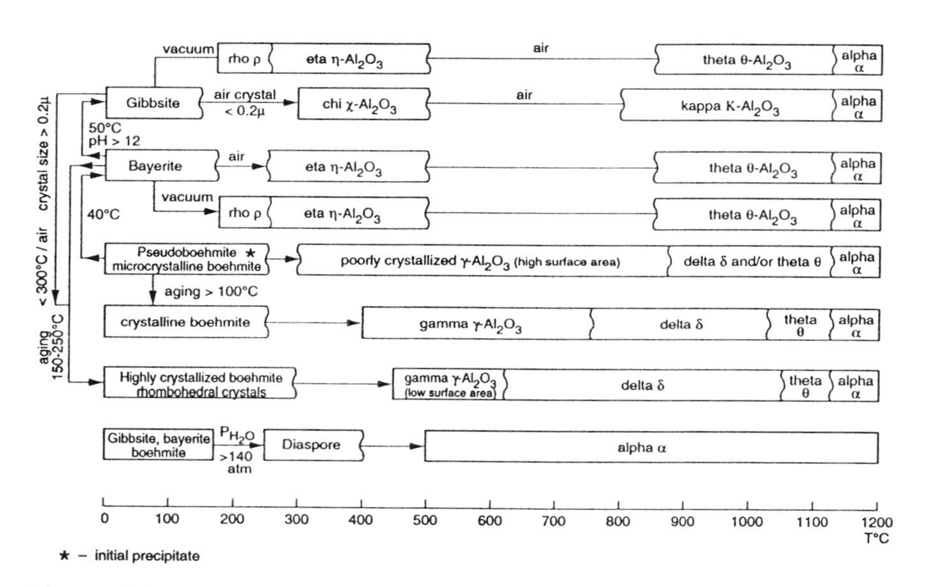

Figure 3.7 Transformation sequence of alumina. Enclosed area indicates range of occurrence, open area indicates range of transition [73].

1. Hydrolysis of Al-alkoxides [74]: $Al(OR)_3 + 3H_2O \rightarrow Al(OH)_3 + 3ROH$, where $R = C_nH_{2n+1}$; this is a common commercial preparation method.
2. Reaction of sodium aluminate with aluminum sulfate [75].
3. Addition of acid to an alkali aluminate [76].
4. Addition of base to a cationic aluminum salt [77].

These reactions are carried out under controlled process conditions. The precipitated alumina is aged and subsequently spray-dried.

The aging conditions of precipitated alumina affect the crystallinity and pore size distribution in the product. For example, an increase in aging temperature, time, and pH results in an increase in pore size (Figure 3.8) [74]. Addition of burnoff additives (e.g., high molecular weight alcohols) during precipitation also affects pore size distribution [78,79]. Pseudoboehmite is often partially peptized with an acid prior to mixing with other catalyst components in order to obtain extrudates with good crushing strength and attrition resistance [79].

Pseudoboehmite is a microcrystalline form of boehmite, which is generally described as an alumina monohydrate or oxyhydroxide [80]. Crystalline boehmite has a cubic structure, shown schematically in Figure 3.9 [80].

Pseudoboehmite is characterized by a broad X-ray peak in the 10–18° range, peaking at 13.50 (2θ). Upon calcination, pseudoboehmite is converted to γ-Al_2O_3 at about 300°C and δ-Al_2O_3 at about 870°C. γ-Al_2O_3 has a spinel-type structure and weakly acidic properties. The acidity is enhanced by fixation of

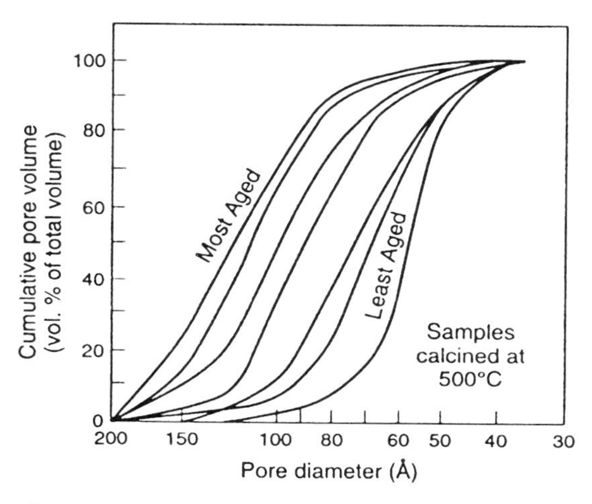

Figure 3.8 Effect of aging on alumina pore size distribution [74].

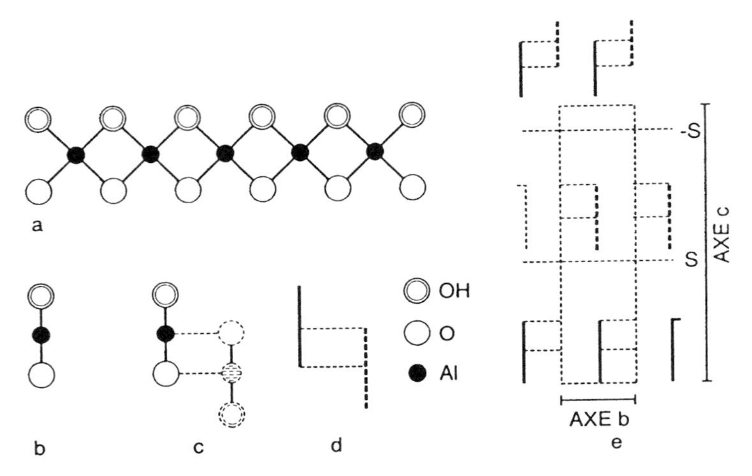

Figure 3.9 Structure of crystalline boehmite. (a) AlOOH chain; (b) side view of AlOOH chain; (c) side view of two antiparallel AlOOH chains; (d) schematic representation of view (c); (e) boehmite structure [80].

halogen (F or Cl) to its surface. It has been reported that fluorinated alumina impregnated with platinum salts has high hydrocracking activity [81].

3.2.2. Silica-Alumina

Amorphous silica-alumina is the acidic support in many commercial amorphous catalysts. It is also a component of many zeolite-based catalysts. Amorphous silica-alumina provides the cracking function of amorphous catalysts and serves as support for the hydrogenation metals. Amorphous silica-alumina also plays a catalytic role in low-zeolite hydrocracking catalysts. In high-zeolite hydrocracking catalysts it acts primarily as support for metals and as binder.

The preparation methods, properties, and catalytic applications (primarily in catalytic cracking) of synthetic silica-alumina gels have been investigated extensively in the 1940s and 1950s, prior to the industrial application of zeolites. A detailed review of the literature published on this subject until the late 1950s is given in [82].

Silica-alumina hydrogels can be prepared by one of the following methods:

1. Precipitation of hydrous alumina on a silica hydrogel [82].
2. Reaction of a silica sol with alumina sol. A silica sol is prepared, for example, by reacting a sodium silicate solution with an acid under controlled conditions. It can also be obtained by contacting a solution of sodium silicate with an ion exchange resin to remove the sodium ions

[83]. An alumina sol can be obtained by reacting aluminum metal with hydrochloric acid [83]. The silica sol may also react with alumina hydrate or an aluminum salt solution [84].

3. Coprecipitation of silica-alumina gel from sodium silicate and aluminum salt solution. In this case a sodium silicate solution is reacted with one or several aluminum salts under controlled conditions. Both cationic and anionic aluminum salts can be used in these processes [85].
4. Hydrolysis of organic derivatives of silicon and aluminum [86].

The silica-alumina hydrogels prepared from sodium silicate contain water-soluble sodium salts, usually sulfates or chlorides. Due to their deleterious effect on both gel and zeolite acidity, such salts are removed by washing. Some of the sodium ions are attached to the hydrogel due to its ion exchange properties and can be removed by ammonium exchange.

A number of preparative variables, such as silica-to-alumina ratio, solid concentration, slurry pH, gelation agent, reaction temperature and time, aging temperature and time, and cationic species present, affect the physical and catalytic properties of the resulting hydrogel [87–90].

Low-alumina gels (e.g., 10–13 wt % Al_2O_3) generally have lower pore volume, lower average pore diameter, higher surface area, and are catalytically less active than higher-alumina gels (~25 wt % Al_2O_3) [82,91]. A comparison between different Ni,W/SiO_2-Al_2O_3 hydrocracking catalysts with the same nickel and tungsten content but different alumina content has shown that the best activity is obtained with a catalyst having 50 wt % Al_2O_3 in the catalyst base [106]. The acidity of the gel varies with the alumina content: Broensted acidity reaches a maximum in cogels with about 70% SiO_2 (Figure 3.10) [92]. The Broensted acid sites have a wide distribution of strengths ($-3 \geqslant H_0 \geqslant -15$) [93].

Both Broensted and Lewis acid sites have been identified in silica-alumina gels. The presence of acid sites is responsible for their catalytic activity [94]. To explain the presence of acid sites, it is assumed that in the reaction between alumina and silica some of the aluminum becomes tetrahedrally coordinated and carries a negative charge, which is compensated by a proton (Broensted acid site) [95]. Dehydration by thermal treatment converts the Broensted acid to a Lewis acid. However, in high-silica SiO_2-Al_2O_3 gel, the ratio of Lewis to Broensted acid sites remains constant over a wide temperature range [96]. The acidity of silica-alumina gels is described in more detail by Tanabe [92].

Different methods designed to prepare amorphous silica-aluminas with controlled pore size have been reported [90,97]. Snell [98,99] has shown that the formation of meso- and macropores in amorphous silica-aluminas is favored by synthesizing the gels at high pH in concentrated solutions containing certain pore-regulating agents.

Peri [100] suggested that different structures, having sites with different

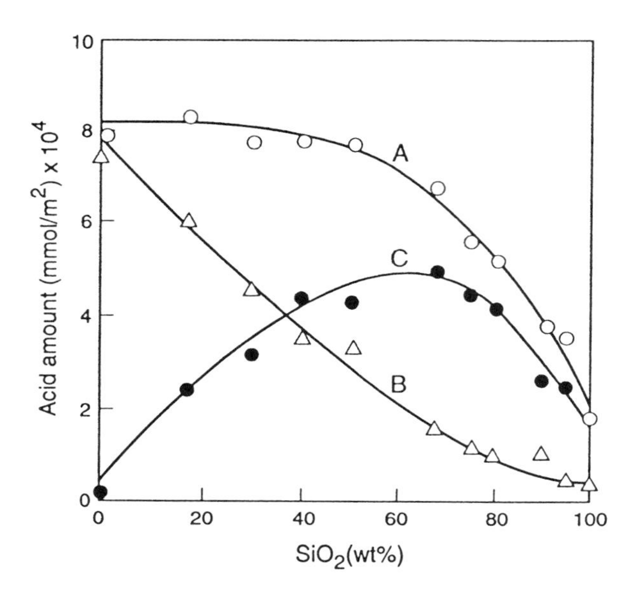

Figure 3.10 Acidity of silica-alumina gel as a function of silica content. A: Total acidity; B: Lewis acidity; C: Broensted acidity (C = A − B) [92].

acidities and catalytic properties, can exist on the surface of amorphous silica-alumina gels. ^{27}Al-MASNMR spectroscopic studies by Gilson et al. [101] have shown that a fresh, amorphous silica-alumina gel with an Si/Al ratio of 2.15 contains predominantly tetrahedral Al, whereas the steamed gel shows the presence of tetrahedral, penta-coordinated, and octahedral Al species (Figure 3.11). The ^{29}Si-MASNMR spectrum of amorphous silica-alumina gel shows a broad band at about 110 ppm, characteristic of silica [102].

Silica-alumina gels consist of aggregates of small, spherical particles and are amorphous to X rays. The discovery of X-ray amorphous zeolites has led to the assumption that the existence of X-ray amorphous zeolite domains in some silica-alumina gels might be the cause of their catalytic activity [103].

The surface area of the gel can vary from 100 to 600 m^2/g, depending on such factors as preparation conditions, type of packing of the gel particles, their size, density, thermal/hydrothermal treatment of the gel, etc. [82]. The same factors affect the pore volume and pore size distribution of the gel. Calcination in the presence of steam reduces the surface area and pore volume, increases the average pore radius, and broadens the distribution of pore radii [82]. The changes are related to the growth of the larger particles in the gel at the expense of smaller ones and the collapse of smaller pores (<30 Å) under these conditions. The changes reflect the "aging" of the gel during hydrothermal treatment and can

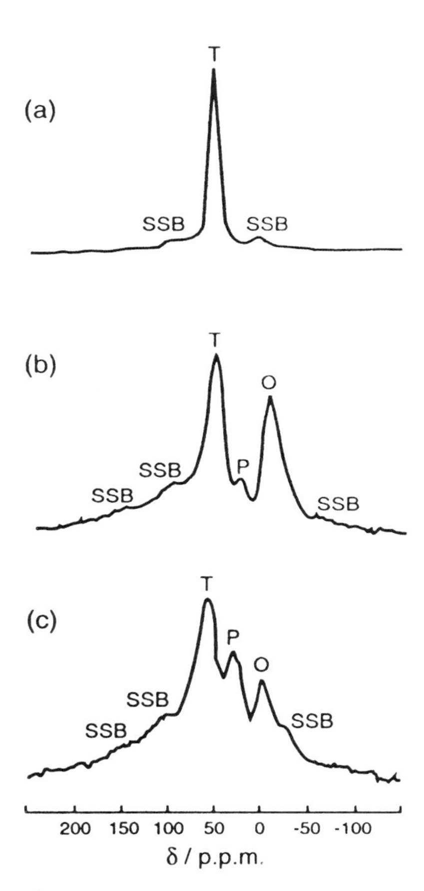

Figure 3.11 ^{27}Al-MASNMR spectra of fresh and steamed SiO_2-Al_2O_3 gel (Si/Al = 2.15). (a) Fresh gel; (b) gel steamed at 427°C/4 h; (c) gel steamed at 760°C/4 h. (T, tetrahedral Al; P, penta-coordinated Al; O, octahedral Al; SSB, spinning sidebands) [101].

result in some blocking of access to the zeolite particles. Aging under these conditions also results in some loss of matrix acidity and catalytic activity. The ion exchange capacity of silica-alumina gels is due to the presence of tetrahedral aluminum on the gel surface.

Different amorphous silica-aluminas with various compositions and properties have been used in hydrocracking catalyst formulations, e.g. [104–106]. Silicon-enriched amorphous silica-alumina, obtained by treatment with $(NH_4)_2$-SiF_6, has also been reported [107]. In this material, the silica-alumina gel is fluorinated and some of the aluminum replaced by silicon.

3.2.3. Other Support Components

Numerous other support components are described in the literature. In addition to the previously mentioned amorphous oxides, the use of acid-treated clays, pillared clays, sepiolite, beidellite, layered silicates, acid metal phosphates, and other solid acids has been reported. These solids differ by their acidity, surface area, pore volume, pore size distribution, and hardness. They therefore convey different physical and catalytic properties to the catalyst (see Section 7.7).

3.3. Metal Component

The metal component of hydrocracking catalysts provides the catalyst with the hydrogenation function. The same metals also have a dehydrogenation function (see Section 6.2). The hydrogenation activity is determined by the following factors:

1. Metal type
2. Ratio between metals (when more than one metal is used)
3. Amount of metal used
4. Degree of metal dispersion
5. Metal–support interaction
6. Location of metal on support

3.3.1. Metal Type

Numerous hydrogenation metals have been evaluated for hydrocracking. Those used in commercial catalysts are noble metals and nonnoble transition metals from group VIA (molybdenum, tungsten) and group VIIIA (cobalt, nickel). Other metals, such as chromium [108], vanadium [109], iron [110], rhodium and ruthenium [111], iridium and niobium [112], have also been recommended.

Among noble metals, platinum is more active than palladium [113]. For different hydrogenation components, the activity decreases in the following order: noble metal > sulfided transition metal > sulfided noble metal [5,114]. For that reason, noble metal catalysts are commonly used in a low-sulfur or sulfur-free environment.

While a sulfur-free environment can be readily obtained under laboratory conditions, commercially it is more difficult to completely remove H_2S from the gases contacting the catalyst. In many commercial operations where noble metal catalysts are used, after removal of H_2S the recycle gas still contains low levels of sulfur (at least several tens of ppm). Sulfur at such levels can react with noble metals and form the corresponding sulfides. Only in units with a separate recycle system for the second-stage operation (see Chapter 10) can the noble metal catalysts used at that stage be maintained practically sulfide-free. In some instances, a certain amount of sulfur is deliberately left in the recycle gas in order to limit the hydrogenation activity of the noble metal catalyst, and thus improve the quality of some products (e.g., gasoline octane) and reduce hydrogen consumption.

Most hydrocracking of industrial feedstocks is carried out in the presence of higher concentrations of hydrogen sulfide and organic sulfur compounds. In that case, catalysts with nonnoble metals are being used. Whereas noble metal catalysts are sulfided by H_2S during the hydrocracking process, the nonnoble hydrogenation metals are sulfided prior to catalyst use.

3.3.2. Metals Ratio

A study of the hydrogenation activity of group VI and VIII nonnoble metals, using toluene as a model compound in the presence of hydrogen sulfide, concluded that the optimum atomic ratio ρ is about 0.25 (Figure 3.12) [5,115]:

$$\rho = \frac{\text{M (gr. VIII)}}{\text{M (gr. VIII)} + \text{M (gr. VI)}} \simeq 0.25$$

where M (group VIII) = Co, Ni and M (group VI) = Mo, W. The hydrogenation activity of various M (group VI) + M (group VIII) couples decreases in the following order:

$$\text{Ni-W} > \text{Ni-Mo} > \text{Co-Mo} > \text{Co-W}$$

The same decrease in hydrogenation activity is also observed when hydrogenating heavier aromatics. For noble metals, the hydrogenation activity decreases in the order Pt > Pd. However, the activity of the corresponding sulfides decreases in the order PdS > PtS.

3.3.3. Metal Amount

In addition to the type of metal and metals ratio, the hydrogenation function is strongly affected by the amount of metal used. This in turn will depend on the desired balance between the cracking and hydrogenation functions. Examples in the patent literature suggest that the noble metal content of hydrocracking catalysts is usually 1 wt % or less, whereas that of nonnoble metals is larger: 3–8 wt % of cobalt or nickel oxides, and 10–30 wt % of molybdenum or tungsten oxides.

3.3.4. Degree of Dispersion

The degree of dispersion of the noble metal on catalyst support plays a key role in determining the hydrogenation activity of the catalyst. During the preparation of noble metal catalysts, the reduction conditions with hydrogen are critical in order to obtain a highly dispersed metal on catalyst (see Section 5.1). As time on-stream

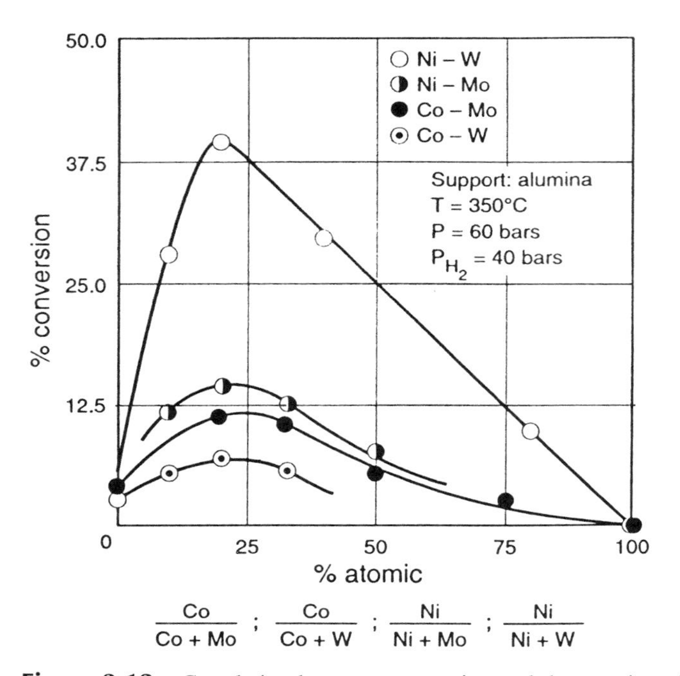

Figure 3.12 Correlation between conversion and the atomic ratio of different hydrogenation components [5].

increases, the hydrogenation activity of the hydrocracking catalyst decreases, due in part to noble metal agglomeration or sulfiding. The hydrogenation activity is restored by redispersion of the metal on catalyst [116,117] (see Section 8.8). In some instances, a dispersion stabilizer metal, such as ionic iron or chromium, is added to the catalyst [118]. Different methods can be used to measure the dispersion of metals on a support (see Chapter 9).

For catalysts whose hydrogenation function is provided by nonnoble metals, dispersion of the mixed metal oxides on catalyst support will also enhance hydrogenation activity [5].

3.3.5. Metal–Support Interaction

The noble metal hydrogenation component is ordinarily added to the zeolite base by ion exchange with an aqueous solution of a suitable compound, such as $[Pd(NH_3)_4](NO_3)_2$ or $[Pt(NH_3)_4](NO_3)_2$ [119]. Upon reduction to a zero valent state, the noble metal dispersed in the zeolite interacts with the support. Such interaction results in an electronic deficiency of the reduced noble metal. For example, in the case of platinum dispersed in Y zeolites, the resistance to sulfur poisoning was attributed to the electronic deficiency of the small platinum particles in the zeolite. Such deficiency results in weakly adsorbed sulfur [120]. Furthermore, zeolite acidity appears to favor a high degree of noble metal dispersion in the zeolite [121] (see also Section 8.8).

Impregnation of ultrastable Y zeolite with ammonium heptamolybdate followed by calcination results in a strong interaction between MoO_3 and most zeolite OH groups. Impregnation with nickel salts leads to interaction only with the most acidic OH groups [122]. Impregnation with nickel and molybdenum salts results in strong interaction between the two metal oxides and regenerates some of the original zeolite acidity.

Nonnoble metals in a hydrocracking catalyst can also interact with the nonzeolitic component of the support. For example, it was shown that nickel reacts with the acid sites of silica-alumina gel, forming a nickel compound and thus reducing the acidity of the support. Catalyst sulfiding leads to the formation of nickel sulfide and regenerates the original active sites [123].

The use of unsupported, finely dispersed catalysts for hydroprocessing has also been described. For example, the use of a colloidal suspension of vanadium, niobium, tantalum, molybdenum, tungsten, iron, cobalt, or nickel sulfides for the hydroconversion of hydrocarbons has been reported [124]. Iron oxide particles are used in hydroprocessing of residue feedstocks.

3.3.6. Location of Metal on Support

The metals may be located in the zeolitic and/or in the nonzeolitic component of the support, depending on the catalyst preparation method and subsequent treatment. When preparing a catalyst by reducing the metal cations exchanged into the zeolite, the following situations may arise [125]:

1. The atoms remain at sites previously occupied by ions.
2. The atoms remain in the same cage but form metal atom clusters.
3. The atoms migrate from smaller cages into larger cages.
4. The atoms agglomerate to form clusters in the zeolite supercages.
5. The atoms migrate outside the zeolite crystals where they sinter, forming metal particles.

In general, as the reduction conditions or the catalytic reaction conditions become more severe, the tendency of the metal atoms to migrate and agglomerate will increase. Progressive agglomeration results in gradual reduction of catalytic activity (see also Section 8.8).

When using the ion exchange method, the metal ions are located primarily at the ion exchange sites of the support. Impregnation allows the introduction of larger amounts of metal than ion exchange would permit. The solution of metal compound used for impregnation fills the pores of the zeolitic and nonzeolitic components of the dehydrated catalyst. The thermal decomposition product (usually a metal oxide when nonnoble metals are used) and the metals obtained upon reduction will therefore be associated not only with the zeolite but with the whole catalyst. When the metal compound used for impregnation contains the metal in cationic form, some of the metal will be exchanged into the zeolite.

References

1. G. A. Mills, H. Heinemann, T. H. Milliken, and A. G. Oblad, *Ind. Eng. Chem. 45*:134 (1953).
2. P. Weisz, *Adv. Catal. 13*:137 (1962).
3. J. H. Sinfeld, *Adv. Chem. Eng. 5*:37 (1964).
4. J. H. Sinfeld, *Bimetallic Catalysts: Discoveries, Concepts and Applications*, Wiley, New York, 1983.
5. J. P. Franck and J. F. LePage, Proc. 7th Intern. Congr. Catal., Tokyo, 1981, p. 792.
6. J. W. Ward, in *Applied Industrial Catalysis*, Vol. 3 (B. E. Leach, ed.), Academic Press, New York, 1984, p. 270.
7. J. W. Scott and A. G. Bridge, *Adv. Chem. Ser. 103*:113 (1971).
8. C. L. Thomas, *Catalytic Processes and Proven Catalysts*, Academic Press, New York, 1970, p. 173.
9. A. G. Bridge, J. Jaffe, and B. E. Powell, Proc. Mtg. Jap. Petr. Inst., Tokyo, Japan, 1982.
10. M. E. Reno, B. L. Schaefer, R. T. Penning, and B. M. Wood, NPRA Ann. Mtg., March 1987, San Antonio, TX, AM-87-60.
11. R. F. Sullivan and J. W. Scott, *Heterogeneous Catalysis: Selected American Histories*, A.C.S. Symposium Series 222, Washington, D.C., 1983, p. 293.
12. M. Adachi, H. Okazaki, and M. Ushio, U.S. Patent No. 4,895,822 (1990).
13. D. W. Breck, *Zeolite Molecular Sieves*, Wiley, New York, 1974.
14. R. Szostak, *Molecular Sieves, Principles of Synthesis and Identification*, Van Nostrand Reinhold, New York, 1989.
15. C. Marcilly and J. P. Franck, in *Catalysis by Zeolites*, (B. Imelik et al., eds.), Elsevier, Amsterdam, 1980, p. 93.
16. J. W. Ward, *Catalysts in Petroleum Refining 1989* (D. L. Trimm et al. eds.), Elsevier, Amsterdam, 1990, p. 417.
17. K. H. Steinberg, K. Becker, and K. H. Nestler, *Acta Phys. Chem. 31*:441 (1985).
18. J. W. Ward, U.S. Patent No. 4,829,040 (1989).
19. J. W. Ward, U.S. Patent No. 5,228,979 (1993).
20. K. L. Hickey, Jr. and R. A. Morrison, U.S. Patent No. 4,812,223 (1989).
21. D. F. Best and V. Nair, U.S. Patent No. 5,230,790 (1993).
22. J. P. Bournonville, P. Dufresne, C. Marcilly, L. Petit, F. Raatz, and C. Travers, U.S. Patent No. 4,909,924 (1990).
23. T. V. Harris, A. Rainis, D. S. Santilli, and S. I. Zones, U.S. Patent No. 5,106,801 (1992).
24. S. I. Zones, U.S. Patent No. 4,589,977 (1986).

25. S. I. Zones, U.S. Patent No. 4,589,976 (1986).
26. M. R. Apelian, T. F. Degnan, Jr., D. O. Marler, and D. N. Mazzone, U.S. Patent No. 5,227,353 (1993).
27. P. Chu and M. K. Rubin, U.S. Patent No. 5,026,943 (1991).
28. S. J. Miller, U.S. Patent No. 4,859,312 (1989).
29. F. P. Gortsema, G. N. Long, R. J. Pellet, and J. A. Rabo, U.S. Patent No. 4,857,495 (1989).
30. F. P. Gortsema, G. N. Long, R. J. Pellet, J. A. Rabo, and A. R. Springer, U.S. Patent No. 4,818,739 (1989).
31. E. M. Flanigen, B. M. Lok, R. L. Patton, and S. T. Wilson, U.S. Patent No. 4,781,814 (1988).
32. M. L. Occelli, U.S. Patent No. 4,946,579 (1990).
33. K. Z. Steigleder, U.S. Patent No. 4,661,239 (1987).
34. J. W. Ward, U.S. Patent No. 4,604,187 (1986).
35. J. Scherzer, *Catalytic Materials: Relationship Between Structure and Reactivity* (T. E. Whyte et al., eds.), A.C.S. Symposium Series 248, Washington, D.C., 1984, p. 157.
36. R. Szostak, *Introduction to Zeolite Science and Practice*, Elsevier, Amsterdam, 1991, p. 153.
37. D. F. Best, G. N. Long, R. J. Pellet, and J. A. Rabo, U.S. Patent No. 5,019,240 (1991).
38. R. D. Bezman and D. R. Cash, U.S. Patent No. 5,277,793 (1994).
39. J. Scherzer, *Octane-Enhancing Zeolitic FCC Catalysts: Scientific and Technical Aspects*, Marcel Dekker, New York, 1990, p. 21.
40. C. V. McDaniel and P. K. Maher, Conf. Mol. Sieves, 1967, Soc. Chem. Ind., London (1968).
41. R. D. Bezman and J. A. Rabo, U.S. Patent No. 4,401,556 (1983).
42. P. Gelin and C. Gueguen, *Appl. Catal. 38*:225 (1988).
43. J. W. Ward, U.S. Patent No. 4,879,019 (1989).
44. F. G. Gortsema, J. R. Pellet, A. R. Springer, and J. A. Rabo, U.S. Patent No. 5,047,139 (1991).
45. G. T. Kerr, *J. Phys. Chem. 72*:2594 (1968).
46. G. W. Skeels and D. W. Breck, *Proc. 6th Int. Zeol. Conf.*, Reno, Nevada, 1983.
47. H. K. Beyer and I. Belenzkaya, *Catalysis by Zeolites* (B. Imelik et al., eds.), Elsevier, Amsterdam, 1980, p. 203.
48. T. L. Carlson, W. S. Millman, and J. W. Ward, U.S. Patent No. 4,826,587 (1989).
49. D. F. Best, G. N. Long, R. J. Pellet, J. A. Rabo, and E. T. Wolynic, U.S. Patent No. 4,735,928 (1988).
50. D. E. Clark, U.S. Patent No. 4,689,137 (1987).
51. J. Scherzer, *J. Catal. 54*:285 (1978).
52. D. F. Best and J. G. Vassilakis, U.S. Patent No. 5,013,699 (1991).
53. G. T. Kerr, *J. Phys. Chem. 71*:4155 (1967).
54. D. W. Breck and G. W. Skeels, Proc. 6th Int. Congr. Catal., Vol. 2, London, 1976, p. 645.
55. G. T. Kerr, J. M. Miale, and R. J. Mikovsky, U.S. Patent No. 3,493,519 (1970).
56. J. Scherzer, U.S. Patent No. 4,477, 336 (1984).

57. J. Scherzer, *Catal. Rev. Sci. Eng. 31*:215 (1989).
58. J. Dwyer, F. R. Fitch, G. Qin, and J. C. Vickerman, *J. Phys. Chem. 86*:4574 (1982).
59. G. Garralon, V. Fornes, and A. Corma, *Zeolites 8*:268 (1988).
60. J. Dwyer, F. R. Fitch, F. Machado, G. Qin, S. M. Smyth, and J. C. Vickerman, *J. Chem. Soc., Chem. Commun. 422* (1981).
61. R. E. Ritter, J. E. Creighton, T. G. Roberie, D. S. Chin, and C. C. Wear, NPRA Annual Mtg., Los Angeles, CA, March 1986, AM-86-45.
62. V. Patzelova and N. I. Jaeger, *Zeolites 7*:240 (1987).
63. A. Hoek, T. Huizinger, and I. E. Maxwell, U.S. Patent No. 4,857,170 (1989).
64. J. Scherzer and J. L Bass, *J. Catal. 28*:101 (1973).
65. R. Beaumont and D. Barthomeuf, *J. Catal. 30*:288 (1973).
66. G. T. Kokotailo, S. L. Lawton, D. H. Olson, and W. M. Meier, *Nature 272*:438 (1978).
67. W. M. Meier and D. H. Olson, *Atlas of Zeolite Structure Types*, 2nd ed., Butterworths, London, 1987.
68. P. Dejaifre, J. C. Vedrine, and E. G. Derouane, *J. Catal. 63*:331 (1980).
69. F. G. Dwyer and E. E. Jenkins, U.S. Patent No. 3,941,871 (1976).
70. P. A. Jacobs and R. von Balmoos, *J. Phys. Chem. 86*:3050 (1982).
71. P. Dufresne and C. Marcilly, U.S. Patent No. 4,766,099 (1988).
72. J. W. Ward, U.S. Patent No. 4,664,776 (1987).
73. J. P. Boitiaux, J. M. Devès, B. Didillon, and C. R. Marcilly, *Catalytic Naphtha Reforming*, Marcel Dekker, New York, 1994, p. 79.
74. R. K. Oberlander, *Applied Industrial Catalysis*, Vol. 3 (B. E. Leach, ed.), Academic Press, New York, 1983, p. 63.
75. T. E. Block and J. Scherzer, U.S. Patent No. 4,332,923 (1982).
76. N. Bell, J. Price, and J. Rigge, U.S. Patent No. 3,630,670 (1971).
77. J. Scherzer and A. T. Liu, U.S. Patent No. 4,332,782 (1982).
78. D. Bosmadjian, G. N. Fulford, B. I. Parsons, and D. S. Montgomery, *J. Catal. 1*:547 (1962).
79. R. E. Tischer, *J. Catal. 72*:255 (1981).
80. B. C. Lippens and J. J. Steggerda, in *Physical and Chemical Aspects of Adsorbents and Catalysis* (B. G. Linsen, ed.), Academic Press, New York, 1970, p. 171.
81. A. B. R. Weber, Thesis, University of Delft, The Netherlands, 1957.
82. L. B. Ryland, M. W. Tamele, and J. N. Wilson, in *Catalysis*, Vol. 7 (P. H. Emmett, ed.), Van Nostrand Reinhold, 1960, p. 1.
83. J. J. Jaffe, U.S. Patent No. 3,637,527 (1972).
84. J. Lim and D. Stamires, U.S. Patent No. 4,142,995 (1979).
85. W. A. Seesee, E. W. Albers, and J. S. Agee, U.S. Patent No. 4,226,743 (1980).
86. T. O. Mitchell and D. D. Whitehurst, U.S. Patent No. 4,003,825 (1977).
87. C. G. Plank and L. C. Drake, *J. Colloid. Sci. 2*:399 (1947).
88. A. G. Oblad, T. H. Milliken, Jr., and G. A. Mills, *Adv. Catal. 3*:208 (1951).
89. M. R. S. Manton and J. C. Davidtz, *J. Catal. 60*:156 (1979).
90. Ph. Courty and Ch. Marcilly, *Preparation of Catalysts III*, Elsevier, Amsterdam, 1983, p. 485.
91. W. C. Cheng and K. Rajagopalan, *Fluid Catalytic Cracking II* (M. L. Occelli, ed.), A.C.S. Symposium Series 452, Washington, D.C., 1991, p. 199.

92. K. Tanabe, *Solid Acids and Bases*, Academic Press, New York, 1970.
93. K. Hashimoto, T. Masuda, and H. Sasaki, *Ind. Eng. Chem. Res. 27*:1792 (1988).
94. M. W. Tamele, *Disc. Faraday Soc. 8*:270 (1950).
95. C. L. Thomas, *Ind. Eng. Chem. 41*:2564 (1949).
96. L. L. Murrell and N. C. Dispenziere, Jr., *Catal. Lett. 2*:329 (1989).
97. D. E. W. Vaughan, P. K. Maher, and E. W. Albers, U.S. Patent No. 3,838,037 (1974).
98. R. Snell, *Appl. Catal. 11*:271 (1984); *12*:189, 347 (1984).
99. R. Snell, *Appl. Catal. 33*:281 (1987); *36*:249 (1988).
100. J. B. Peri, *J. Catal. 41*:227 (1976).
101. J. P. Gilson, G. E. Edwards, A. W. Peters, K. Rajagopalan, R. F. Wormsbecker, T. G. Roberie, and M. P. Shatlock, *J. Chem. Soc., Chem. Commun. 91* (1987).
102. J. Sanz, V. Fornés, and A. Corma, *J. Chem. Soc., Faraday Trans. I, 84*(9), 3113 (1988).
103. G. E. Derouane, S. Detremmerie, Z. Gabelica, and N. Blom, *Appl. Catal. 1*:201 (1981).
104. Mark O'Hara, U.S. Patent No. 3,216,922 (1965).
105. L. Hilfman, U.S. Patent No. 3,825,504 (1974).
106. L. Hilfman, U.S. Patent No. 4,061,563 (1977).
107. S. L. Lambert and M. W. Schoonover, U.S. Patent No. 5,259,948 (1993).
108. A. J. Hensley, Jr., T. D. Nevitt, and M. A. Tait, U.S. Patent No. 4,476,011 (1984).
109. R. J. Lawson, R. W. Johnson, and L. Hilfman, U.S. Patent No. 4,422,959 (1983).
110. J. R. Anderson and N. R. Avery, *J. Catal. 8*:48 (1967).
111. H. F. Wallace and K. E. Hayes, *J. Catal. 29*:83 (1973).
112. S. F. Abdo and J. W. Koepke, U.S. Patent No. 4,871,445 (1989).
113. S. Sivasanker, K. M. Reddy, and P. Ratnasamy, *Catalysts in Petr. Refining, 1989* (D. L. Trimm et al., eds.), Elsevier, Amsterdam, 1990, p. 335.
114. I. E. Maxwell, *Catal. Today 1*:385 (1987).
115. J. F. LePage, Applied Heterogeneous Catalysis, Edition Technip, Paris, 1987, p. 112.
116. J. W. Ward, U.S. Patent No. 3,835,028 (1974).
117. O. Feeley and W. M. H. Sachtler, *Appl. Catal. 67*:141 (1990).
118. W. M. H. Sachtler, M-S. Tzou, and H-J. Jiang, U.S. Patent No. 4,654,317 (1987).
119. J. A. Rabo and J. E. Boyle, U.S. Patent No. 3,236,762 (1966).
120. R. A. Dalla Betta and M. Boudart, Proc. 5th Int. Congr. Catal., North Holland, Amsterdam, 1973, *2*, p. 1329.
121. G. V. Echevskii and K. G. Ione, *Catalysis by Zeolites*, Elsevier, Amsterdam, 1980, p. 273.
122. V. Fornes, M. I. Vazquez, and A. Corma, *Zeolites 6*:125 (1986).
123. G. E. Langlois, R. F. Sullivan, and C. J. Egan, *J. Phys. Chem. 70*:3666 (1966).
124. W. K. T. Gleim and J. G. Gatsis, U.S. Patent No. 3,165,463 (1965).
125. S. T. Homeyer and W. M. H. Sachtler, *Zeolites: Facts, Figures, Future* (P. A. Jacobs and van Santen, eds.), Elsevier, Amsterdam, 1989, p. 975.
126. E. Gorin, U.S. Patent No. 4,247,385 (1981).
127. G. H. Bagshaw and C. W. Zielke, U.S. Patent No. 4,257,873 (1981).

4

CATALYST PREPARATION AND MANUFACTURING

4.1. General Methods and Concepts

Hydrocracking catalysts, like most industrial catalysts, can be prepared by a variety of methods. The manufacturing method chosen usually represents a balance between manufacturing cost and the degree to which the desired chemical and physical properties are achieved. Although there is a relationship between catalyst formulation, preparation procedure, and catalytic properties, the details of that relationship are not always well understood due to the complex nature of catalyst systems. For industrial catalysts, the chemical composition plays a decisive role in their performance, although physical and mechanical properties are also of major importance. Catalyst preparation methods are described in more detail in several reviews and textbooks [1–5].

The preparation of supported metal catalysts usually involves the following major steps:

1. Precipitation
2. Filtration (decantation, centrifugation)
3. Washing
4. Drying
5. Forming
6. Calcining
7. Impregnation
8. Activation

Other steps, such as kneading or mulling, grinding, and sieving, may also be required. Depending on the preparation method used, some of these steps may be eliminated, whereas other steps may be added. For example, kneading or comulling of the wet solid precursors is used in some processes instead of precipitation. When the metal precursors are precipitated or comulled together with the support precursors, the impregnation step can be eliminated. A general description of the different preparation steps follows.

4.1.1. Precipitation and Comulling

Precipitation involves the mixing of solutions or suspensions of materials, resulting in formation of a precipitate. The precipitate may be crystalline or amorphous. It is often submitted to aging or ripening in solution prior to filtration in order to make desirable textural or structural changes in the solid. Ripening may increase the crystal size, increase the size of amorphous particles, convert amorphous to crystalline material, create a certain pore structure in an amorphous precipitate, or change the original crystal structure of the material.

Mulling or kneading of wet solid materials usually leads to the formation of a paste that is subsequently formed and dried. The mulled or kneaded product is submitted to thermal treatment in order to obtain more intimate contact between components and better homogeneity by thermal diffusion and solid state reactions.

Precipitation or mulling is often used to prepare the support for a metal catalyst, whereas the metal component is subsequently added by impregnation. Zeolites, amorphous silica-alumina, and alumina are the major components used in the preparation of hydrocracking catalyst supports (see Chapter 3). Both precipitation and mulling methods are used in the preparation of these catalysts (see below).

The support determines the mechanical properties of the catalyst, such as attrition resistance, hardness, and crushing strength. High surface area and proper pore size distribution are generally required. An essential requirement for any support is resistance to sintering under reaction and regeneration conditions. In supports with a catalytic function (e.g., acid supports in hydrocracking catalysts), that function has to be carefully controlled to meet the requirements of the catalytic process (see Chapter 7).

When preparing the catalyst support from its components or their precursors, binders, die lubricants, and pore-forming additives may also be added to the mixture. For example, a silica sol suspension such as Ludox or an alumina suspension is often used as an effective binder. Although silica is less reactive, alumina may also play a catalytic role. Die lubricants, such as graphite or stearic acid, are added to facilitate subsequent forming. Pore-forming additives, such as wood flower, starch, organic polymers, or carbon fibers, are sometimes used in the preparation of catalyst supports. These additives are burned off during subsequent calcination and form a network of large pores.

The pore size distribution and other physical properties of a catalyst support prepared by the precipitation method are also affected by the precipitation and aging conditions of the precipitate as well as by subsequent drying, forming, and calcining. For example, the properties of silica-alumina gel prepared by reacting a sodium silicate solution with an aluminum salt solution can vary significantly with precipitation conditions (pH, temperature, solution concentration, aging temperature and time), as well as with thermal treatment and forming conditions (see Section 3.2.2).

Gelatinous precipitates are very difficult to filter and usually require an aging step prior to filtration. When filtration of such precipitates is too difficult, decanting or centrifugation can be used. The gels can be coagulated by electrolytes, but washing to remove the electrolyte impurities may cause peptization of the gels. To avoid this, the wash water of filtered gels usually contains electrolytes that subsequently can be readily removed from the washed gel by calcination (e.g., ammonium salts).

4.1.2. Forming

The final shape and size of catalyst particles is determined in the forming step. When preparing a catalyst support, forming and thermal treatment take place prior to addition of the metal component. Catalysts and catalyst supports are commonly formed into extrudates, spheres (beads), or pellets. Shell catalysts, in which the catalyst is deposited in the form of a skin on the outside portion of a support, are less common. The optimum shape and size of the catalyst particle is determined by the need for high activity, acceptable mechanical strength, and type of catalytic reactor used. Pressure drop in fixed-bed reactors is strongly affected by catalyst shape and size. For fixed-bed processes, such as hydrocracking, primarily extrudates or spherical catalyst particles are used.

Extrudates are obtained by extruding a thick paste through a die with perforations. In addition to the catalyst components or their precursors, the paste may contain a peptizing agent, die lubricant, and binder. The spaghetti-like extrudate is usually dried and then broken into short pieces, whose length is two to four times their diameter. The extrudate is then dried and/or calcined. The water content of the paste submitted to extrusion is critical because it determines the density, pore size distribution, and mechanical strength of the product. The water content of the paste is usually kept close to the minimum at which extrusion is still possible.

The form of the extrudates may vary. The simplest form is cylindrical, but other forms, such as trilobes, twisted trilobes, or tetralobes, are also found in commercial products. Catalysts with multilobal cross-sections have a higher surface-to-volume ratio than cylindrical extrudates. When used in a fixed bed, these shaped catalyst particles help reduce diffusional resistance, create a more open bed, and reduce pressure drop. Extrusion is a forming process widely used in the manufacture of hydrocracking catalysts (see Section 4.3.2).

Spherical catalyst particles can be obtained by several methods. One method consists of spray-drying a slurry, as in the manufacture of fluid cracking catalysts.

Another method of obtaining spherical catalyst particles consists of mixing a powder with a spray of liquid in a tilted, rotating pan. Through a snowballing effect large catalyst spheres are formed. The spheres are screened and only those in the proper size range are used. Such spheres generally have a high pore volume and low crushing strength.

Spheres can be obtained by the "oil drop" method, whereby precipitation or coagulation occurs upon the pouring of a liquid into a second immiscible liquid. Spherical bead catalysts are obtained by this process. For example, by pouring an aqueous colloidal solution of silica-alumina into a hot oil bath, spherical silica-alumina beads can be obtained (for more details, see Section 4.3.2).

Pellets (or rings) are obtained by compression of powder in a die. A binder (clay, alumina gel) and 1% or 2% of a die lubricant (e.g., graphite) are added to the mixture. This forming process can be applied only to free-flowing powder mixtures that cohere upon pressing. It is a forming process that is more expensive than extrusion.

For use in fixed beds, catalysts generally range from about 1.5 to 10 mm in diameter and have length-to-diameter ratios of about 1 for pelleted catalysts and up to about 3 or 4 for extrudates [1]. With larger particle sizes, diffusion resistance may reduce the rate of reaction in the center of the particles and therefore decreases catalyst activity per unit mass. For that reason larger particles often have holes or various shapes that increase the surface-to-volume ratio. Particle size, shape, and packing affects the void fraction in a catalyst bed and therefore has a strong impact on pressure drop in the reactor (see Section 10.5).

4.1.3. Drying and Calcining

Thermal treatment is generally applied before and after impregnation of the formed catalyst. For catalysts prepared by precipitation or comulling of all components (including the metal components), only drying may be required prior to forming, with subsequent calcination of the formed product.

Thermal treatment of the catalyst or support eliminates water, volatile and unstable cations or anions (e.g., NH_4^+, CO_3^{2-}, NO_3^-), as well as organic compounds (e.g., die lubricants, pore-forming additives, residual organic solvents). These components were introduced during preparation but have to be eliminated from the final product. Thermal treatment also decomposes metal precursors, such as $Pt(NH_3)_4^{2+}$ or $Pd(NH_3)_4^{2+}$. Furthermore, calcination increases the strength of the final catalyst particle, by causing crystal phase or compound formation through thermal diffusion, solid state reactions, and beginning sintering.

The drying and calcination conditions are of critical importance in determining the physical as well as catalytic properties of the product. Surface area, pore size distribution, stability, attrition resistance, crushing resistance, as well as catalytic activity are affected by the drying and calcination conditions. These conditions are drying and calcination temperature, time, environment (air, vacuum, inert gas), steam partial pressure, and rate of temperature increase. For example, excessive calcination conditions may result in structural collapse of the support, loss in surface area, loss of smaller pores, and destruction of active sites.

The calcination conditions of the support are often different from those of the impregnated catalyst. In the presence of a dispersed metal on the support, excessive heating (or heating in the presence of water vapors) may cause sintering of the metal. At high temperatures, metal oxides that are active catalyst components can react with the support and form inactive compounds or solid solutions. For example, divalent metal oxides (of Ni, Co, Cu, Mn) can react with γ-Al_2O_3 to form unreactive $MeAl_2O_4$. In general, heavy metals are more readily incorporated into the support when the catalyst is prepared by coprecipitation than when the metal is added by impregnation. However, the calcination temperature of the catalyst should be at least as high as that encountered in the reactor in order to avoid undesirable changes and decompositions during the catalytic process.

4.1.4. Impregnation

Impregnation is used to incorporate a metal component into a preformed catalyst support. The method can be used in the preparation of supported noble metal catalysts because it allows the deposition of very small amounts of metal on the surface of a support in finely dispersed form. Impregnation also avoids trapping of the metal at inaccessible sites in the support, as may happen with the precipitation method. Supported nonnoble metal catalysts, requiring fairly large amounts of metal, can also be prepared by impregnation. Specifically, the method is used in the preparation and manufacture of hydrocracking catalysts (see below).

The support used in this process is usually thermally treated before impregnation in order to open the pores to the impregnating solution. Capillary forces ensure that the liquid is drawn into the entire pore system of the support. Diffusion is rapid and may take from a few seconds to a few minutes.

Several impregnation methods may be used for catalyst preparation: (a) impregnation by the immersion method (dipping), (b) impregnation to incipient wetness, and (c) diffusional impregnation. In the first method, the calcined support is immersed in an excess of solution containing the metal compound. The solution fills the pores and is also adsorbed on the support surface. The excess solution is drained off. Impregnation to incipient wetness is carried out by tumbling or spraying the activated support with a volume of solution having the proper concentration of metal compound, and equal to or slightly less than the pore volume of the support. The impregnated support is dried and calcined. Because metal oxides are formed in the process, this calcination step is also called oxidation.

Impregnation to incipient wetness allows a more precise control of the amount of metal incorporated into the catalyst. However, the maximum loading that is obtainable by a single impregnation is limited by the solubility of the metal compound. The solubility of a compound can sometimes be enhanced by some chemical modification. For example, addition of ammonia to a solution of nickel sulfate or chloride enhances its solubility due to the formation of the more soluble nickel ammine complex. Multiple impregnation with the same or different metal compound solution can also be carried out using an intermediate drying step [4,5]. The method may be used when more than one metal is deposited on the support. In some instances, prior to drying, the impregnated support is immersed into a solution that causes precipitation of the active component in the catalyst pores.

Another, less common impregnation method is diffusional impregnation. In that case, the support is saturated with water or with an acid solution, and immersed into the aqueous solution containing the metal compound. That compound subsequently diffuses into the pores of the support through the aqueous phase. For example, the method is applied in the preparation of Pt/Al_2O_3 catalysts, using hydrated alumina and a solution of H_2PtCl_6 as reagent. Such a diffusion is slow and may require several hours to ensure good Pt distribution in the catalyst [6].

Impregnation methods can also be characterized by the presence or absence of strong interaction between active metal and support. In the first case, a chemical bond is established between metal and support. In the second case, there is no affinity between metal and support. A typical example of impregnation with strong interaction is the impregnation of an acidic support with a solution in which the metal is in cationic form. Ionic exchange takes place and the metal is chemically bound to the support at the exchange sites. It has been demonstrated that impregnation with interaction is far superior to that without interaction in terms of metal dispersion and catalyst performance [7].

Impregnation of a dry support may be accompanied by liberation of heat [8]. If uncontrolled, it can lead to insufficient penetration of the metallic component into the support. When using the immersion method, the heat is rapidly dissipated in the excess solution used for impregnation. When the incipient wetness method is used, the impregnation is often done gradually to allow the generated heat to dissipate.

The degree of metal dispersion in the extrudate or pellet is determined by the nature of the support surface (e.g., acidic in amorphous silica-alumina or zeolites, amphoteric in alumina), type of metal compound used, method of impregnation, strength of interaction, and the chemical reactions that occur during drying and calcining. Acidic supports will bind cations by ion exchange, whereas amphoteric supports can bind cations or anions, depending on the pH of the impregnating solution. A pH that is too high or too low should be avoided in order to prevent partial dissolution of the support.

When preparing a bifunctional noble metal catalyst with an acid support, optimum performance is achieved when (a) a small amount (usually <1 wt %) of noble metal (Pt, Pd) is uniformly distributed in the catalyst particle with maximum dispersion (i.e., atomic dispersion) and maximum accessibility; (b) acid sites are in close vicinity of the noble metal [9]; and (c) the physical and mechanical properties of the catalyst are adequate.

The compound used for impregnation may contain the noble metal in cationic or anionic form, e.g., $Pt(NH_3)_4^{2+}$ or $PtCl_6^{2-}$. When using an acid support such as amorphous silica-alumina or an acid zeolite, impregnation with a solution containing the metal as a cationic ammine complex leads to ion exchange and a higher degree of metal dispersion. However, a strong interaction between metal cation and support may deplete the initial solution before it reaches the center of the extrudate or pellet. Additional diffusion of the metal compound from solution into the pores results in a homogeneous metal distribution in the support.

Chloroplatinic acid, H_2PtCl_6, is a common platinum reagent used in catalyst preparation [10]. It is strongly adsorbed on alumina but not on silica gel. By contrast, $[Pt(NH_3)_4]Cl_2$ is strongly adsorbed on silica gel but less on alumina.

Another method of preparing supported catalysts is precipitation of an active precursor in the presence of suspended support particles. After precipitation, the solids are filtered, dried, formed, and thermally treated. However, the distribution of the active component throughout the support may not be uniform. For that reason, impregnation remains the most widely used method to deposit metal precursors on supports.

Drying the impregnated material can affect the distribution of the active metal in the support. Initially, evaporation occurs at the outer surface of the catalyst particle, but liquid evaporated from small pores is replaced by liquid drawn from larger pores by capillarity. The place where crystallization of the metal compound begins and its ultimate distribution depends on the initial concentration of the impregnating solution, rate of nucleation, rate of heating, and the possibility of surface migration [11].

4.2. Preparation of Hydrocracking Catalysts

Bench scale and pilot plant preparation methods of hydrocracking catalysts are described in this section. Some of the methods described are also used in catalyst manufacturing. Selected commercial manufacturing processes are described in Section 4.3.

The preparation methods of hydrocracking catalysts can be divided into two categories:

1. Methods in which all catalyst components (or their precursors) are combined in one step, by comulling or cogelling
2. Methods in which the catalyst support is prepared first, followed by the addition of the hydrogenation components

4.2.1. Comulling

The comulling method of category (1) involves comulling of all catalyst components (or their precursors), followed by extrusion, drying, and calcination in air [12]. In a typical preparation, calculated amounts of the following compounds are comulled: (a) a nickel or cobalt compound (e.g., nitrate, carbonate, acetate); (b) a molybdenum or tungsten compound (e.g., ammonium heptamolybdate, molybdenum oxide, ammonium metatungstate, tungsten oxide); (c) an acid zeolite; (d) a binder (e.g., pseudoboehmite); and (e) possibly another catalytic component (e.g., amorphous silica-alumina). The zeolite type and content can vary within a wide range. The binder alumina is partially acid-peptized prior to comulling. Water is added during comulling in order to obtain a homogeneous paste suitable for extrusion. The paste is extruded through dies with 1/8- or 1/16-in. openings. The extrudate is dried, cut into short pieces ($<1/2$ in. length), and calcined in air between 480°C and 650°C. Upon calcination, pseudoboehmite is converted to γ-Al_2O_3, and the precursors of the hydrogenation components are decomposed to the corresponding metal oxides.

The methods of category (2) involve the comulling of the cracking components, such as a zeolite (and/or silica-alumina gel) with a binder, such as pseudoboehmite, and extruding the mixture. The extrudate is then dried and calcined in air. The calcined extrudate is subsequently combined with the hydrogenation components or their precursors, usually by impregnation.

4.2.2. Cogelling

This method is used primarily for the preparation of amorphous catalysts. "Cogel" catalysts are made by coprecipitating most [13] or all [14,15] components from solution, resulting in a homogeneous hydrogel. The gel is subsequently washed, dried, and calcined [16]. Cogelling is also used to prepare amorphous catalyst supports, such as silica-alumina gels [13]. The catalyst support is then mixed with the hydrogenation metal precursors. The hydrogenation metals may be dispersed in an insoluble form throughout the gel, adsorbed by the gel from solution, or added by impregnation [17,18]. Cogelled catalysts may be used in the processing of heavier feedstocks primarily to middle distillates [19–21]. The activity of cogel catalysts can be enhanced by addition of small amounts of zeolite [15,19].

4.2.3. Catalyst Forming

The catalyst-forming operation is designed to provide a product suitable for use in the hydrocracking process. The product should have satisfactory crush strength, an open-pore structure, and high activity.

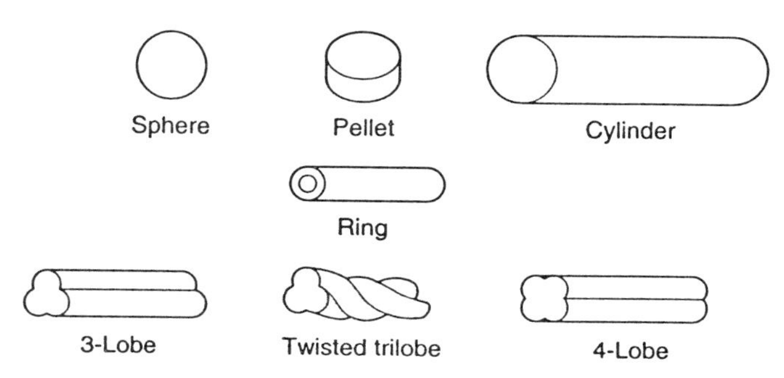

Figure 4.1 Particle shapes of commercial hydrocracking catalysts.

Hydrocracking catalysts may be prepared in different forms. They are made in the form of pellets, extrudates, or beads (Figure 4.1). Pellets are made by compacting in a die the catalyst components (with or without hydrogenation metals) in the presence of an alumina binder and about 2% graphite as a lubricant. The pellets are cylindrical and about ⅛-in. in diameter.

A more common form of hydrocracking catalysts is extrudates. In this case, the catalyst support components, with or without hydrogenation components, are mixed with a small amount of water until a paste is formed. The paste is forced under pressure through an orifice of ⅛, 1/16, or 1/32 in. diameter. The extrudate is broken up into short cylinders (<½ in.), dried, and calcined at about 480–650°C [12]. If the extrudates are made without hydrogenation components, they are subsequently impregnated with a solution of these components, dried, and again calcined.

The most common form of extrudates is cylindrical, although trilobe and tetralobe hydrocracking catalysts have also been described [22]. Hydrocracking catalysts shaped in the form of beads are obtained by slowly pouring the slurry or sol of catalyst components into a bath of hot oil (oil drop method). The metal component may then be added by impregnation. However, the use of this method is quite limited and is applied mostly to amorphous catalysts (see Section 4.3.2).

4.2.4. Impregnation

The method consists of addition of the hydrogenation metal precursors in solution to the formed and dried catalyst support. It may be used for addition of noble metals, of nonnoble metals when large concentrations of metal are required, when the metal cannot be obtained readily in cationic form (e.g., molybdenum, tungsten), and when the ion exchange capacity of the zeolite is low (e.g., ZSM-5), or nonexistent (e.g., aluminum phosphates, silicalite).

Solutions of both noble and nonnoble metal compounds can be used for impregnation. The noble metal is usually in the form of a water-soluble metal ammine salt. The nonnoble metal compounds used for impregnation are water-soluble salts such as nickel or cobalt nitrate, ammonium heptamolybdate, and ammonium metatungstate.

Two procedures are commonly used for impregnation of supports: impregnation to incipient wetness (pore saturation) and impregnation by immersion (dipping). Aqueous solutions of the corresponding metal salts are used in the process. The impregnated support is dried and calcined. Impregnation with nonaqueous, organic solutions of group VI and VIII metal compounds has also been reported [23].

Both impregnation procedures, when properly implemented, provide a uniform metal distribution in the finished catalyst. The impregnation can be carried out in one or two steps, i.e., all of the metals can be added at the same time, or the solution of one metal is added first, followed by drying and impregnation with a solution of the second metal [4,5]. The flow diagram for a typical catalyst preparation method, involving comulling and impregnation, is shown in Figure 4.2.

In those instances where diffusion becomes critical to catalyst performance, a surface impregnation (coating) procedure is used. In this case, the metal-containing solution is sprayed on the catalyst support. The impregnated support is then dried and calcined. By using this procedure, most of the metals are deposited on the outer surface of the catalyst support [4].

In some preparations, the solubility of the nickel salts is enhanced by addition of ammonia to the solution. Addition of phosphoric acid increases the solubility of ammonium heptamolybdate due to the formation of a complex phosphomolybdate. Phosphate also converts some of the Al_2O_3 surface to a basic aluminum phosphate [24] and may decrease coke formation.

By incorporating the metals into the catalyst through impregnation by the incipient wetness or dipping method, the possibility exists of a nonuniform metal distribution in the catalyst support, with higher metal concentrations near the support surface leading to rind formation [4,25]. Therefore, the metal distribution in the catalyst extrudate or pellet has to be monitored (e.g., by electron microprobe analysis) and a uniform metal distribution secured by careful impregnation.

4.2.5. Ion Exchange

Due to the high ion exchange capacity of zeolites, metal ion exchange is a method often used to introduce hydrogenation metals into hydrocracking catalysts. The application of this method also results in a high degree of metal dispersion in the catalyst. Since the zeolite framework carries a negative charge, the metals used for exchange are in cationic form. In the case of noble metal catalysts, the compound most frequently used for ion exchange is a complex tetrammine salt, such as

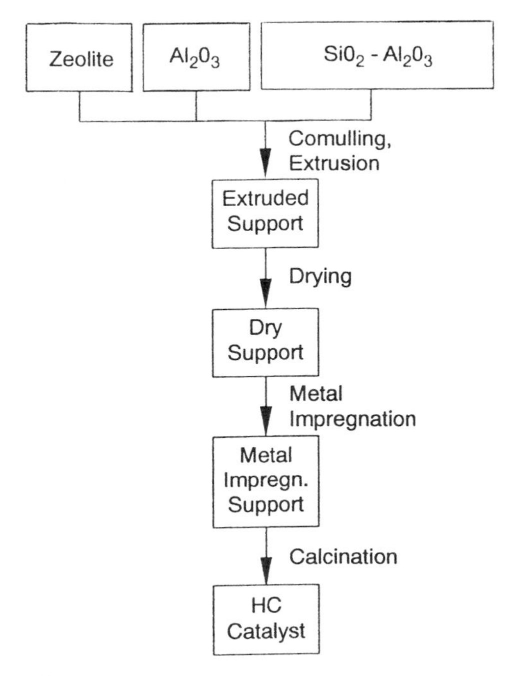

Figure 4.2 Flow diagram for a typical hydrocracking catalyst preparation using comulling and impregnation.

$[Pd(NH_3)_4](NO_3)_2$. The complex ammine is stable in a wide pH range and therefore suitable for ionic exchange [26,27]. This involves slow addition of a dilute solution of the metal ammine complex to a dilute zeolite slurry and allowing the mixture to equilibrate. Ion exchange of zeolite powder by this method results in a uniform distribution of cations in the zeolite crystals. For that reason it is a common method used to prepare noble metal–loaded zeolite catalysts. The exchange can also be carried out on zeolite-containing extrudates or pellets. In that case, strong calcination of the zeolite-containing support prior to ion exchange should be avoided in order to prevent destruction of ion exchange sites. When exchanging a zeolite-containing support, there is a possibility of preferential concentration of the metal near the surface, leading to rind formation.

Noble metal–loaded catalysts, containing the metal in zero valent state, are obtained by calcination in flowing air or oxygen of the material containing the metal ammine complex, followed by reduction in hydrogen. Reduction is usually carried out in the process unit. The degree of metal dispersion obtained is very sensitive to the calcination and reduction conditions; small changes in these conditions can result in either metal dispersion near 100% or the presence of large metal particles on the external surface [28] (see Chapters 5 and 8).

Nonnoble metals, such as nickel and cobalt, can be introduced into the zeolite by ion exchange [29]. A solution of a water-soluble salt of the metal (e.g., nitrate) is used for the exchange. The ionic exchange is carried out under controlled conditions (pH, concentration, temperature) in order to avoid precipitation of basic metal salts or hydroxides. The use of zeolites exchanged with rare earths and nickel (or cobalt) ions in hydrocracking catalysts has also been reported [30]. However, molybdenum and tungsten cannot be readily ion-xchanged into zeolites because they occur mostly in anionic form. In order to incorporate both types of metal (cationic and anionic), ion exchange can be combined with metal impregnation. One of the hydrogenation components (e.g., nickel, cobalt) is introduced into zeolite powder by ion exchange prior to its combination with other catalyst components. The second hydrogenation component (molybdenum, tungsten) is introduced by impregnation of the extruded or pelleted support. The metal-exchanged zeolite powder can also be impregnated with the second hydrogenation component prior to its combination with a binder and extrusion or pelleting [31]. It should be noted that the exchange of cationic molybdenyl ions into zeolites and the use of these zeolites in hydrocracking catalyst preparations has been reported [32].

The metal-containing support is finally dried and/or calcined (oxidized) in air. In this process, moisture and other volatile components are removed from the catalyst and the nonnoble metals are converted to metal oxides. The calcined catalyst is converted to an active form by sulfiding (see Chapter 5).

4.3. Manufacturing Processes of Hydrocracking Catalysts

The catalyst manufacturing processes can be divided into two major categories, similar to the categories of preparation methods previously described. In some processes, all catalyst components (or their precursors) are combined in one step by comulling or cogelling, followed by extrusion, drying, and calcination in air (see Section 4.2). This route eliminates the impregnation step and saves a calcination step. When the manufacturing of catalyst pellets is the objective, the catalyst components are mixed as dry powders and compressed in dies usually in the presence of a lubricant.

In other processes, the catalyst support is manufactured first, followed by addition of the hydrogenation component, usually by impregnation, and finishing by calcination. The manufacturing of catalyst supports is usually carried out by (a) comulling followed by extrusion or by (b) the oil drop method. These processes are used in the manufacture of both conventional and mild hydrocracking catalyst, as well as for manufacturing of some dewaxing and residue hydroprocessing catalysts.

The selected manufacturing process usually depends on composition, form, and required catalytic properties of the final product. Each process involves the following steps: manufacturing of the major catalyst components (zeolite, alumina); combining the catalyst components; forming the catalyst; and finishing the product. The catalyst-manufacturing plant has usually several production units for the different stages of the manufacturing process. Catalyst support production and catalyst finishing are usually carried out in different units. The manufacture of zeolite-containing catalysts requires a separate zeolite-manufacturing unit.

The details of different manufacturing processes are proprietary information. In this section, only a general outline of typical manufacturing processes is provided.

4.3.1. Zeolite Crystallization and Modification

Many commercial hydrocracking catalysts contain Y-type zeolites. The catalytically active forms of this zeolite are obtained from NaY. The preparation of NaY zeolites has been described in detail in the literature [33–37]. The zeolites are prepared from a silica source (e.g., colloidal silica, fumed silica, sodium silicate), alumina source (e.g., hydrated alumina, sodium aluminate), sodium hydroxide, and nucleating agents (see Figure 3.3). The nucleating agents consist of amorphous sodium aluminosilicate and are added to initiate the crystallization process [38]. Calcined clay can also be used as raw material for zeolite synthesis.

The crystallization is usually carried out in large tanks, at temperatures between 80°C and 100°C. It can last from 8 hours to several days, depending on raw materials, crystallization conditions, degree of zeolite crystallinity, and composition desired. The crystallization process can be monitored by X-ray diffraction of control samples. Once completed, the crystallized zeolite is separated from solution ("mother liquor") by filtration over rotary vacuum filters or horizontal belt filters, and water-washed. A typical NaY zeolite has an SiO_2/Al_2O_3 mole ratio of 5.0 or higher, a surface area of over 800 m^2/g, high crystallinity, only minor X-ray-etectable impurities, and a particle size in the range of 0.5–3 μm.

The NaY zeolite is submitted to a series of treatments in order to convert the zeolite into a catalytically active material (Figure 4.3) [39]. The NaY zeolite is first exchanged with ammonium ions to remove most of the sodium ions. The ionic exchange is carried out commercially in ionic exchange tanks or on horizontal belt filters. In the former case, the zeolite is slurried with a solution containing the exchange ions under specific exchange conditions, such as solid concentration, pH, temperature, and time. The exchanged zeolite is then filtered. When using belt filters for the exchange, the zeolite is placed on the filters and sprayed from spray bars with the exchange solution. In both cases the exchanged zeolite is subsequently washed in order to remove the excess salts. When carrying out the ammonium exchange by one of these procedures, a partially ammonium-

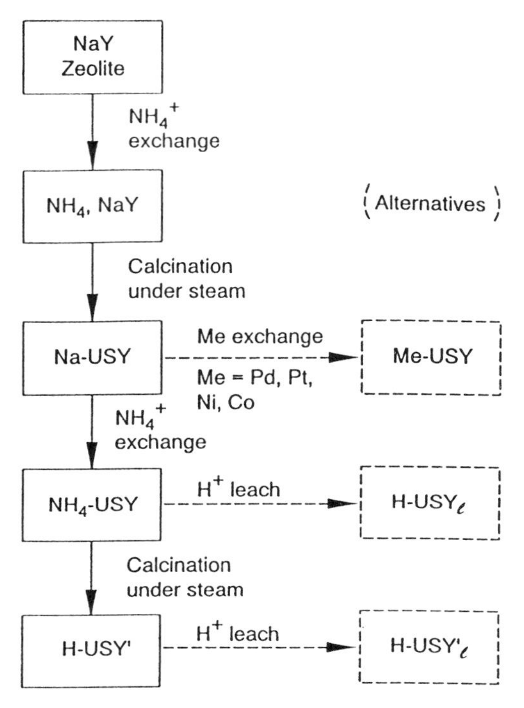

Figure 4.3 Conversion of NaY zeolite into different active forms used in hydrocracking catalyst manufacturing.

exchanged Y zeolite is obtained with about 2–4 wt % Na_2O still left in the zeolite [40].

To obtain the dealuminated form of the zeolite, the most common commercial procedure consists of hydrothermal treatment of ammonium-exchanged Y zeolite. The partially ammonium-xchanged zeolite is calcined at elevated temperature (over 540°C; 1000°F) in the presence of steam in rotary calciners or kilns. Multistage moving belt furnaces may also be used. While moving in the calciner from one end to another, the material passes through several closely controlled heating zones. During the hydrothermal treatment the zeolite is partially dealuminated and stabilized, while the remaining sodium ions are moved to exchangeable positions [33]. The calcination conditions determine the degree of dealumination and unit cell size of the product (Na-USY). Noble metals, nickel/cobalt [41], or other ions [42–44] may be exchanged into the zeolite before or after hydrothermal treatment.

Subsequent ammonium exchanges of Na-USY zeolite remove most of the remaining sodium ions. The product dealuminated zeolite (NH_4-USY) contains less than 1 wt % Na_2O, has good stability, and has a smaller unit cell size than the original zeolite [40]. This material can be used as is in the manufacture of hydrocracking catalysts or can be submitted to further treatment. In the latter case the material (a) may be submitted to acid leaching in order to remove nonframework aluminum or (b) can be further hydrothermally treated to further reduce the zeolite unit cell size. Calcination is done under steam in a rotary calciner [39]. The product zeolite (H-USY′) can be used in catalyst manufacturing as is or be submitted to acid leaching for removal of nonframework aluminum.

Acid leaching is usually carried out in a leaching tank with buffered acid solutions [45]. The severity of the treatment is controlled by varying acid concentration, temperature, and reaction time. In general, the higher the severity, the lower the alumina and sodium content of the zeolite. Mild acid leaching, leading to a SiO_2/Al_2O_3 ratio between 6.5 and 12, is generally preferred in order to obtain good catalytic activity.

The medium-pore zeolite ZSM-5, commonly used in the manufacture of dewaxing catalysts, can be prepared by one of several methods described in the literature [46,47]. It can be prepared by reacting SiO_2, partially dissolved in a quaternary ammonium hydroxide, with a solution of sodium aluminate in water. The resulting gel is heated in an autoclave at about 150°C for 5–8 days, leading to the crystallization of ZSM-5. The organic cation is decomposed by calcination and Na,H-ZSM-5 is formed. Amines, alcohols, ethers, or certain transition metal complexes can be used as templates during synthesis instead of quaternary ammonium cations. ZSM-5 zeolites can also be prepared in the absence of organic templates. Ammonium exchange, followed by drying and calcination, converts the zeolite to H-ZSM-5.

4.3.2. Catalyst Support Manufacturing

The catalyst support for many commercial catalysts consists of an acidic component (zeolite and/or amorphous silica-alumina) and a binder (alumina). It has both physical and catalytic functions. It defines catalyst particle size and shape, as well as the physical/mechanical properties of these particles (attrition and abrasion resistance, crushing strength, hardness, density). Its porosity controls the diffusion of feedstock and product molecules in the catalyst particle. Its acidity defines the cracking function of the catalyst. Therefore, the production of a catalyst support with well-defined physical and chemical properties is of crucial importance in the catalyst manufacturing process.

There are several processes used in the manufacture of catalyst supports, depending on the desired product. With regard to size and shape, it was already shown that three forms of catalyst supports are encountered among commercial hydrocracking and mild hydrocracking catalysts: extrudates, spheres, and pellets. Each of these requires a different manufacturing process. The production of extrudates is the most economical and is widely used. Typical manufacturing processes of extruded catalyst supports and spherical catalyst supports are described in this section.

a. Extruded catalyst support. The extruded catalyst support is manufactured in the catalyst support production unit of the plant. The extruded, zeolite-based catalyst support is often made from modified zeolite powder and alumina powder. In some instances, amorphous silica-alumina powder is added. Peptizers, lubricants, pore-forming agents, and binders may also be added during the comulling and forming process. The most common shapes of commercial hydrocracking catalysts are cylinders of 1/16 or 1/8 in. diameter. The crystallization and modification of Y zeolite has been described in the previous section. Alumina for catalyst manufacturing is obtained in spray-dried form from the alumina process unit or from an outside supplier. Pseudoboehmite can be manufactured by one of the procedures described in Section 3.2.1. It is also used in the manufacture of mild hydrocracking catalyst supports.

Amorphous silica-alumina powder is prepared in a separate manufacturing unit. One of the methods used is reacting diluted solutions of sodium silicate (waterglass) and aluminum salts in precipitation tanks under controlled reaction conditions. Reactant concentration, reaction pH, temperature, and gel aging conditions are closely controlled. The precipitated gel is filtered, washed to remove excess salts, reslurried in water, and spray-dried. Another method is based on formation of a silica and alumina sol, followed by gradual precipitation from homogeneous solution with urea or hexamethylene tetramine (both compounds hydrolyze gradually in solution and generate ammonia). This method has the advantage of producing a more uniform gel [48]. The resulting silica-alumina gel is filtered, washed, reslurried in water, and spray-dried. The dry silica-alumina powder is used in catalyst support manufacturing.

A simplified flow diagram of the extruded catalyst support manufacturing process is shown in Figure 4.4. Weighed amounts of zeolite, alumina, and, if required, silica-alumina powder are transferred via a transfer bottle to a commercial blender, where the dry powders are intimately blended. The resulting blend is transferred to a mixer (muller), where specified amounts of additive solutions are added. The additive solutions consist of water, nitric acid, acetic acid, or ammonia, or combinations of these materials. Their role is to peptize the alumina in the mixture and to facilitate the formation of a wet, thick paste in the mixer. Too much paste fluidity would result in extrudates with poor physical/mechanical properties. Alumina may also be peptized prior to addition of other components. The amount of additive used and residence time in the mixer are critical parameters.

The product from the mixer is extruded, as described in Section 4.2.3. The

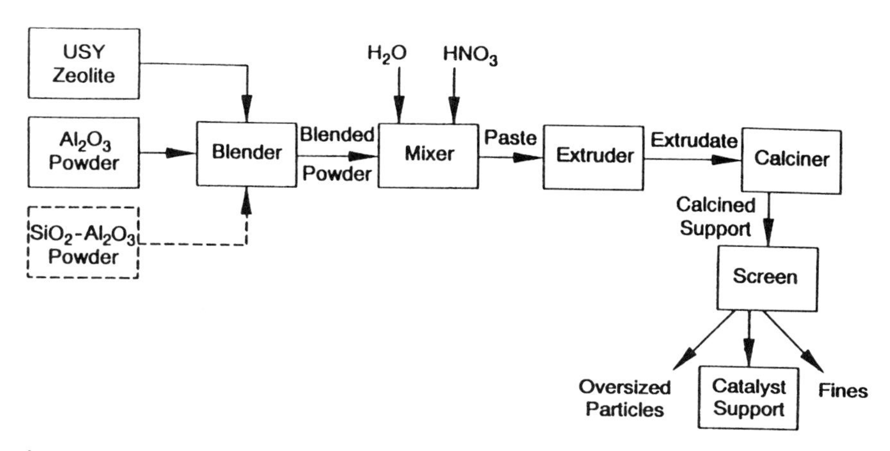

Figure 4.4 Flow diagram of extruded catalyst support manufacturing process.

sized extrudate is then transferred to the calciner. In some processes, the extrudate is dried prior to calcining.

The calciner has a continuous metal belt that moves the extrudates through a prescribed heating process. The calciner usually has two heating zones, with temperatures in the first zone between 150 and 315°C (300 and 600°F) and in the second zone between 650 and 700°C (1200 and 1300°F). In the first zone physically adsorbed water is removed. In the second zone, pseudoboehmite is converted to γ-alumina. If a zeolite in ammonium-exchanged form is present, deammoniation and some dehydroxylation takes place. Calcination conditions (temperature and residence time) also define the physical/mechanical properties of the catalyst support. In some manufacturing processes, belt calciners have been replaced by rotary calciners. The calciner product is screened in order to separate oversized particles and fines from the calcined catalyst support. The catalyst support is then transferred to the catalyst finishing section.

b. Spherical catalyst support. Catalyst supports containing primarily amorphous silica-alumina are sometimes manufactured in spherical form, using the oil drop process [13]. Such a process is used by Universal Oil Products (UOP). In this process, an alumina sol is obtained by digesting aluminum pellets or gibbsite with hydrochloric acid and adding ammonia and urea (or hexamethylene tetramine) for neutralization. A silica sol is obtained separately by reacting a sodium silicate solution (waterglass) with hydrochloric acid in the presence of urea. The urea added to these aqueous sols plays a role in the subsequent processing steps. The neutralized alumina sol is mixed with the silica sol, and the aqueous mixture is fed to the top of a forming tower filled with circulating hot oil. Upon contact between the two immiscible liquids, silica-alumina spheres are formed. The spheres fall to the bottom of the forming tower and are fed to an aging tank filled with hot oil. At the high temperature in the forming tower and aging tank (over 95°C), urea is decomposed and the freed ammonia helps set the gel [49,50]. Aging can be carried out at atmospheric or higher pressure. In the latter case, a higher aging temperature and shorter aging time can be used. After aging in hot oil, the formed spheres are further aged in an aqueous ammonia solution. During the aging step the formed spheres acquire the necessary physical/mechanical properties. Once aging is completed, the formed spheres are washed with an aqueous solution of ammonium nitrate to help remove ammonium chloride and sodium ions from the silica-alumina gel. The washed spheres are transferred first to a dryer and then to a calciner. The resulting spherical catalyst support is screened and then transferred to the catalyst finishing section. A simplified flow diagram of the oil drop process is shown in Figure 4.5.

Spherical alumina supports for mild hydrocracking catalysts can also be manufactured by the oil drop method. Hexamethylene tetramine is added to an aluminum oxychloride sol and the mixture fed to the top of a column holding oil at

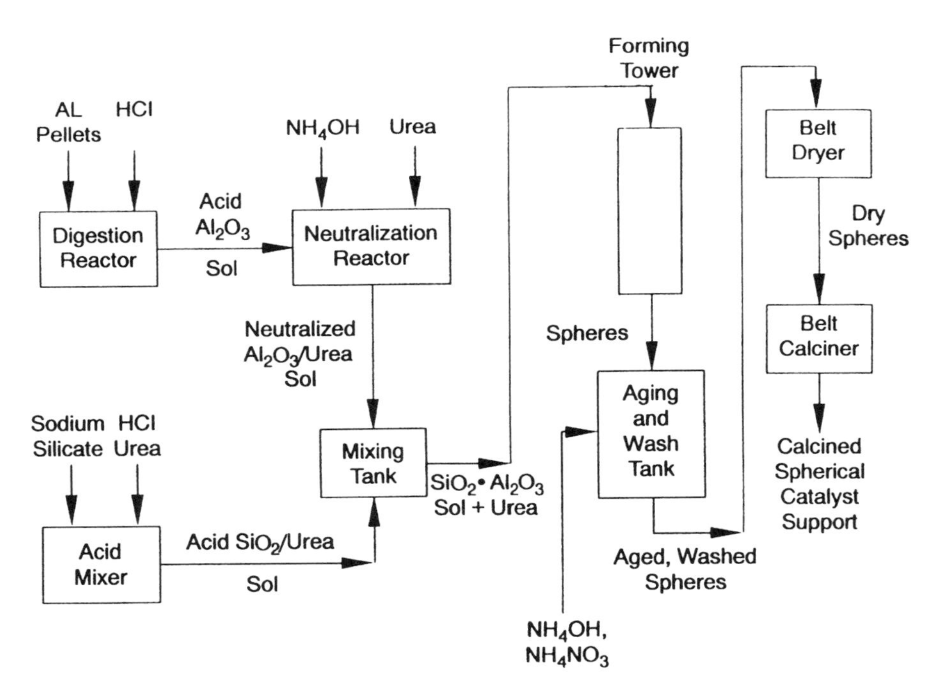

Figure 4.5 Flow diagram of spherical catalyst support manufacturing process using the "oil drop" method.

100°C [4]. At that temperature, the tetramine is decomposed with liberation of ammonia, which reacts with the oxychloride. The resulting spheres of alumina gel are aged, dried, calcined, and impregnated with the appropriate metal salt solution (e.g., Co-Mo or Ni-W salts). The impregnated spheres are subsequently dried and calcined.

4.3.3. Catalyst Finishing

Catalyst finishing consists of conversion of the catalyst support to the final catalyst. It involves two major steps: impregnation and calcination. Impregnation with solutions of hydrogenation metal compounds is typically done commercially by the pore saturation/evaporation method or by the dipping method. With the pore saturation/evaporation method, the metal solution is added to the catalyst support in an amount equal to or in slight excess of the absorption capacity of the support. The volume of solution added to the catalyst by this procedure should contain the amount of hydrogenation metals required in the finished catalyst. Water and ammonia present in solution are evaporated, leaving a metal deposit on the catalyst. By "dipping" the catalyst, an equilibrium metals solution is absorbed by the catalyst support and the excess drained off. Flow diagrams of the two catalyst finishing processes are shown in Figure 4.6.

a. Pore saturation/evaporation. The pore volume of the calcined catalyst support is determined by "titration" with water of a catalyst sample to the point of incipient wetness. A weighed amount of calcined catalyst support is then transferred to a rotating, steam-jacketed evaporator. A measured amount of impregnating solution, made up in the metals mix tank, is slowly drained into the rotating evaporator. The rotation assures that all of the catalyst support is saturated with the impregnating solution. Once the saturation is completed, steam is injected into the evaporator jacket. The steam provides the heat necessary to evaporate the free water and ammonia. Preheated air is used to remove water and ammonia vapors from the evaporator. The catalyst obtained from the evaporator usually contains 30–35% moisture. The impregnated catalyst support is then transferred to the calciner feedhopper.

b. Dip impregnation. If the catalyst is dip-impregnated, a weighed amount of calcined catalyst support is transferred to a dip basket. The basket is then lowered into a dip tank to which impregnating solution is added from the metals mix tank. The catalyst bed is completely covered with impregnating solution and the solution circulated through the basket by a circulating pump. When impregnation is completed, the basket with impregnated catalyst support is lifted from the dip tank and placed on a drain table to drain the excess solution from the basket. The impregnated material is then transferred to the calciner feedhopper.

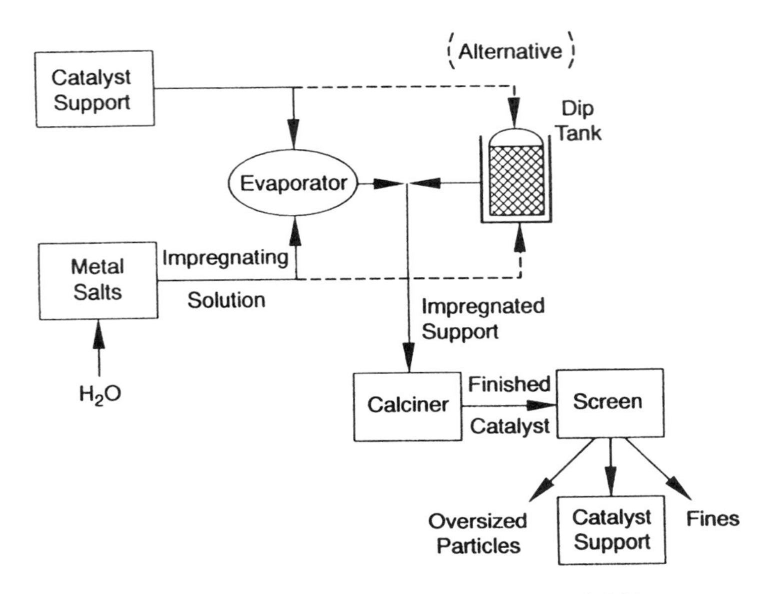

Figure 4.6 Flow diagram of hydrocracking catalyst finishing process.

c. Calcination ("Oxidation"). Rotary calciners or belt calciners with multiple heating zones are used in the process. Calcination is carried out in an oxidizing atmosphere, usually in air. Since the hydrogenation metals are converted to metal oxides, this calcination step is also called oxidation, in order to differentiate it from calcination of the support. In the first heating zone, free water is removed. In the subsequent heating zones with higher temperatures, the metals are converted to oxides. The calcination temperature of the impregnated catalyst is generally lower than that of the catalyst support. The calcined material usually has less than 2% moisture.

The finished catalyst is screened in order to remove fines and oversized particles. These particles are usually recycled back in the process. The on-size catalyst particles are loaded into drums and weighed. Supersacks or sling bins may also be used for packaging.

4.3.4. Environmental Control

In order to operate in a healthy and safe environment, environmental control equipment is installed in most catalyst plants, next to production equipment. A dust collection system controls the dust generated during dry powder handling and mixing. Another dust collection system is installed at the calciners. Such a system consists of a cover for the dust-generating equipment, a duct connecting the cover to a filter/receiver, and a suction-generating fan. The collected dust is usually recycled.

Noxious fumes formed during the manufacturing process are controlled, often with scrubbers. Since the metal salt solutions used for impregnation contain ammonia, fumes of ammonia are generated during calcination of impregnated catalyst support. The equipment where ammonia fumes are produced is covered with hoods that are connected by ducts to the fume exhaust. The fumes are sent to scrubbers or to a stack for disposal. NO_x generated during calcination of the extruded catalyst support or during calcination ("oxidation") of the impregnated catalyst support can be controlled, e.g., by selective catalytic reduction with NH_3 in the presence of a vanadia-titania catalyst. Waste solutions from impregnation are collected and either are sent for disposal or are drummed for recycling.

References

1. Ch. N. Satterfield, *Heterogeneous Catalysis in Industrial Practice*, McGraw-Hill, New York, 1991, p. 87.
2. A. B. Stiles, *Catalyst Manufacture. Laboratory and Commercial Preparations*, Marcel Dekker, New York, 1983.
3. M. V. Twigg (ed.), *Catalyst Handbook*, Wolfe, 1989, p. 34.
4. J. F. LePage, *Applied Heterogeneous Catalysis*, Editions Technip, Paris, 1987.
5. N. Pernicone and F. Traina, *Applied Industrial Catalysis*, Vol. 3, Academic Press, New York, 1984, p. 1.
6. P. B. Weisz, *Trans. Faraday Soc. 63*:1801, 1807, 1815 (1967).
7. T. A. Dorling, B. W. J. Lynch, and R. L. Moss, *J. Catal. 20*:190 (1971).
8. C. Marcilly and J. P. Franck, *Rev. Inst. Fr. Petr. 39*:337 (1984).
9. J. P. Franck and J. F. LePage, Proc. 7th Int. Congr. Catal., Tokyo, 1981, p. 792.
10. J. W. Geus, *Preparation of Catalysts*, Vol. 3 (G. Poncelet, P. Grange, and P. A. Jacobs, eds.), Elsevier, Amsterdam, 1983, p. 1.
11. J. R. Anderson, *Structure of Metallic Catalysts*, Academic Press, New York, 1985.
12. J. W. Ward, in *Preparation of Catalysts III* (G. Poncelet, P. Grange, and P. A. Jacobs, eds.), Elsevier, Amsterdam, 1983, p. 587.
13. L. Hilfman, U.S. Patent No. 4,061,563 (1977).
14. J. J. Jaffe, U.S. Patent No. 3,280,040 (1966).
15. J. J. Jaffe, U.S. Patent No. 3,637,527 (1972).
16. A. G. Bridge, D. R. Cash, and J. F. Mayer, NPRA Ann. Mtg., March 1993, San Antonio, TX, AM-93-60.
17. J. A. Meyer, U.S. Patent No. 4,186,111 (1980).
18. T. J. Gray, U.S. Patent No. 3,210,296 (1965).
19. R. L. Howell, R. F. Sullivan, C. W. Hung, and D. S. Laity, Paper presented at the Japan Petroleum Institute, Petroleum Refining Conference, Tokyo, Oct. 1988.
20. R. F. Sullivan and J. A. Meyer, *Hydrotreating and Hydrocracking*, A.C.S. Symposium Series 20, Washington, D.C., 1975, p. 28.
21. J. W. Ward, U. S. Patent No. 4,097,365 (1978).

22. W. R. Gustafson, U.S. Patent No. 3,966,644 (1976).
23. R. W. Johnson and M. J. O'Hare, U.S. Patent No. 4,497,704 (1985).
24. M. J. Baird, Ch. L. Gutberlet, and J. T. Miller, U.S. Patent No. 4,431,516 (1984).
25. J. W. Ward, *Applied Industrial Catalysis*, Vol. 3 (B. E. Leach, ed.), Academic Press, New York, 1984, p. 272.
26. C. Naccache, J. F. Dutel, and M. Che, *J. Catal. 29*:179 (1973).
27. P. Fletcher and R. P. Townsend, *Zeolites 3*:129 (1983).
28. M. S. Tzou, B. K. Teo, and W. M. H. Sachtler, *J. Catal. 113*:220 (1988).
29. R. J. Bertolacini, H. M. Brennan, and L. C. Gutberlet, U.S. Patent No. 3,597,349 (1971).
30. J. W. Ward, U.S. Patent No. 4,565,621 (1986).
31. H. W. Haynes, J. F. Parcher, and N. E. Helmer, *Ind. Eng. Chem. Proc. Des. Dev. 22*:401 (1983).
32. E. L. Moorehead, U.S. Patent No. 4,297,243 (1981).
33. D. W. Breck, *Zeolite Molecular Sieves*, Wiley, New York, 1974.
34. D. W. Breck, U.S. Patent No. 3,130,007 (1964).
35. J. S. Magee and J. J. Blazek, *Zeolite Chemistry and Catalysis*, A.C.S. Monograph 171 (J. A. Rabo, ed.), Washington, D.C., 1976, p. 615.
36. R. Szostak, *Molecular Sieves*, Van Nostrand Reinhold, New York, 1989.
37. H. Van Bekkum, E. M. Flanigen, and J. C. Jansen (eds.), *Introduction to Zeolite Science and Technology*, Elsevier, New York, 1991.
38. D. E. W. Vaughan, G. C. Edwards, M. G. Barrett, U.S. Patent No. 4,340,573 (1982).
39. J. Scherzer, *Octane-Enhancing Zeolitic FCC Catalysts*, Marcel Dekker, New York, 1990, p. 67.
40. J. Scherzer, *Catal. Rev. Sci. Eng. 31*(3):215 (1989).
41. R. J. Bertolacini, H. M. Brennan, and L. C. Gutberlet, U.S. Patent No. 3,597,349 (1971).
42. J. W. Ward, U.S. Patent No. 4,429,053 (1984).
43. L. Hilfman, R. W. Johnson, and M. J. O'Hara, U.S. Patent No. 4,198,286 (1980).
44. W. A. Welsh, U.S. Patent No. 4,456,693 (1984).
45. J. Scherzer, U.S. Patent No. 4,477,336 (1984).
46. P. A. Jacobs and J. A. Martens, *Synthesis of High-Silica Aluminosilicate Zeolites*, Elsevier, New York, 1987.
47. R. Szostak, *Molecular Sieves*, Van Nostrand Reinhold, New York, 1989.
48. L. Gordon, M. L. Salutsky, and H. H. Willard, *Precipitation from Homogeneous Solutions*, Wiley, New York, 1959.
49. M. W. Schoonevea, U.S. Patent No. 4,318,896 (1982).
50. J. C. Hayer, U.S. Patent No. 3,887,493 (1975).

5

CATALYST ACTIVATION

5.1. General Concepts

The formed, calcined catalyst is converted to a catalytically active form by a process called activation. When the catalyst is of the noble metal type, activation consists of hydrogen reduction of the calcined catalyst, in which the metal is usually present as oxide. Nonnoble metal catalysts are activated by transforming the catalytically inactive metal oxides into active metal sulfides. In some publications, the term "activation" covers both oxidation and reduction (or sulfiding) steps.

In the case of noble metal catalysts, calcination in air prior to reduction is necessary to avoid metal sintering. The presence of water vapors is generally avoided, in order to prevent metal sintering. By using an excess of hydrogen, the water formed during reduction can be swept away. The noble metal oxides can be readily reduced with hydrogen between 300°C and 500°C [1]. A variety of chemical reagents may also be used for reduction (e.g., hydrazine, formaldehyde).

The reduction mechanism with hydrogen involves several steps. Initially, hydrogen reacts with O^{2-} ions of the metal oxide, leading to the formation of metal nuclei and water. Subsequently, dissociative chemisorption of hydrogen on the metal takes place, and the resulting atomic hydrogen migrates to the metal/metal oxide interface, where further reduction takes place. The reduction with atomic hydrogen is much more rapid. The reduction of the oxide takes place sometimes at some distance from the metal due to migration of hydrogen along the surface ("hydrogen spillover" effect) [2,3].

The initial reduction temperature should be sufficiently high to lead to the formation of a large number of metal nuclei without causing sintering. The optimum reduction temperature also depends on the support. The reducibility decreases in the order Pt > Pd > Ni > Co > Fe. Sintering and dispersion of supported metals is discussed in more detail in Section 8.8.

Both noble and nonnoble metal hydrocracking catalysts are calcined in the catalyst-manufacturing plant and therefore are in metal oxide form when supplied to the refiner. The catalysts are commonly activated in situ in the hydrocracking unit prior to contact with feedstock. Nonnoble metal catalysts may also be activated outside the unit (ex situ) and then transferred into the unit.

5.2. Activation of Noble Metal–Zeolite Catalysts

The method that converts the metal–zeolite precursor into a zeolite-supported metal catalyst generally involves two steps: calcination in air or oxygen (oxidation), and reduction in hydrogen (activation) [4–6]. Calcination prior to reduction is necessary to obtain a well-dispersed metal catalyst. Calcination is usually carried out in flowing air or oxygen, whereas reduction is usually done in flowing hydrogen. Calcination removes the physisorbed water from the zeolite and, in the presence of metal ammine complexes, decomposes the complex ions. Calcination leads to the formation of metal cations and, in some instances, to metal oxides. Reduction results in formally zero valent metal. The severity of the calcination and reduction step (calcination and reduction temperature, time of treatment, partial pressure of oxygen or hydrogen) will affect the location and dispersion of the metal and therefore affect the catalytic properties of the zeolite-supported metal catalyst. Direct reduction of the noble metal ammine complex in hydrogen at 370°C results in very low dispersion [7]. However, gradual heating in oxygen up to 450°C prior to hydrogen reduction results in highly dispersed metals. In commercial operations, the reduction is usually carried out at high pressure under flowing hydrogen (up to 50 atm). Metal agglomeration due to water generated during reduction is avoided by slowly raising the temperature to about 370°C and maintaining it until reduction is completed [8]. Higher temperatures would result in the migration of the metal from the interior of the zeolite to the external surface, with formation of large metal crystals (for more details, see Section 8.8.1).

The reduction of platinum group metal ammine complexes in zeolites has been the subject of numerous publications. Most of these studies center on Pt–zeolite systems, while the study of Pd–zeolite systems is more limited. Although platinum is not used commonly in commercial hydrocracking catalysts, these studies have been used to establish general correlations between catalyst oxidation–reduction conditions and some key properties of the reduced catalysts, such as degree of metal dispersion and location of dispersed metal. Furthermore, these studies have led to the development of reaction mechanisms for catalyst oxidation and reduction.

Direct reduction of platinum ammine complexes in zeolites by hydrogen at 350°C occurs according to the following mechanism:

$$Pt(NH_3)_4^{2+} + 2H_2 \rightarrow Pt(NH_3)_2H_2 + 2NH_3 + 2H^+$$

$$Pt(NH_3)_2H_2 \rightarrow Pt^0 + 2NH_3 + H_2$$

According to Dalla Beta and Boudart [7], reduction causes the formation of a mobile platinum hydride intermediate, which leads to platinum agglomeration. Calcination under vacuum or in an inert atmosphere also leads to the formation of Pt^0 according to the following reaction:

$$Pt(NH_3)_4{}^{2+}Y \xrightarrow{T} Pt^0 + 2HY + 2/3N + \tfrac{10}{3}\,NH_3$$

The calcination and reduction of platinum and palladium ammine complexes in NaY zeolites has been studied in detail by Sachtler and co-workers [9]. It was shown that calcination in oxygen, in addition to destroying ammine ligands from the exchanged metal ions and removing physisorbed water from the zeolite, affects the location of the metal ions [10,11]. They remain in the supercages at low temperatures [12,13], but at higher temperatures the ions migrate to smaller cages (i.e., hexagonal prisms and sodalite cages) where they displace the sodium ions. Such migration is explained by the tendency of transition metal ions to move to regions with high negative charge density in the zeolite, i.e., to the smaller zeolite cages, which provide greater charge stability. Pt^{2+} and Pd^{2+} ions occupy positions in the sodalite cage, next to the hexagonal prism [9,11,13,14].

The reduction reaction, e.g., $Pt^{2+} + H_2 \rightleftharpoons Pt^0 + 2H^+$, results in the formation of protons attached to framework oxygen. These acidic OH groups can be identified by infrared spectroscopy. The intensity of the OH band increases with progressive metal reduction [9].

The location of metal ions after calcination determines the reduction temperature and strongly influences the particle formation and growth during the reduction step. Reduction of ions located in the sodalite cages occurs at higher temperatures than those required to reduce metal ions located in supercages [15]. The metal atoms formed in the sodalite cages may subsequently migrate to the supercage. According to Sachtler [9], the reduction and migration of platinum ions (bare or complexed) located in the supercages occurs under reducing conditions (flowing H_2, $T \leqslant 200°C$) according to the following sequence:

1. Formation of a neutral nucleus Pt_n ($n = 1, 2, 3, \ldots$)
2. Migration of the complexed or bare ions to a supercage containing a nucleus
3. Adsorption of the cation on the nucleus
4. Release of the ligands
5. Dissociative adsorption of dihydrogen
6. Release of two protons which become attached to oxide ions on the cage walls
7. Migration of the primary particles and their coalescence to larger particles

If the calcination is mild enough to leave the metal in the supercage in predominantly tetrammine form, reduction will result in low dispersion due to the high mobility of the complex ion. The reduced metal particles can reach a size larger than the supercage (whose diameter is about 13 Å) and extend into adjacent cages, forming grape-shaped aggregates [11,16]. Such aggregates have also been reported for Pd/NaY [11]. Further agglomeration can lead to the formation of metal particles up to 10 nm in the bulk of the zeolite crystal [16]. With a partially decomposed complex ion, such as the Pd diammine ion that shows less mobility, higher dispersion can be achieved. The bare metal ions (Pt^{2+} or Pd^{2+}), obtained upon more severe calcination, have lower mobility than the ammine complexes due to stronger interaction with the negatively charged zeolite framework. Upon reduction, metal particles are formed that are smaller than the supercage. When all metal ions are located in small cages, the reduced metal atoms remain in the small cages at low reduction temperatures but migrate to the supercage or even to the external surface at high reduction temperatures. This mechanism has been found to prevail in zeolites containing Pt, Pd, or Ni. These observations lead to the conclusion that for faujasite-type zeolites containing noble metal ammine complexes, the calcination and reduction conditions play a crucial role in determining the location and dispersion of the metal in the zeolite. The same applies to other noble metal zeolite types, e.g., $Pt(NH_3)_4^{2+}$/ZSM-5 [17]. In that case, the optimum calcination temperature is about 300°C, similar to that established for faujasite-type zeolites.

In addition to the calcination and reduction temperature, the degree of dispersion also depends on the complex ion exchanged into the zeolite. For example, it was shown that the degree of dispersion decreases in the order $Pt(NH_3)_4^{2+} > Pt(NH_3)_2(H_2O)_4^{2+} > Pt(H_2O)_4^{2+}$ [18]. The presence of water favors metal agglomeration [19] (see also Section 8.8.2). The presence of other metal ions may also affect metal catalyst dispersion [20].

5.3. Activation of Nonnoble Metal Catalysts

Air-calcined, nonnoble metal catalysts are activated by sulfiding, i.e., by converting the group VI and VIII metal oxides into metal sulfides. The sulfiding methods used for hydrocracking catalysts are very similar to those used for hydrotreating catalysts. If the catalyst is sulfided prior to its use, the process is called "presulfiding" (or "presulfurizing"). Interaction between sulfur compounds and metal oxides in the catalyst is strong, and oxysulfides can be formed even at room temperature [21,22]. Upon heating in the reactor the oxysulfides are converted into the active metal sulfide form. Sulfiding reactions are highly exothermic. According to some authors, maximum catalyst activity is often reached when the amount of sulfur supplied to the catalyst is in excess of the stoichiometric amount [23].

Several sulfiding procedures have been described in the literature [24].

These procedures can be classified in three groups based on the sulfiding agent used:

1. Presulfiding with H_2/H_2S gas mixtures
2. Presulfiding with nonspiked feedstock
3. Presulfiding with spiked feedstock

Each of these procedures can be applied in situ, with the catalyst in the reactor during sulfiding; or ex situ, when catalyst sulfiding is carried out outside the reactor.

a. *Presulfiding with H_2S* is carried out by passing through the reactor an H_2/H_2S gas mixture containing 2–5 mol % H_2S. The reactor temperature is gradually raised until sulfiding starts, as indicated by exothermicity in the catalyst bed around 150°C (300°F) and the formation of water vapor. The temperature is held until H_2S is evolved from the reactor, then slowly increased (e.g., to 350°C, 660°F) while continuing the flow of H_2/H_2S gas. The maximum sulfiding temperature generally does not exceed the expected maximum bed temperature anticipated for start-of-run. The completion of the presulfiding process is determined by analyzing the recycle gas for H_2S. It can take from several hours to several days, depending on reactor temperature and on amount of sulfur to be deposited on the catalyst. The sulfided catalyst is cooled and is then ready for contact with the feedstock. Presulfiding with H_2/H_2S gas mixtures is often applied in laboratory experiments [24].

b. *Presulfiding with nonspiked feedstock* is a wet phase process, in which sulfiding is achieved with the sulfur from a regular feedstock, such as a VGO. In this case, some catalyst manufacturers recommend that presulfiding be preceded by a drying and soaking step [24].

Drying under hydrogen is necessary in order to remove the moisture from the hygroscopic catalyst. Otherwise subsequent heating of the wet catalyst with oil may cause mechanical damage to the catalyst and result in pressure drop problems in the unit. The catalyst is dried by slowly raising the temperature in order to avoid any reduction of the metals to lower valence states and to vaporize the water gradually. Some authors recommend a maximum drying temperature of about 130°C (270°F) [23].

Soaking of the catalyst with feedstock prior to presulfiding is done to wet catalyst particles with oil and to eliminate dry areas in the catalyst bed. The presence of dry areas results in a reduction in overall activity of the catalyst bed. When soaking with a sulfur-containing feedstock, some sulfur adsorption may occur. The soaking is usually done at 150°C (300°F) [24].

After soaking has been completed, the temperature is increased in order to decompose the sulfur compounds and free H_2S that will react with metal oxides. Presulfiding with a nonspiked feedstock is usually applied at relatively low temperatures, i.e., ~300°C (575°F) or less. The lower temperature is required in order to avoid the competing reduction reactions of the trioxide phase to the dioxide phase (e.g., MoO_3 to MoO_2), which would prevent the subsequent formation of the sulfide phase [25,26]. Under these conditions, the sulfiding process is slow and takes a long time to reach completion.

c. *Presulfiding with spiked feedstock* is another wet phase process, in which sulfiding is achieved by using a feedstock spiked with a sulfur-containing organic compound [23,24,27]. Such compounds release H_2S at a much lower temperature than the sulfur compounds present in regular feedstocks. The spiking agents are usually added to a straight run feedstock, such as a light gas oil. A list of spiking agents, as well as their sulfur content, decomposition temperature, and required presulfiding temperature, is shown in Table 5.1. The presulfiding temperature with a specific spiking agent may vary depending on the degree of sulfiding required. The spiking agents most frequently used are dimethyl sulfide (DMS) and dimethyl disulfide (DMDS). Carbon disulfide is now rarely used due to environmental restrictions. The application of mercaptans is also limited due to their bad odor [24].

The effectiveness of a spiking agent increases as its decomposition temperature decreases. Cost, environmental restriction, and availability are also factors affecting the choice of spiking agent for commercial application.

Spiked presulfiding is a common practice in many refineries. It is more advantageous than nonspiked presulfiding because it requires shorter time off-stream, lower presulfiding temperature, and gives higher sulfur deposition on catalyst [27]. The maximum activity of catalysts prepared with spiked presulfiding is usually 7–8% higher than that obtained with nonspiked presulfiding [24].

Spiked presulfiding can also be done in gas phase by vaporizing a low-boiling-range feedstock (e.g. light gas oil) together with a spiking agent and passing the vapors through the catalyst bed. It may also be done by injecting a spiking agent into the recycle hydrogen streams. However, in liquid phase sulfiding, the liquid feedstock acts as an effective heat sink and aids in dissipating the heat generated during sulfiding.

Table 5.1 Spiking Agents Used for Presulfiding

Spiking Agent	Sulfur, wt %	Decomposition temp. °C (°F)	Presulfiding temp. °C (°F)
Carbon disulfide (CS_2)	84.2	175 (347)	240–260 (470–500)
Dimethyl disulfide (DMDS)	68.1	200 (390)	230–250 (450–480)
Dimethyl sulfide (DMS)	51.1	250 (480)	270–280 (520–535)
n-Butylmercaptan (NBM)	34.7	225 (436)	260 (500)
Di-tertiary nonyl polysulfide (TNPS)	~37	160 (320)	200 (390)

The spiked presulfiding process may be carried out in situ or ex situ. In situ presulfiding takes place during the startup of the unit. It involves unit operating time loss while completing the sulfiding. In ex situ presulfiding, the catalyst is usually impregnated with organic sulfur compounds that convert the metal oxides to oxysulfide species. The amount of sulfur is adjusted to the metal content of the catalyst. Commercial presulfided catalysts are stable and have little or no odor of sulfur compounds. Upon heating in the reactor, the oxysulfides are converted to active metal sulfides. Ex situ presulfiding has the advantage of shorter time off-stream, avoids handling of toxic and flammable materials such as DMS and DMDS, and provides the refiner with a catalyst having the proper sulfur content.

Most presulfiding processes were first developed for hydrotreating catalysts and subsequently have been adapted to hydrocracking. The changes required are due primarily to the stronger acidity of hydrocracking catalysts. For in situ liquid presulfiding, treatment with gaseous ammonia or other nitrogen-containing compounds is used to passivate the acid sites of the catalyst. Such passivation prevents the occurrence of cracking reactions and coke deposition during presulfiding.

An ex situ process adapted to hydrocracking is the 4A-CAT process developed by Eurecat for presulfiding zeolite-containing hydrocracking catalysts [28]. This process is a modified version of Eurecat's "Sulficat" process, developed for ex situ presulfiding of hydrotreating catalysts [29]. The key of the 4A-CAT (*A*ctivity *A*djustment by *A*mmonia *A*dsorption) process is the passivation of zeolitic acid sites by ammonia. A measured amount of nitrogen compound that is readily transformed into ammonia is added to the spiked feedstock. Cracking reactions are thus avoided during presulfiding. Because the passivation is reversible, catalyst activity is readily restored upon heating the sulfided catalyst to normal operating temperature. A presulfiding process applicable to hydrotreating and hydrocracking catalysts has also been developed by CRI.

During the hydrocracking process, the nonnoble metal catalysts must be kept in a sulfided state; therefore, a minimum concentration of H_2S must be maintained in the reacting gases. In the absence of H_2S, some nickel (cobalt) sulfide may be reduced to metal, which in turn may sinter.

5.4. Composition and Structure of Sulfided Catalysts

Conventional sulfided catalysts used in hydroprocessing have been the subject of numerous studies and reviews [30–38]. Most of these studies have focused on hydrotreating catalysts, which often contain the nonnoble metals at similar ratios as hydrocracking catalysts. Despite extensive research, there is still considerable uncertainty with regard to the nature of catalytically active sites, their structure, and their mechanism of action.

Prior to sulfiding, the thermally treated catalyst contains the catalytically active metals in oxidic form, i.e., MoO_3/WO_3 and CoO/NiO. Upon sulfiding in the presence of H_2, MoO_3 and WO_3 are converted to sulfides having the metal in 4-valent state: MoS_2 and WS_2, respectively. Cobalt and nickel can form different sulfides depending on the amount of metal present and on sulfiding temperature. For example, high cobalt/molybdenum ratio and high sulfiding temperature favor formation of Co_9S_8. In nickel/molybdenum catalysts, Ni_3S_2 has been identified.

MoS_2 has a layered structure with weak interaction between the sulfur atoms in adjacent layers. MoS_2 exhibits hexagonal morphology. At low Mo concentrations, the MoS_2 hexagons can exist as two-dimensional clusters, one layer thick, on the catalyst support. Their size can vary from 20 Å to over 60 Å. At higher Mo concentrations, multilayers (crystallites) of MoS_2 are formed. In $CoMo/Al_2O_3$ catalysts, most of the cobalt is adsorbed on the edge of MoS_2 crystallites, forming a structure usually referred to as the Co-Mo-S phase. This is believed to be the main active site, in which Co acts as a promoter [34–36]. In the presence of larger amounts of cobalt, some of the cobalt may also be present as Co_9S_8 crystallites on the support. Some Co may enter the alumina lattice. A schematic representation of different phases present in a typical $CoMo/Al_2O_3$ catalyst is shown in Figure 5.1 [34]. In supported NiMo catalysts, an Ni-Mo-S phase is formed, and Ni acts as a promoter.

Adsorption of organic nitrogen compounds on these catalysts has shown the presence of Lewis acid centers on both the alumina support and the metal sulfides.

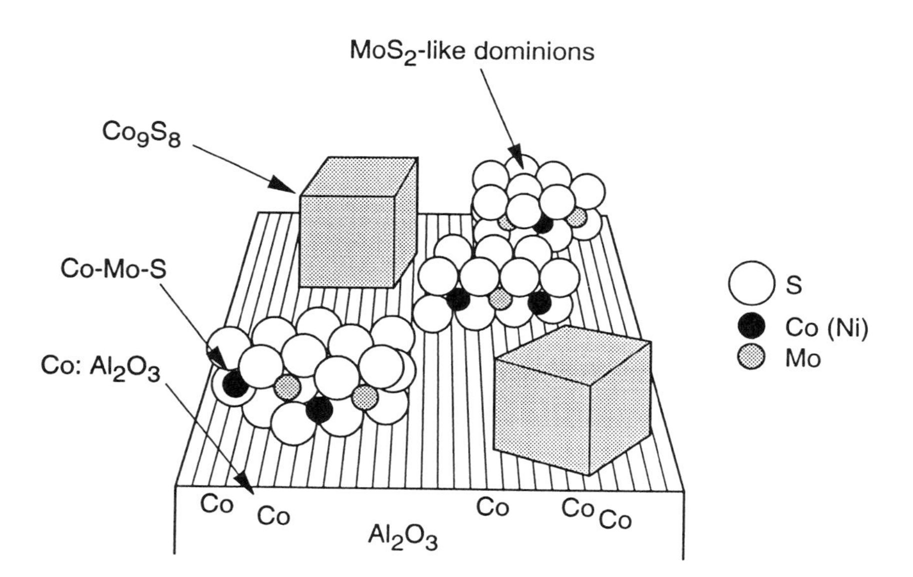

Figure 5.1 Schematic representation of different phases present in a sulfided, typical $Co,Mo/Al_2O_3$ catalyst [34].

It was concluded that the latter are due to sulfur anion vacancies [39], i.e., coordinatively unsaturated sites, which are located at the edges of the crystallites, where Mo atoms are incompletely coordinated by S anions [40]. These vacancies are sites for adsorption and reaction of reactant molecules. The Lewis acid sites on Al_2O_3 are not active in hydrogenation reactions. A combination of several sulfur anion vacancies in a specific geometrical arrangement is considered a reaction center [41,42]. Different geometrical arrangements of these vacancies are required for different types of reaction, e.g., for hydrogenation vs. C-S hydrogenolysis. Uncoordinated Co and Ni also have enhanced activities, explaining their promoting effect. Hydrogen is adsorbed at sulfur sites, forming SH sites, that presumably are active for hydrogen transfer to reactant molecules.

No systematic studies have been reported with regard to active sites and reaction mechanisms on nonnoble metal catalysts with supports containing several components, e.g., zeolite/amorphous component or SiO_2-Al_2O_3/Al_2O_3. Such systems are more complex and different metal sulfides are likely to be present, depending on the preparation method used.

References

1. J. R. Anderson, *Structure of Metallic Catalyst*, Academic Press, New York, 1975.
2. S. Khoobiar, *Catalyst Supports and Supported Catalysts* (A. B. Stiles, ed.), Butterworths, London, 1987, p. 201.
3. G. C. Bond, in *Spillover of Adsorbed Species* (G. M. Pajonk, S. J. Teichner, and J. E. Germain, eds.), Elsevier, Amsterdam, 1988, p. 1.
4. P. Gallezot, *Catal. Rev. Sci. Eng. 20*:121 (1979).
5. W. J. Reagan, A. W. Chester, and G. T. Kerr, *J. Catal. 69*:89 (1981).
6. T. Kubo, H. Arai, H. Tominaga, and T. Kunuzi, *Bull. Soc. Chem. Soc. Jpn. 45*:607, 613 (1972).
7. K. A. Dalla Betta and M. Boudart, in *Proc. 5th Int. Congr. Catal.*, Vol. 1 (J. W. Hightower, ed.), Elsevier, New York, 1973, p. 329.
8. J. W. Ward, *Prep. of Catalysts III*, Elsevier, Amsterdam, 1983, p. 587.
9. S. T. Homeyer and W. M. H. Sachtler, in *Zeolites: Facts, Figures, Future* (P. A. Jacobs and R. A. van Santen, eds.), Elsevier, Amsterdam, 1989, p. 975.
10. P. A. Jacobs, in *Metal Microstructures in Zeolites* (P. A. Jacobs et al., eds.), Elsevier, Amsterdam, 1982, p. 71.
11. G. Bergerat, P. Gallezot, and B. Imelik, *J. Phys. Chem. 85*:411 (1981).
12. M. S. Tzou, B. K. Teo, and W. M. H. Sachtler, *J. Catal. 113*:220 (1988).
13. P. Gallezot, A. Alarcon Diaz J. A. Dalmon, A. I. Renouprez, and B. Imelik, *J. Catal. 39*:334 (1975).
14. P. Gallezot, in *Metal Microstructure in Zeolites* (P. Jacobs, et al., eds.), Elsevier, Amsterdam, 1982, p. 167.
15. S. H. Park, M. S. Tzou, and W. M. H. Sachtler, *Appl. Catal. 24*:85 (1986).
16. D. Exner, N. Jaeger, and G. Schulz-Ekloff, *Chem. Ing. Techn. 52*:734 (1980).

17. G. Gianetto, G. Perot, and M. Guisnet, in *Catalysis by Acids and Bases* (B. Imelik et al., eds.), Elsevier, Amsterdam, 1985, p. 265.
18. M. Guerin, C. Kappenstein, F. Alvarez, G. Gianetto, and M. Guisnet, *Appl. Catal. 45*:325 (1988).
19. R. C. Hansford and R. H. Hass, U.S. Patent No. 3,287,257 (1966).
20. W. M. H. Sachtler, M-S. Tzou, and H. J. Jiang, U.S. Patent No. 4,654,317 (1987).
21. R. Scheffer, J. C. M. De Jonge, P. Arnoldy, and J. A. Moulijn, *Bull. Soc. Chim. Belg. 93*:8 (1984).
22. E. Payen, S. Kasztelan, J. Grimblot, and J. P. Bonnelle, *Catal. Today 4*, 57 (1988).
23. J. F. Martin and F. Plantenga, Paper presented at Akzo Chem. Symposium during NPRA Annual Meeting, October 1988, San Francisco.
24. H. Hallie, *Oil Gas J.*, Dec. 20, 1982, p. 69.
25. F. E. Massoth, *J. Catal. 36*:164 (1975).
26. F. E. Massoth and C. L. Kibby, *J. Catal. 47*:300 (1977).
27. Ch. Bryson and B. Christman, *Oil Gas J.*, Nov. 2, 1987, p. 35.
28. P. Dufresne, G. Berrebi, and J. Wilson, AICHE Annual Meeting, November 1989, San Francisco.
29. G. Berrebi and R. Roumieu, *Bull. Soc. Chim. Belg. 96*:11 (1987).
30. B. Delmon, Proc. 4th Int. Congr. on Uses of Molybdenum, 1979, p. 73.
31. B. C. Gates, J. R. Katzer, and G. C. A. Schuit, *Chemistry of Catalytic Processes*, McGraw-Hill, New York, 1979, Ch. 5.
32. P. Grange, *Cat. Rev. Sci. Eng. 21*:135 (1980).
33. P. Ratnasamy and S. Sivashanker, *Cat. Rev. Sci. Eng. 22*:401 (1980).
34. H. Topsoe and B. S. Clausen, *Catal. Rev. Sci. Eng. 26*(3,4): 395 (1984).
35. H. Topsoe, B. S. Clausen, N. Y. Topsoe, and E. Pedersen, *Ind. Eng. Chem. Fund. 25*:25 (1986).
36. R. Prins, V. H. J. DeBeer, and G. A. Somorjai, *Catal. Rev. Sci. Eng. 31*:1 (1989).
37. H. Topsoe, B. S. Clausen, N-Y Topsoe, and P. Zeuthen, in *Catalysts in Petroleum Refining 1989* (D. L. Trimm et al., eds.), Elsevier, Amsterdam, 1990, p. 77.
38. F. E. Massoth, *Adv. Catal. 27*:265 (1978).
39. C. Vogdt, T. Butz, A. Lerf, and H. Knotzinger, *Polyhedron 5* (1986).
40. C. B. Roxlo, M. Daage, A. F. Ruppert, and R. R. Chianelli, *J. Catal. 100*:176 (1986).
41. F. E. Massoth and G. Muralidhar, *Proc. 4th Int. Conf. on Uses of Molybdenum*, 1982, p. 343.
42. E. Furimsky and F. E. Massoth, *Catal. Today 17*(4): 536 (1993).

6

REACTIONS AND REACTION PATHWAYS

6.1. Reactions

The reactions that occur during hydrocracking have been studied extensively. These studies have shown that hydrocracking can take place by three major routes [1]:

1. Noncatalytic, thermal cleavage of C-C bonds via hydrocarbon radicals, with hydrogen addition (hydropyrolysis).
2. Monofunctional C-C bond cleavage with hydrogen addition over hydrogenation components consisting of metals (Pt, Pd, Ni), oxides, or sulfides (hydrogenolysis).
3. Bifunctional C-C bond cleavage with hydrogen addition over bifunctional catalysts consisting of a hydrogenation component dispersed on a porous, acidic support. In petroleum refining, most hydrocracking reactions follow this route.

In addition to hydrocracking reactions, a number of other reactions take place during the hydrocracking process (Table 6.1). In most hydrocracking processes, the feedstock is submitted to hydrotreating prior to hydrocracking, in the same or in different reactors. Hydrotreating is a pretreatment that partially hydrogenates unsaturated hydrocarbons and removes hetero atoms from the feedstock. Compounds containing these hetero atoms have a deleterious effect on the hydrocracking catalyst and must be removed. The initial reactions in the hydrocracking process are therefore hydrotreating reactions and occur at a relatively high rate over a hydrotreating catalyst. These reactions include hydrodesulfurization (HDS), hydrodenitrogenation (HDN), hydrodeoxygenation (HDO), olefin hydrogenation, and partial aromatics hydrogenation. A comparison of the rates of different hydrotreating reactions shows that the reaction rate decreases in the order HDS > HDN > olefin saturation > aromatic saturation.

Table 6.1 Typical Hydroprocessing Reactions

A. Reactions Occurring Mostly During Hydrotreating

Reaction	
Hydrodesulfurization:	R, S $+ 6H_2 \rightarrow$ R $+ H_2S$
Hydrodenitrogenation:	R, N $+ 7H_2 \rightarrow$ R $+ NH_3$
Hydrodemetallation:	M-porphyrin $\xrightarrow[(H_2S)]{H_2}$ M_xS_y + H-porphyrin
Hydrodeoxygenation:	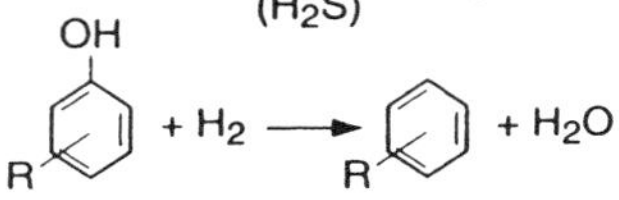 OH, R $+ H_2 \rightarrow$ R $+ H_2O$
Olefin hydrogenation:	$+ H_2 \rightleftharpoons$

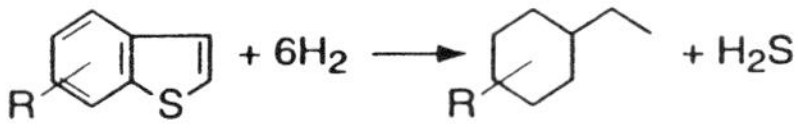

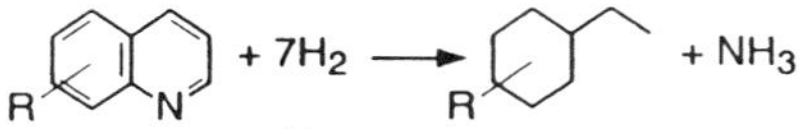

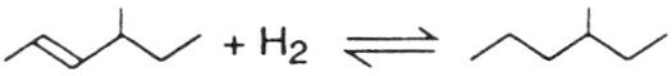

B. Reactions Occurring Mostly During Hydrocracking

Reaction	
Monoaromatics hydrogenation:	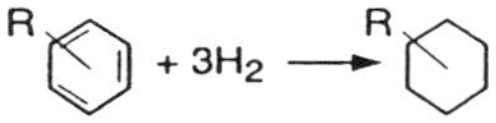 R $+ 3H_2 \rightarrow$ R
Hydrodealkylation:	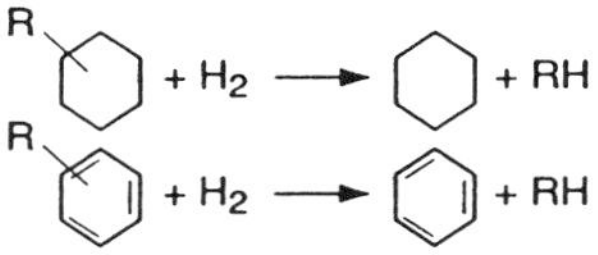 R $+ H_2 \rightarrow$ + RH
	R $+ H_2 \rightarrow$ + RH
Hydrodecyclization:	R $+ 2H_2 \rightarrow$ R $+ C_2H_6$
Hydrocracking:	$C_nH_{2n+2} + H_2 \rightarrow C_aH_{2a+2} + C_bH_{2b+2}$ (a+b = n)
Isomerization of paraffins:	$\rightleftharpoons$

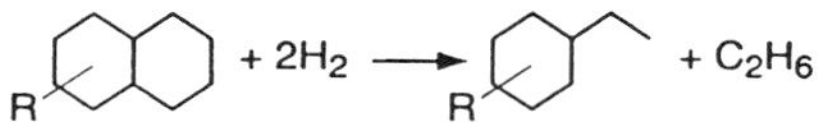

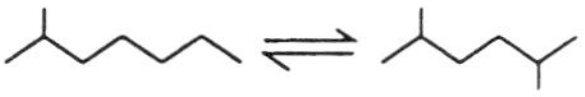

C. Reactions Occurring During Hydrotreating and Hydrocracking

Reaction	
Polyaromatics hydrogenation:	R $+ 2H_2 \rightarrow$ R $+ 3H_2 \rightarrow$ R
Coking:	Polyaromatics $\xrightarrow{+\text{ Olefins}}$ Alkylation $\xrightarrow{-H_2}$ Cyclization $\xrightarrow{-H_2}$ Coke precursors

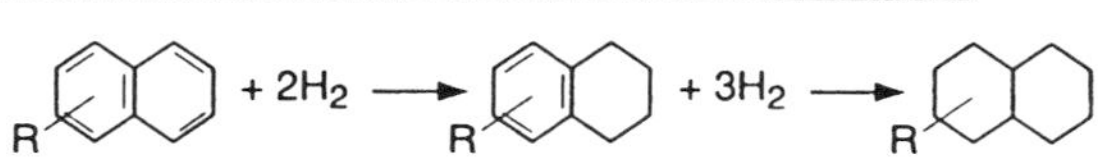

In the presence of a hydrocracking catalyst, further hydrogenation of aromatics takes place, followed by naphthene ring opening (hydrodecyclization). The side chains of cyclic compounds are cracked (hydrodealkylation), along with long-chain paraffins. Isomerization of paraffins followed by cracking also takes place. Coke is formed gradually on the catalyst during the hydrocracking process.

Hydrogenation reactions that occur during the hydrocracking process lead to partial or total saturation of olefinic and aromatic hydrocarbons. Both aliphatic and aromatic hydrocarbons can be hydrocracked. In the case of aromatics, the side chain is relatively easy to remove (dealkylation). Cracking the aromatic ring is much more difficult and requires hydrogenation prior to cracking. Isomerization of paraffins results in branched molecules, whereas isomerization of naphthenes results in changes in ring size.

6.2. Reaction Mechanism and Reaction Pathways

The mechanism of hydrocracking reactions over bifunctional catalysts has been investigated extensively. Most studies were carried out using model compounds, primarily paraffins [2–14], and, to a lesser extent, naphthenes, alkylaromatics, and polyaromatics [15–17]. The reaction pathway for conversion of some industrial feedstocks [18–20], as well as that of heterocyclic compounds [21,22], has also been investigated.

The mechanisms of hydrocracking reactions are essentially the carbenium ion mechanism of catalytic cracking reactions coupled with that of hydrogenation and isomerization reactions. Although the initial reactions in hydrocracking are similar to those in catalytic cracking, the presence of excess hydrogen and of a hydrogenation component in the catalyst results in hydrogenated products and inhibits some of the secondary reactions, such as coke formation and secondary cracking.

Details of earlier work (until 1970) regarding the chemistry of hydrocracking are given in the comprehensive review by Langlais and Sullivan [3]. The reaction mechanism and reaction pathways in hydrocracking over zeolite-based catalysts are described mostly by J. Weitkamp, P. Jacobs, M. Guisnet, and their coworkers in numerous publications (see below).

6.2.1. Hydroconversion of Paraffins

The mechanism of paraffin hydroconversion over bifunctional amorphous catalysts was studied in detail in the 1960s (see, for example, [24–26]). Those and later studies were carried out mostly by using model compounds. Based on the pioneering work of Mills et al. [27] and Weisz [28], a carbenium ion mechanism was proposed, similar to that previously described for catalytic cracking, with additional hydrogenation and skeletal isomerization. More recent studies of *n*-paraffin hydroconversion over noble metal–loaded, zeolite-based catalysts have concluded that the reaction mechanism is similar to that proposed earlier for amorphous, bifunctional hydrocracking catalysts [3,5,23–25].

Hydrocracking of *n*-paraffins over a bifunctional catalyst goes through the following steps:

1. Adsorption of *n*-paraffins on metal sites

2. Dehydrogenation with formation of *n*-olefins
3. Desorption from metal sites and diffusion to acid sites
4. Skeletal isomerization and/or cracking of olefins on the acid sites through carbenium ion intermediates
5. Desorption of formed olefins from acid sites and diffusion to metal sites
6. Hydrogenation of these olefins (*n*- and iso-) on metal sites
7. Desorption of resulting paraffins

The elementary reactions corresponding to the reaction path described are shown in Table 6.2. Product analysis has shown that whenever several reaction pathways are possible, the one leading to the formation and subsequent cracking of a tertiary carbenium ion is preferred (reactions (d) and (e) in Table 6.2) [5]. While the hydrogenation/dehydrogenation and isomerization reactions are reversible, the cracking reactions are irreversible.

a. Types of isomerization and β-scission mechanisms. The rearrangement of secondary alkylcarbenium ions may lead to another secondary carbenium ion through an alkyl shift (type A isomerization), or it may lead to a tertiary alkylcarbenium ion (branching) via a protonated cyclopropane (PCP) intermediary (type B isomerization) (Table 6.3) [7,30,31]. The rate of type A isomerization is usually higher than that of type B.

Beta-scission can lead to the formation of tertiary and secondary carbenium ions, but no primary carbenium ions are formed. Several ß-scission mechanisms have been suggested for the cracking of branched secondary and tertiary carbenium ions (Figure 6.1) [6]. Type A β scission, which converts a tertiary carbenium ion to another tertiary carbenium ion, has the highest reaction rate and is the most likely to occur. The reaction rates decrease in the following order: $A \gg B_1 > B_2 > C$. One should note that each type of reaction requires a minimum number of carbon atoms in the molecule and a certain type of branching in order to take place.

The proposed β-scission mechanisms suggest that the *n*-paraffins submitted to hydrocracking may undergo several isomerizations until a configuration is attained that is favorable to β scission.

The cracking of isomers occurs preferentially near the center of the hydrocarbon chain and practically no methane or ethane formation is observed. For large carbenium ions, the cracking by β scission is more likely to occur with dibranched and tribranched isomers than with monobranched ones [6]. Furthermore, the cracking of lower molecular weight paraffins via β scission is less likely to occur, which explains their high yields even at high conversions.

The rate of hydroconversion of individual paraffins over both amorphous [33,34] and zeolite-based catalysts, such as Pt/CaY and Pt/USY [1,5], increases with increasing chain length. Iso-to-normal ratio for paraffins in the product is high. This is due primarily to the isomerization of secondary carbenium ions to

Table 6.2 Typical Reaction Path in Hydrocracking of *n*-Paraffins

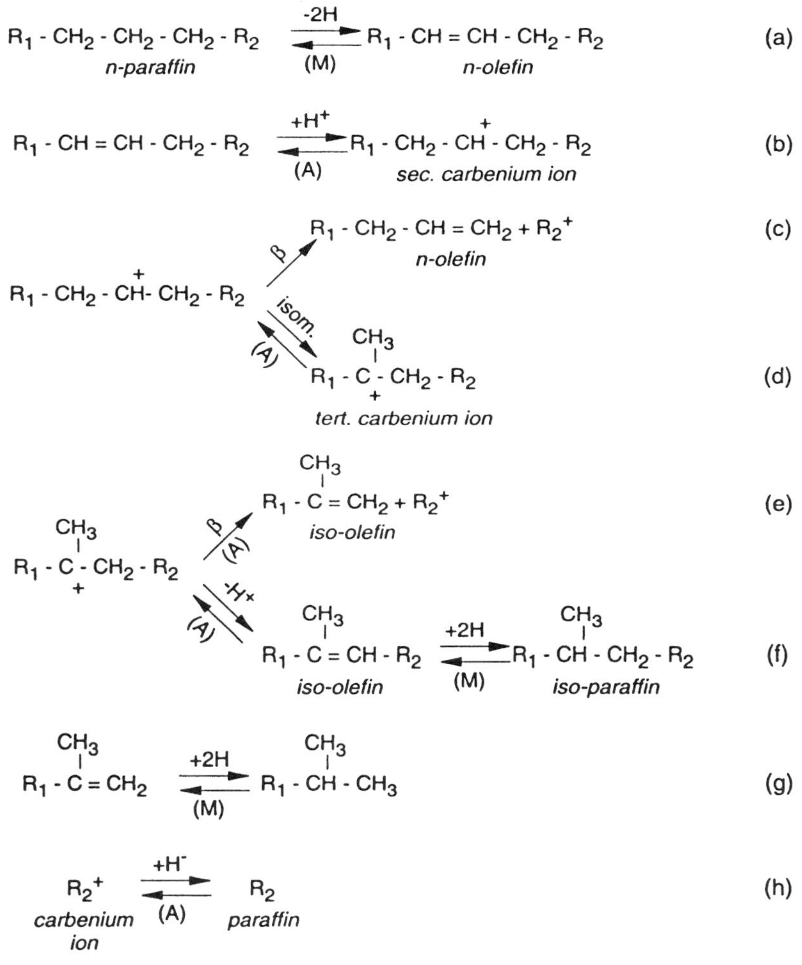

$$R_1 - CH_2 - CH_2 - CH_2 - R_2 \underset{(M)}{\overset{-2H}{\rightleftharpoons}} R_1 - CH = CH - CH_2 - R_2 \qquad (a)$$

n-paraffin → *n-olefin*

$$R_1 - CH = CH - CH_2 - R_2 \underset{(A)}{\overset{+H^+}{\rightleftharpoons}} R_1 - CH_2 - \overset{+}{C}H - CH_2 - R_2 \qquad (b)$$

sec. carbenium ion

$$R_1 - CH_2 - \overset{+}{C}H - CH_2 - R_2 \xrightarrow{\beta} R_1 - CH_2 - CH = CH_2 + R_2^+ \qquad (c)$$

n-olefin

$$R_1 - CH_2 - \overset{+}{C}H - CH_2 - R_2 \underset{(A)}{\overset{isom.}{\rightleftharpoons}} R_1 - \underset{+}{\overset{CH_3}{\overset{|}{C}}} - CH_2 - R_2 \qquad (d)$$

tert. carbenium ion

$$R_1 - \underset{+}{\overset{CH_3}{\overset{|}{C}}} - CH_2 - R_2 \xrightarrow[(A)]{\beta} R_1 - \overset{CH_3}{\overset{|}{C}} = CH_2 + R_2^+ \qquad (e)$$

iso-olefin

$$R_1 - \underset{+}{\overset{CH_3}{\overset{|}{C}}} - CH_2 - R_2 \underset{(A)}{\overset{-H^+}{\rightleftharpoons}} R_1 - \overset{CH_3}{\overset{|}{C}} = CH - R_2 \underset{(M)}{\overset{+2H}{\rightleftharpoons}} R_1 - \overset{CH_3}{\overset{|}{C}H} - CH_2 - R_2 \qquad (f)$$

iso-olefin → *iso-paraffin*

$$R_1 - \overset{CH_3}{\overset{|}{C}} = CH_2 \underset{(M)}{\overset{+2H}{\rightleftharpoons}} R_1 - \overset{CH_3}{\overset{|}{C}H} - CH_3 \qquad (g)$$

$$R_2^+ \underset{(A)}{\overset{+H^-}{\rightleftharpoons}} R_2 \qquad (h)$$

carbenium ion → *paraffin*

β: beta scission, A: acid site, M: metal site

Table 6.3 Skeletal Isomerization Mechanisms

Type A

sec. carbenium ion *sec. carbenium ion*

Type B

sec. carbenium ion *PCP intermediary* *tert. carbenium ions*

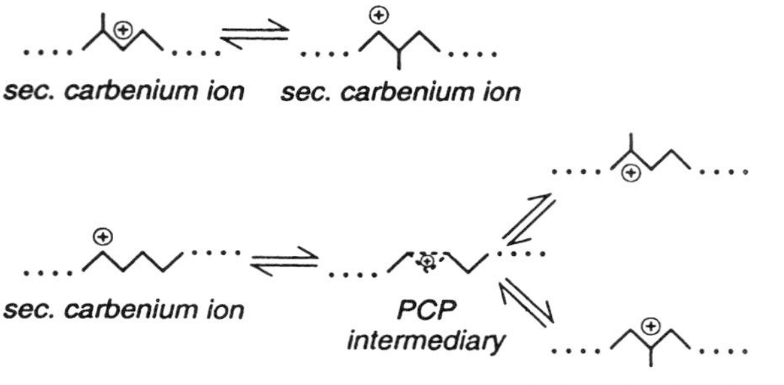

β-Scission mechanisms

Type	Minimum carbon number	Ions involved	Rearrangement
A	8	tert → tert	
B_1	7	sec → tert	
B_2	7	tert → sec	
C	6	sec → sec	

Isomerization mechanisms

Type	Minimum carbon number	Ions involved	Rearrangement
A	6[a]	sec → sec	
B	5[a]	PCP → sec	

a, to observe a chemically different product

Figure 6.1 Possible isomerization and β-scission mechanisms for secondary and tertiary carbenium ion conversion over bifunctional Pt/zeolite catalyst [6].

more stable tertiary ions prior to cracking, as well as to the high rate of hydride ion transfer to the tertiary carbenium ion [3].

b. Effect of hydrogenation-to-acidity ratio and pore geometry. The iso-to-normal ratio in product paraffins increases with decreasing reaction temperature because at higher temperatures the cracking rate of isoparaffins increases faster than that of *n*-paraffins. This is illustrated by the hydrocracking of *n*-decane (Figure 6.2) [35]. The iso-to-normal ratio also increases when the catalyst contains a weak hydrogenation component and a strong acid component. The higher iso-to-normal ratio is attributed to a higher rate of isomerization of the olefinic intermediates at the strong acidic sites. Conversely, partial neutralization of acid sites by ammonia during hydrocracking reduces not only cracking activity but also the iso-to-normal ratio in product paraffins.

The product distribution obtained for *n*-hexadecane hydrocracking with catalysts having different hydrogenation components and different supports is shown in Figure 6.3: a higher hydrogenation-to-acidity ratio in the catalyst (e.g., Pt/CaY, Pt/USY) results in a wider spread of products [1]. Such hydrocracking is sometimes called "ideal hydrocracking" and often results in higher liquid yields. In "ideal hydrocracking," the rate-determining events (isomerization and β scission) occur at the acid sites, whereas the metal sites serve only for rapid hydrogenation and dehydrogenation [7].

The wide spread of products also suggests a high rate of desorption and hydrogenation of the primary cracking products before secondary cracking can occur. The high rate of carbenium ion desorption is due to their displacement by

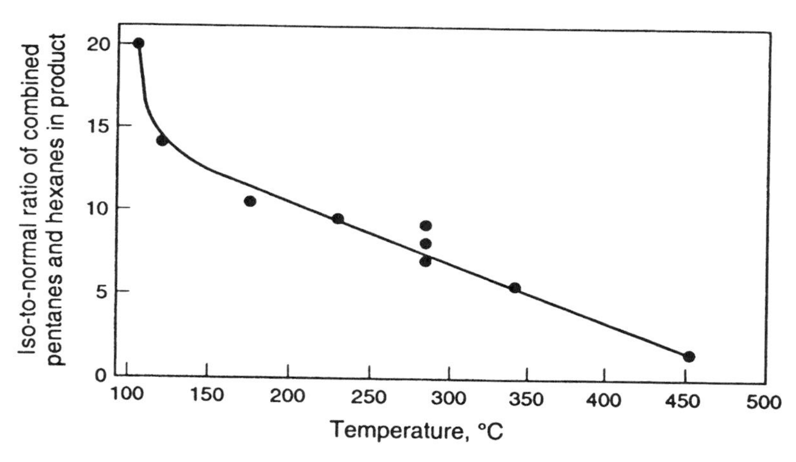

Figure 6.2 Effect of reaction temperature on iso-to-normal paraffin ratio in products obtained from hydrocracking of *n*-decane over strongly acidic catalyst [35].

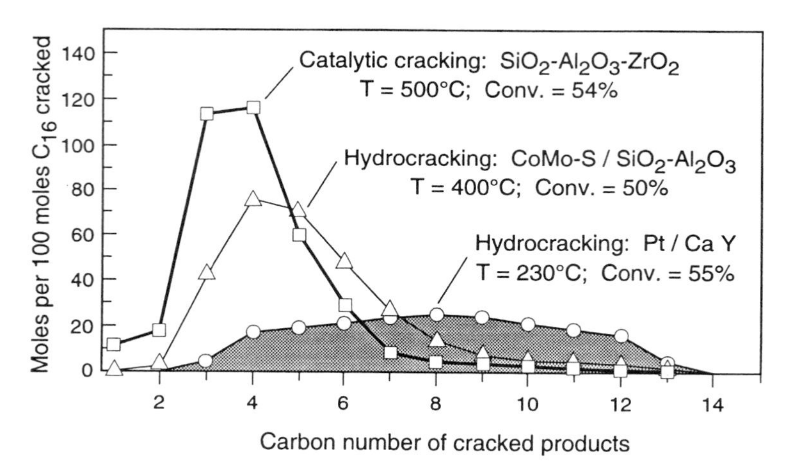

Figure 6.3 Molar carbon number distributions in catalytic cracking and hydrocracking of *n*-hexadecane at 50% conversion [1].

n-olefins, whose steady state concentration is higher in the presence of a strong hydrogenation/dehydrogenation component (competitive sorption/desorption) [1]. The strength of this component can therefore influence the rate of desorption of tertiary carbenium ions and affect product distribution. The data in Figure 6.3 also show that long-chain molecules tend to crack in or near the center because no C_1 or C_2 hydrocarbons are found in the product.

By contrast, on catalysts with low hydrogenation-to-acidity ratios (e.g., Co-Mo-S/SiO_2-Al_2O_3), the moieties of primary cracking reactions remain longer adsorbed at acid sites and undergo secondary cracking. This results in higher yields in low molecular weight products (C_2–C_6). In the absence of a hydrogenation component, as in cracking catalysts, secondary cracking reactions become even more important and lead to high yields of low molecular weight products (Figure 6.3).

Hydrocracking over a catalyst consisting of a strong hydrogenation component (e.g., Pt) and a weak acidic or nonacidic component proceeds via a hydrogenolysis mechanism on the metal [1]. This results in high yields of C_1 and C_2 hydrocarbons, along with *n*-paraffins, and a near absence of isoparaffins.

Using *n*-heptane as a probe molecule and hydrocracking catalysts containing a variety of zeolites, Guisnet et al. [11–13,36] investigated the effect of hydrogenation-to-acidity ratio and of pore geometry on catalyst activity and selectivity. For PtHY and PtHZSM-5 catalysts, the authors found that activity increases with the increase in hydrogenation-to-acidity ratio before reaching a plateau. Pt,H-mordenite showed an increase followed by a decrease in activity for increasing hydrogenation-to-acidity ratios. The observed differences in activity were attributed to differences in zeolite pore geometry: whereas PtHY and PtHZSM-5 have a three-dimensional pore network that facilitates the diffusion of feed and product molecules, mordenite has one-dimensional pores. In mordenite, the pores can be easily blocked by platinum or coke, reducing catalyst activity and leading to rapid deactivation [37].

Catalyst selectivity is also affected by the hydrogenation-to-acidity ratio. The ratio of isomerized-to-cracked *n*-heptane increases with increasing hydrogenation-to-acidity ratio [36]. The presence of a strong hydrogenation component increases the rate of hydrogenation of isoolefin moieties formed at acid sites from the initial feed molecules resulting in higher yields of isomerized products.

At low temperatures and low degrees of conversion, the hydroisomerization of *n*-paraffins predominates. As the temperature increases, the degree of hydroisomerization goes through a maximum and starts declining, whereas the degree of hydrocracking increases (Figure 6.4) [1]. The decrease in hydroisomerization at higher temperatures is due to the consumption of branched isomers by hydrocracking. These results suggest that skeletal isomerization precedes C-C bond cleavage. An increase in the chain length of the *n*-paraffin results in a decrease in required reaction temperature for both hydroisomerization and hydrocracking. The number of branched isomers, and the corresponding number of cracking products, increases significantly with increasing chain length.

At high cracking severity, the primary cracking products undergo secondary isomerization and cracking. The rate of secondary hydroconversion increases with the chain length of the fragment. Other secondary reactions, such as disproportionation, cyclization, and coke formation, may also take place.

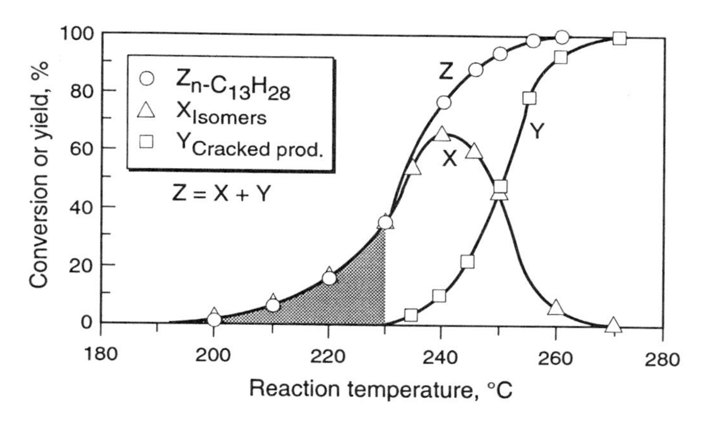

Figure 6.4 Effect of reaction temperature on isomerization and hydrocracking of *n*-tridecane over a Pt/CaY zeolite catalyst [1,31].

6.2.2. Hydroconversion of Naphthenes

The hydrocracking reactions of naphthenes have been described in numerous reports [e.g., 3,17,38–40]. As in the case of paraffins, most of the studies regarding naphthene conversion were carried out with model compounds. These studies have shown that the main reactions of naphthenes with a single five-membered or six-membered ring on bifunctional, hydrocracking catalysts are skeletal isomerization and hydrocracking, similar to those observed for *n*-paraffins. Furthermore, naphthenes have a strong tendency to disproportionate [17].

The most important difference between paraffinic and naphthenic carbenium ions is the difficulty of cracking the naphthenic ring. Only under severe hydrocracking conditions does β scission occur in naphthenic carbenium ions.

a. Reaction mechanisms. Several explanations have been provided to account for the low cracking rates of C-C bonds inside a naphthenic ring. One explanation claims that the noncyclic carbenium ion formed by β scission of a naphthenic carbenium ion has a high tendency to return to the cyclic form [38], e.g.:

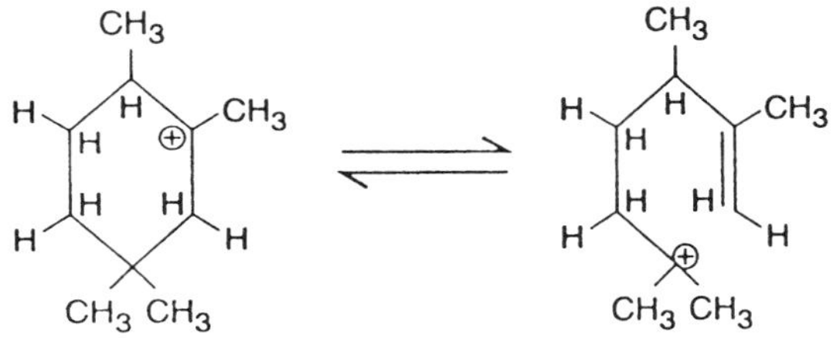

Another explanation, based on the chemical bond theory, was advanced by Brouwer and Hogeveen [41]. According to these authors, the low rate of β scission in naphthenic carbenium ions is due to an unfavorable orientation of the p orbital at the positively charged carbon atom and the β bond that is to be broken. A coplanar orientation of the vacant p orbital and of the β bond, which exists in paraffinic carbenium ions, facilitates β scission. A near-perpendicular orientation of the p orbital vs. the β bond, which exists in naphthenic carbenium ions, is energetically unfavorable to β scission.

A third explanation was provided by Brandenberger et al. [42]. From ring opening experiments with methylcyclopentane, the authors concluded that a so-called direct mechanism of ring opening via nonclassical carbonium ions is involved. This mechanism provides for a direct attack of the acidic proton at a carbon–carbon σ bond, with formation of a pentacoordinated carbon atom and two-electron–three-center bonds (Figure 6.5, I). The carbonium ion opens to form a noncyclic carbenium ion (Figure 6.5, II) that is subsequently stabilized by the mechanism described for paraffins. Data obtained by other authors support this

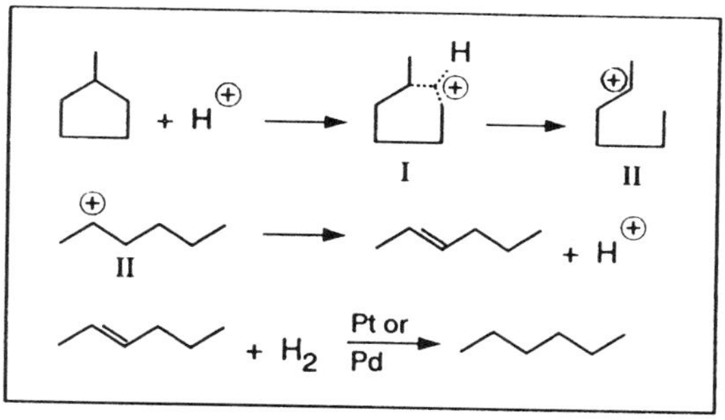

Figure 6.5 Mechanism of direct ring opening of metylcyclopentane via a nonclassical carbonium ion [17,42].

mechanism [43]. More recently, Haag and Dessau [32] showed that at high temperatures this mechanism is valid even for cracking of paraffins.

b. The paring reaction. The paring reaction was discovered in the early 1960s by a group from Chevron [3,35,38]. The authors found that alkylated cyclohexanes, with a total of 10–12 carbon atoms, hydrocrack in a highly selective manner. The alkyl groups are peeled ("pared") from the feed molecule. The principal products are isobutane and a cyclic product with four carbon atoms less than the original naphthene. The product contains very little methane and a high iso-to-normal paraffin ratio. Ring contraction also takes place. The proposed mechanism for hydrocracking tetramethylcyclohexane is shown in Figure 6.6 [35].

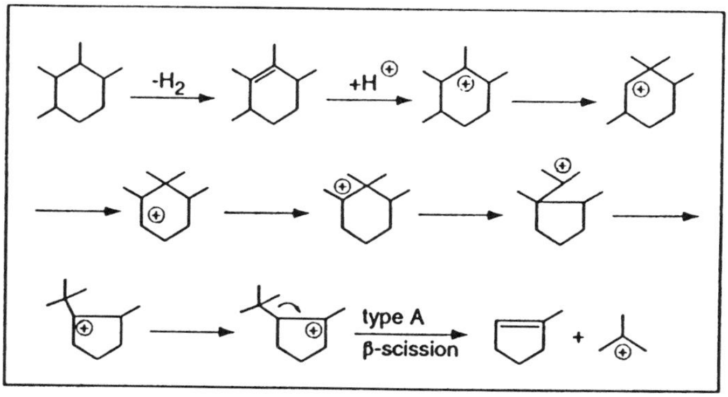

Figure 6.6 Mechanism of the paring reaction [17].

The high concentration of isobutane and ring compounds in the product, as well as the near absence of methane, can be understood if one considers two basic characteristics of naphthene hydrocracking: (a) extensive skeletal isomerization precedes β scission and (b) the cracking of C-C bonds inside the ring is slow [17]. Figure 6.7 shows that the skeletal rearrangement proceeds in several steps until a configuration favorable to type A β scission of a bond outside the ring is achieved. It leads to methylcyclopentene and a tertiary butyl cation, which are stabilized as saturated hydrocarbons via the usual bifunctional mechanism. For substituted naphthenes, the mechanism requires a minimum of 10 carbon atoms in order to make possible a type A β scission (formation of two tertiary moieties; see Figure 6.1). This explains why the cracking rate and selectivity decrease drastically (over 100 times) when a C_{10} naphthene is replaced by a C_9 naphthene. Ring stability has also been observed for larger rings, such as cyclododecane [3].

Fewer studies have been reported on hydrocracking of multiple-ring naphthenes. For example, decalin, a two-ring naphthene, is hydrocracked to light paraffins with high iso-to-normal ratios and single-ring naphthenes with high methylcyclopentane-to-cyclohexane ratios. The product distribution indicates opening of one of the two rings, with subsequent conversion as described for alkylated single-ring naphthenes.

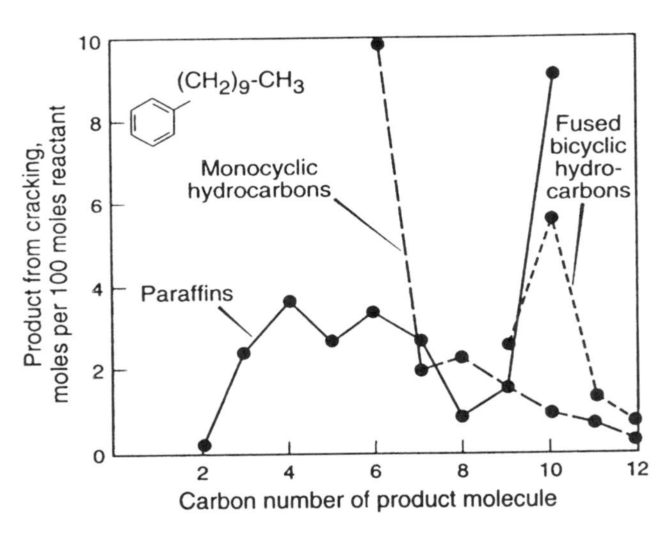

Figure 6.7 Product distribution obtained from hydrocracking of *n*-decylbenzene at 288°C and 82 atm [3].

6.2.3. Hydroconversion of Alkylaromatics

The hydrocracking of a number of alkylaromatics has been investigated, e.g., [3,44–46]. The reactions observed are isomerization, dealkylation, alkyl transfer, paring, and cyclization. This results in a wide variety of reaction products.

Hydrocracking of alkylbenzenes with side chains of three to five carbon atoms gives relatively simple products. For example, hydrocracking of *n*-butylbenzene results primarily in benzene and *n*-butane. Isomerization to isobutane, as well as alkyl group transfer to form benzene and dibutylbenzene, also takes place. The larger the alkyl side chain, the more complex the product distribution. In the latter case, cyclization may also take place. This is illustrated by the hydrocracking of *n*-decylbenzene over an NiS/silica-alumina catalyst (Figure 6.7) [3]. Simple dealkylation to benzene and decane is still the most important reaction, but many other reactions also occur, including cyclization. A considerable amount of C_9–C_{12} polycyclic hydrocarbons, such as tetralins and indanes, is found in the product.

Hydrocracking of polyalkylbenzenes with short side chains, such as hexamethylbenzene, results in light isoparaffins and C_{10} plus C_{11} methylbenzenes as the principal products (Figure 6.8) [35]. Ring cleavage is almost absent. Different reaction mechanisms have been proposed [44–47]. One mechanism, proposed by Sullivan, is similar to that proposed for the paring reaction of polymethylcyclohexane (see Figure 6.6) [35]. If weakly acidic catalysts are used, such as hydrogenation metals supported on alumina, successive removal of methyl groups from the side chain is the principal reaction (hydrogenolysis). Isomerization is minimal in this case.

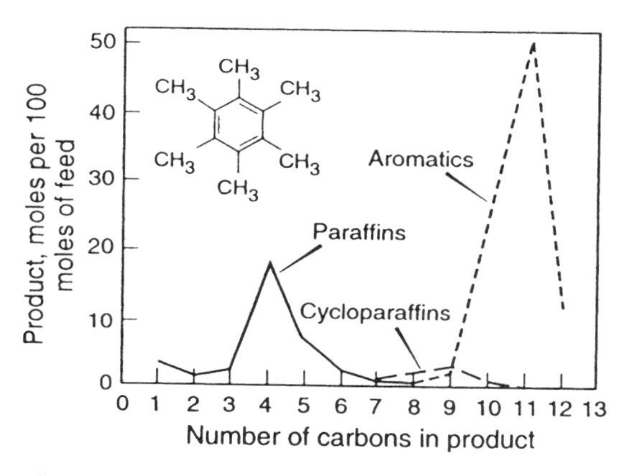

Figure 6.8 Product distribution obtained from hydrocracking of hexamethylbenzene at 349°C and 14 atm [35].

6.2.4. Hydroconversion of Polycyclic Aromatics

This subject has been discussed in a number of publications, e.g., [3,23,46,48–50]. The reaction mechanism of hydrocracking polycyclic aromatics is fairly complex and involves hydrogenation, isomerization, alkylation, and cracking. For example, hydrocracking of phenanthrene over an acidic, amorphous silica-alumina catalyst results in tetralin and methylcyclohexane as the principal products [3]. Only traces of paraffins are formed, although one would expect 1 mol of butane for each mole of tetralin formed. Langlois and Sullivan proposed a mechanism involving partial hydrogenation, ring opening, alkyl transfer, ring closure, and cracking [3]. Another mechanism, involving hydrogenation, isomerization, and cracking, was proposed by Lamberton and Guisnet [52].

Studying the hydrocracking of polycyclic aromatics over silica-alumina-based dual-function catalysts, Qader [48] observed the sequential occurrence of hydrogenation, isomerization, and cracking reactions. Hydrocracking reactions were found to follow first-order kinetics with respect to hydrocarbon concentration at constant hydrogen pressure (Figure 6.9).

When using zeolite-based hydrocracking catalysts, the reaction mechanism for hydrocracking polycyclic aromatics is, in general, similar to the one suggested for amorphous catalysts. However, the major reaction products may differ. Hydrocracking of polycyclic aromatics, such as naphthalene, phenanthrene, and pyrene, over zeolite catalysts involves hydrogenation, ring opening, and dealkylation. For example, hydrocracking of tetralin over an Ni,W/USY catalyst results in a variety of products, in which C_4, C_6, and C_{10} hydrocarbons are dominant. Product analysis indicates that the main reaction path is ring opening followed by dealkylation of *n*-butylbenzene [50].

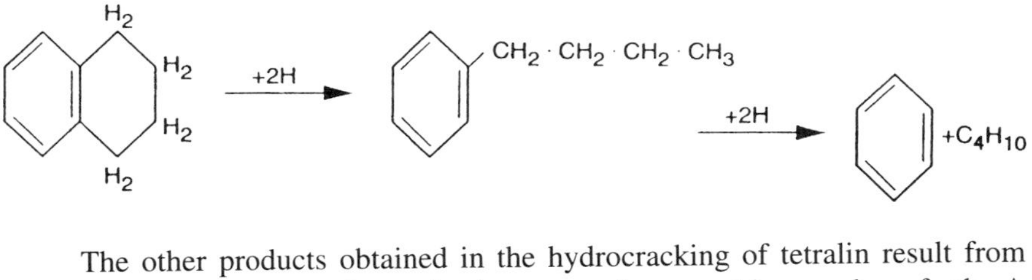

H_2 H_2 H_2 H_2 $\xrightarrow{+2H}$ $CH_2 \cdot CH_2 \cdot CH_2 \cdot CH_3$ $\xrightarrow{+2H}$ $+C_4H_{10}$

The other products obtained in the hydrocracking of tetralin result from secondary reactions, i.e., isomerization, secondary cracking, and so forth. A similar product distribution has been obtained from tetralin hydrocracking over an amorphous Ni,W/SiO_2-Al_2O_3 catalyst.

Hydrocracking of hydrogenated phenanthrene leads to products that indicate a major reaction path similar to the one observed for tetralin hydrocracking, i.e., ring opening followed by dealkylation [23,50]. Substantial amounts of butyltetralins (C_{14}), tetralin (C_{10}), benzene (C_6), and butanes (C_4) are present in the products. Smaller amounts of other products result from secondary reactions. One

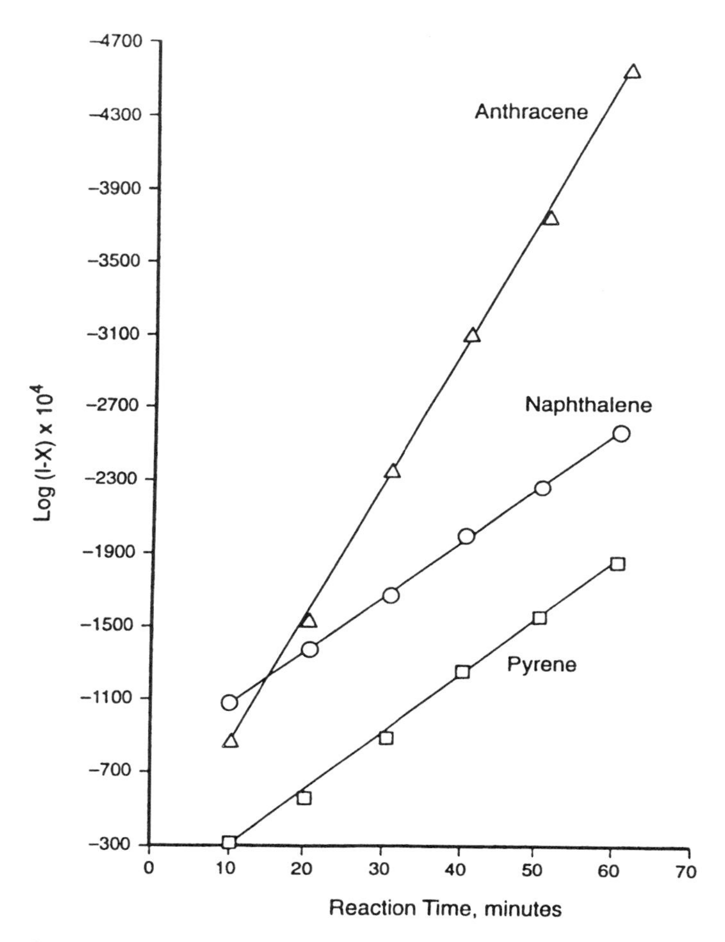

Figure 6.9 First-order reaction plot for hydrocracking polycyclic aromatics. The "x" represents mole fraction of converted hydrocarbons [48].

should note that the major product of phenanthrene hydrocracking over a zeolite catalyst is a mixture of butanes, whereas the products obtained over an amorphous catalyst contain only traces of paraffins. The difference may be due to the difference in strength of the acidic function in the two catalysts.

The effect of hydrogen pressure on aromatics conversion from a thermodynamic and kinetic point of view was investigated by Dufresne et al. [54]. The authors showed that upon hydrogenation of different aromatics at constant temperature, the naphthenic molar fraction in the product increases with hydrogen

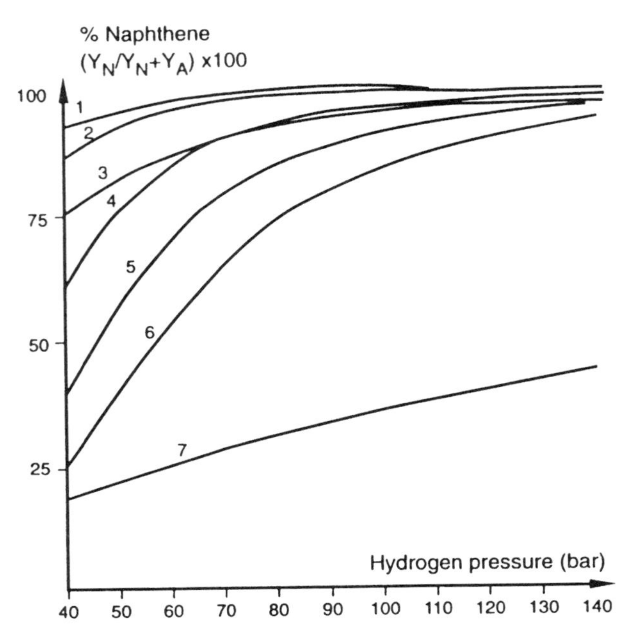

Figure 6.10 Thermodynamic equilibrium for hydrogenation of aromatics at 400°C and different hydrogen pressures [54]: 1, benzene; 2, toluene; 3, naphthalene; 4, decylbenzene; 5, *p*-xylene; 6, mesitylene; 7, phenanthrene.

pressure (Figure 6.10) [54]. Furthermore, the benzene molecules can be completely hydrogenated under normal hydrocracking conditions (e.g., 400°C and 100 bars), whereas the heavier molecules are more difficult to hydrogenate from a thermodynamic point of view.

Conversely, at constant hydrogen pressure, the naphthenic molar fraction in the product decreases with increasing temperature (Figure 6.11) [54]. Therefore, the hydrogenation reaction is favored thermodynamically by high hydrogen pressure and low temperature.

The kinetics and mechanism of catalytic hydrocracking of 1-methylnaphthalene and phenanthrene over Ni,W/USY catalyst has been investigated by Landau et al. [51]. 1-Methylnaphthalene hydrocracking led to 2-methylnaphthalene, methyltetralins, methyldecalins, pentylbenzene, and tetralin. Phenanthrene hydrocracking led to dihydro-, tetrahydro-, and octahydrophenanthrene, butylnaphthalene, tetralin, butyltetralin, and dibutylbenzene.

The hydroconversion of phenanthrene has also been investigated by Lamberton and Guisnet [52], whereas the hydrocracking of fluoranthrene was studied

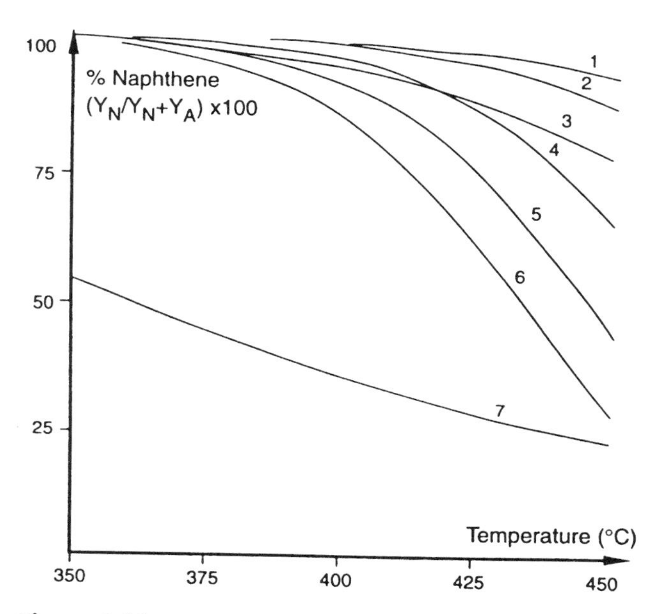

Figure 6.11 Thermodynamic equilibrium for hydrogenation of aromatics at 100 bars hydrogen pressure and different temperatures [54]: 1, benzene; 2, toluene; 3, naphthalene; 4, decylbenzene; 5, paraxylene; 6, mesitylene; 7, phenanthrene.

by Lapinas et al. [53]. The hydroconversion of methylcyclohexane over Pt/USY has been described by Mignard et al. [55].

The partial hydrogenation of polycyclic aromatics, followed by extensive splitting of the saturated rings to form high-octane, monocycle aromatics, plays an important role in enhancing the octane rating of gasoline.

6.2.5. Hydroconversion of Industrial Feedstocks

The reaction scheme for hydrocracking industrial feedstocks is complex. In addition to paraffins, naphthenes, and monoaromatics, such feedstocks contain polyaromatics, organosulfur and organonitrogen compounds, some olefins, as well as small amounts of resins and asphaltenes. Most of these species are readily adsorbed at both acid and metal sites on the catalyst, thus reducing catalyst activity. To preserve catalyst activity, these species have to undergo rapid hydrogenation, followed by hydrodecyclization, hydroisomerization, and hydrocracking. The heterocompounds undergo hydrodesulfurization (HDS) and hydrodenitrogenation (HDN), with H_2S and NH_3 resulting from these reactions. Due to the deleterious effect of heterocompounds on catalyst activity, the amount of these compounds in the feedstock has to be minimized. This is usually accomplished by hydrotreating the feedstock prior to hydrocracking.

Resins, asphaltenes, and other polynuclear aromatics that may be present in high-boiling feedstocks have a strong coking effect on the catalyst. The processing of such feedstocks requires high hydrogen pressures (90–150 bars) in order to reduce coke deposition and to hydrogenate the polyaromatics and heterocycles present (see Chapter 13).

Heavy polynaphthenes and partially hydrogenated polycyclic aromatics are converted on a large-pore amorphous catalyst or on the acidic, amorphous component of a zeolite-based catalyst to lower molecular weight products. The bulky polycyclic hydrocarbon molecules are unable to diffuse into the zeolite pores. Paraffins and alkylaromatics are converted primarily by the zeolite component of the catalyst.

The composition of the catalyst used in the process strongly affects the product distribution resulting from hydrocracking industrial feedstocks (see Chapter 7). For example, a decrease in catalyst acidity (i.e., fewer strong acid sites) results in lower gas and light gasoline yields, whereas jet fuel and diesel yields increase.

The products obtained in hydrocracking of industrial feedstocks are, in general, of superior quality. They have the following characteristics [56]:

1. Light gasoline ($C_5 + C_6$) has a high research octane number due to its high isoparaffinic character.
2. Heavy gasoline is a very good feedstock for catalytic reforming due to its high naphthene content.
3. The burning qualities of jet fuel and diesel, measured by smoke point and diesel index, respectively, are very good due to the low aromatic content of these fuels.
4. The pour point and freezing point of jet and diesel fuel are low due to their low content of linear paraffins.

These properties can be further optimized by the proper selection of feedstock, process parameters, and catalyst (see Chapter 11).

6.2.6 Kinetics

A series of kinetic studies have been carried out and kinetic models have been developed for hydrocracking reactions of individual hydrocarbons. The kinetic study of reactions occurring during hydrocracking of petroleum feedstocks is considerably more difficult because the process involves a network of complex reactions and numerous components. Several reviews of this subject have been published [57–59].

Kinetic models have been developed for hydroisomerization and hydrocracking of model compounds, such as *n*-heptane [60,61], *n*-octane [62], *n*-decane and *n*-dodecane [63,64] over zeolite or amorphous catalysts. Bernardo and Trimm [65] postulated a Langmuir–Hinshelwood type of kinetic model for hydrocracking of light hydrocarbons. Froment et al. [62,63] also assume a Langmuir–Hinshelwood mechanism for chemisorption and use a Langmuir isotherm to express the hydrocarbon concentration on the catalyst surface.

The kinetics of hydrocracking of polycyclic aromatics over amorphous catalysts was investigated by Qader [48] (Figure 6.9). He concluded that the kinetic data obtained were compatible with the dual-site mechanism of Langmuir–Hinshelwood.

Kinetic data for the hydrogenation reaction of different aromatics over different catalysts were obtained by Dufresne et al. [54]. The relative rate constants for hydrogenation reactions carried out over catalysts containing sulfided group VI metals show that polyaromatic molecules are hydrogenated rapidly if the thermodynamic conditions are satisfied. However, when catalysts with group VIII metals (Pt, Ni) are being used, monoaromatics are more easily hydrogenated.

Several studies have been published regarding the kinetics of hydrocracking of gas oil and vacuum distillate [66–69]. From these studies it was concluded that the rate of hydrocracking is of first order with respect to feed concentration. In the kinetic study made by Nasution [68] of vacuum distillate hydrocracking over a Ni,Mo/SiO_2-Al_2O_3 catalyst to produce middle distillate, the author determined the apparent activation energy for hydrocracking to be 108 MJ $kmol^{-1}$.

From the kinetic studies described, one can conclude that although the kinetics of hydrocracking of model paraffins can be described satisfactorily by using Langmuir–Hinshelwood models, such models are difficult to apply to petroleum fractions. The kinetics of hydrocracking of petroleum fractions and its practical applications is described in more detail in Section 10.6.

6.3. Comparison Between Catalytic Cracking and Hydrocracking

Both processes involve C-C bond cleavage and formation of lower molecular weight products. In both processes, the cracking of hydrocarbons occurs according to a carbenium ion mechanism. However, there are significant differences between the two processes and the quality of the resulting products.

The main advantage of hydrocracking over catalytic cracking is the production of a more hydrogenated product and a high liquid yield. Furthermore, in hydrocracking essentially no carbon from feedstock is lost, whereas in catalytic cracking several percentage points are lost as carbon on catalyst (carbon rejection).

Hydrocracking studies with model compounds (Section 6.2.1) have shown that the presence of excess hydrogen and of a hydrogenation/dehydrogenation component in the hydrocracking catalyst affects catalyst selectivity, not only by forming hydrogenated products but by suppressing secondary cracking reactions. This results in only small amounts of C_1 and C_2 hydrocarbons in the products obtained from hydrocracking paraffins, whereas significant amounts of these and other light hydrocarbons are formed during the cracking of the same paraffins (Figure 6.3).

The difference between products obtained by cracking and hydrocracking is even more pronounced when comparing the conversions of polycyclic aromatics. For example, catalytic cracking of phenanthrene over an acidic, amorphous catalyst produces only coke and small amounts of gases, whereas hydrocracking over an amorphous hydrocracking catalyst yields low molecular weight cyclic products [3]. The difference in product distribution is due to the presence of a hydrogenation/dehydrogenation component and of an excess of hydrogen in the hydrocracking reaction.

When hydrocracking industrial feedstocks, the use of higher partial pressures of hydrogen and of relatively low temperatures decreases the rate of coke formation and favors the hydrogenation of olefins and aromatic compounds. The low rate of coke formation allows the use of a fixed-bed process. The hydrogenation of unsaturated compounds is a very exothermic reaction and involves a significant consumption of hydrogen. As a result, almost no olefinic and few aromatic products are formed.

By contrast, catalytic cracking of industrial feedstocks leads to rapid coke buildup on the catalyst and requires continuous regeneration. This is accomplished in a fluid catalytic cracking unit, where the catalyst cycles continuously between reactor and regenerator. Fluid catalytic cracking (FCC) is therefore a carbon rejection process. Furthermore, the cracking reactions taking place in the reactor are endothermic and the necessary energy is provided in part by burning off the coke on spent catalyst in the regenerator. Thus, coke formation and coke combustion play a key role in the FCC process. The absence of a high partial pressure of hydrogen in the FCC unit not only makes possible the rapid buildup of coke on the catalyst but results in products containing a significant amount of olefinic and aromatic compounds. This accounts for the high-octane rating of FCC gasoline, but also for the poor quality of middle distillates obtained in the FCC process.

A key advantage of hydrocracking over cracking is the flexibility of the hydrocracking process. Such flexibility allows the treatment of heavy feeds rich in polynuclear aromatics, which are resistant to catalytic cracking. Furthermore, the flexibility of the hydrocracking process allows the optimization of yields and quality of a variety of products. This is illustrated in Table 6.4 [70]. The data show product yields and quality obtained by converting a California vacuum gas oil (VGO) by fluid catalytic cracking and by hydrocracking. The hydrocracking process is designed to operate in the middle-distillate mode. The FCC process

Table 6.4 Comparison of Conversion Processes [70]: FCC vs. Hydrocracking

Feedstock	*California vacuum gas oil*	
Boiling range, °C	238–562	
Specific gravity, g/cm^3	0.9224	
Nitrogen, wppm	2660	
Sulfur, wt %	1.18	
Aniline point, °C	72	
Hydrogen, wt %	12.11	
Aromatics, wt %	32	
Product yields and properties		
Process	FCC	Hydrocracking
Gasoline		
Boiling range, °C	C5-220	C5-180
Yield, vol %	55	30
Gas oil	LCO	Diesel
Boiling range, °C	220–360	180–370
Yield, vol %	20	80
Specific gravity, g/cm^3	0.9806	0.8324.
Sulfur, ppm	15,000	<5
Aromatics, wt %	70	20
Cetane index	21	56

generates higher gasoline yields, whereas the hydrocracking process generates higher middle-distillate yields. The data show that the quality of the gas oil fraction obtained by hydrocracking is superior to that obtained by FCC. The diesel fraction produced by hydrocracking has less aromatics, less sulfur, and a higher cetane index.

Hydrocracking is a more expensive process than FCC. It requires costly high-pressure equipment and consumes large quantities of hydrogen. The price of hydrocracking catalysts is higher than that of cracking catalysts. However, the former are kept much longer in the unit. Whereas most FCC catalysts can be used in a variety of cracking processes regardless of unit design, hydrocracking catalysts are often tied to a specific process (see Chapter 10).

References

1. J. Weitkamp and S. Ernst, in *Guidelines for Mastering the Properties of Molecular Sieves* (D. Barthomeuf et al., eds.), Plenum Press, New York, 1990, p. 343.
2. H. L. Conradt and W. E. Garwood, *Ind. Eng. Chem. Proc. Des. Dev. 3*:38 (1964).
3. G. E. Langlois and R. F. Sullivan, *Adv. Chem. Ser. 97*, A.C.S., Washington, D.C., 1970, p. 38.

4. G. F. Froment, *Catal. Today 1*:455 (1987).
5. J. Weitkamp, *Hydrocracking and Hydrotreating*, A.C.S. Symp. Series 20 (J. W. Ward and S. A. Qader, eds.), Washington, D.C., 1975, p. 1.
6. J. A. Martens, P. A. Jacobs, and J. Weitkamp, *Appl. Catal. 20*:239 (1986).
7. J. A. Martens, M. Tielen, and P. A. Jacobs, *Catal. Today 1*:435 (1987).
8. J. A. Martens, P. A. Jacobs, and J. Weitkamp, *Appl. Catal. 20*:283 (1986).
9. M. Steijns, G. F. Froment, P. Jacobs, J. Uyterhoeven, and J. Weitkamp, *Ind. Eng. Chem., Prod. Des. Dev. 20* (4), 654 (1981).
10. J. Weitkamp and H. Schulz, *J. Catal. 29*:361 (1973).
11. M. Guisnet, F. Alvarez, G. Giannetto, and G. Perot, *Catal. Today 1*:415 (1987).
12. C. Thomazeau, C. Canaff, J. L. Lemberton, M. Guisnet, and S. Mignard, *App. Catal. A 103*(1): 163 (1993).
13. G. Perot, P. Hilaireau, and M. Guisnet, in *Proc. 6th Int. Zeol. Conf.* (D. Olson and A. Bisio, eds.), Butterworth, London, 1984, p. 427.
14. A. Montes, G. Perot, and M. Guisnet, *React. Kinet. Catal. Lett. 13*(1): 77 (1980).
15. H. W. Haynes, Jr., J. P. Parcher, and N. E. Heimer, *Ind. Eng. Chem. Proc. Des. Dev. 22*:401 (1983).
16. R. F. Sullivan, M. M. Boduszynski, and J. C. Fetzer, *Energ. Fuel 3*:603 (1989).
17. J. Weitkamp, S. Ernst, and H. G. Karge, *Erdoel Kohle-Erdgas-Petrochem. 37*(10): 457 (1984).
18. S. A. Qader and G. R. Hill, *Ind. Eng. Chem. Proc. Des. Dev. 8*:98 (1969).
19. R. F. Sullivan and J. A. Meyer, *Hydrotreating and Hydrocracking*, A.C.S. Symp. Series 20, Washington, D.C., 1975, p. 28.
20. W. Schneider, E. Mueller, R. Zschocke, and E. Onderka, *Chem. Techn. (Leipzig) 33*:508 (1981).
21. G. C. Hadjiloizou, J. B. Butt, and J. S. Dranoff, *J. Catal. 135*(1): 27 (1992).
22. G. C. Hadjiloizou, J. B. Butt, and J. S. Dranoff, *J. Catal. 135*(2): 481 (1992).
23. K. H. Steinberg, K. Becker, and K. H. Nestler, *Acta Phys. Chem. 31*:441 (1985).
24. H. Beuther and O. A. Larson, *Ind. Eng. Chem. Proc. Des. Dev. 4*:177 (1965).
25. G. E. Langlois, R. F. Sullivan, and C. J. Egan, *J. Phys. Chem. 70*:3666 (1966).
26. H. L. Coonradt and W. E. Garwood, *Ind. Eng. Chem., Proc. Des. Dev. 3*:38 (1964).
27. G. A. Mills, H. Heinemann, T. H. Milliken, and A. G. Oblad, *Ind. Eng. Chem. 45*:134 (1953).
28. D. B. Weisz, *Adv. Catal. 13*:137 (1962).
29. J. P. Franck and J. F. LePage, *Proc. 7th Int. Congr. Catal.*, Tokyo, 1981, p. 792.
30. D. M. Brouwer, *Rec. Trav. Chim. 87*:1435 (1968).
31. J. Weitkamp, *Ind. Eng. Chem., Prod. Res. Dev. 21*:550 (1982).
32. W. O. Haag and R. M. Dessau, *Proc. 8th Int. Congr. Catal.*, Vol. 2, Verlag Chemie, Weinheim, 1984, p. 305.
33. R. C. Archibald, B. S. Greensfelder, G. Holzman, and D. H. Rowe, *Ind. Eng. Chem. 52*:745 (1960).
34. R. A. Flinn, O. A. Larson, and H. Beuther, *Ind. Eng. Chem. 52:153* (1960).
35. R. F. Sullivan and J. W. Scott, *Heterogeneous Catalysis: Selected American Histories*, A.C.S. Symp. Series 222 (B. H. Davis and W. P. Hettinger, eds.), Washington, D.C., 1983, p. 293.
36. G. Giannetto, F. Alvarez, F. R. Ribeiro, G. Perot, and M. Guisnet, *Guidelines for*

Mastering the Properties of Molecular Sieves (D. Barthomeuf et al., eds.), Plenum Press, New York, 1990, p. 355.
37. M. M. Otten, M. J. Clayton, and H. H. Lamb, *J. Catal. 149*(1): 211 (1994).
38. C. J. Egan, G. E. Langlois, and R. J. White, *J. Am. Chem. Soc. 84*:1204 (1962).
39. M. G. Luzzaraga and A. Voorhies, Jr., *Ind. Eng. Chem., Prod. Res. Dev. 12*:194 (1973).
40. J. M. Beelen, V. Ponec, and W. M. H. Sachtler, *J. Catal. 28*(3): 376 (1973).
41. D. M. Brouwer and H. Hogeveen, *Rec. Trav. Chim. 89*:211 (1970).
42. S. G. Brandenberger, W. L. Callender, and W. K. Meerbott, *J. Catal. 42*:282 (1976).
43. E. G. Christoffel and K. H. Röbschläger, *Ind. Eng. Chem., Prod. Res. Dev. 17*:331 (1978).
44. R. F. Childs and S. Winstein, *J. Am. Chem. Soc. 90*:7144 (1968).
45. R. F. Childs and S. Winstein, *J. Am. Chem. Soc. 90*:7146 (1968).
46. R. F. Sullivan, C. J. Egan, and G. E. Langlois, *J. Catal. 3*:183 (1964).
47. R. F. Sullivan, C. J. Egan, G. E. Langlois, and R. P. Sieg, *J. Am. Chem. Soc. 83* (5), 1156 (1961).
48. S. A. Qader, *J. Inst. Pet. (London) 59*, No. 568, 178 (1973).
49. S. A. Qader and G. R. Hill, *Ind. Eng. Chem. Proc. Des. Dev. 8*:98 (1969).
50. H. W. Haynes, Jr., J. F. Parcher, and W. E. Helmer, *Ind. Eng. Chem. Proc. Des. Dev. 22*:401 (1983).
51. R. N. Landau, S. C. Korré, M. Neurock, and M. T. Klein, in *Catalytic Hydroprocessing of Petroleum and Distillates* (M. C. Oballa and S. S. Shih, eds.), Marcel Dekker, New York, 1994, p. 421.
52. J. L. Lemberton and M. Guisnet, *Appl. Catal. 13*:181 (1984).
53. A. T. Lapinas, W. T. Klein, B. C. Gates, A. Macris, and J. E. Lyons, *Ind. Eng. Chem. Res. 26*:1026 (1987).
54. P. Dufresne, P. H. Bigeard, and A. Billon, *Catal. Today 1*(4): 367 (1987).
55. S. Mignard, Ph. Caillette, and N. Marchal, in *Catalytic Hydroprocessing of Petroleum and Distillates* (M. C. Oballa and S. S. Shih, eds.), Marcel Dekker, New York, 1994, p. 447.
56. Ch. Marcilly and J. P. Frank, in *Catalysis by Zeolites* (B. Imelik et al., eds.), Elsevier, New York, 1980, p. 93.
57. S. Mohanty, D. Kunzru, and D. N. Saraf, *Fuel 69*:1467 (1990).
58. H. Sue and H. Sugiyama, *Petrotech (Tokyo) 5*:942 (1982).
59. K. H. Steinberg, K. Becker, and K. H. Nestles, *Acta Phys. Chem. 31*:441 (1985).
60. F. Y. A. El-Kady, M. F. Menoufy, and H. A. Hassan, *Ind. J. Tech. 21*:300 (1983).
61. V. P. Sokolov and N. M. Zaidman, *Kinet. Katal. 24(4)*:898 (1983).
62. M. A. Baltanas, H. Vansina, and G. F. Froment, *Ind. Eng. Chem. Prod. Res. Dev. 22*:531 (1983).
63. M. Steijns and G. F. Froment, *Ind. Eng. Chem. Prod. Res. Dev. 20*:660 (1980).
64. Y. Y. Goldfarb, Y. R. Katsobashvili, and A. L. Rozental, *Kinet. Katal. 22*:668 (1981).
65. C. A. Bernardo and D. L. Trimm, *Rev. Port. Quim 19*:369 (1977).
66. S. A. Qader and G. R. Hill, *Ind. Eng. Chem. Proc. Des. Dev. 8(1)*:98 (1969).
67. F. Y. A. El-Kady, *Ind. J. Tech. 17*:176 (1979).
68. A. S. Nasution, *Proc. 5th Int. Semin. New Dev. Engine Oils, Ind. Oils, Fuels Addit.*, 1985.
69. B. E. Stangeland, *Ind. Eng. Chem. Proc. Des. Dev. 13*:71 (1974).
70. Unocal Technology Seminar, Kyoto, Japan, 1991.

7

CORRELATIONS BETWEEN CATALYST COMPOSITION AND CATALYST PERFORMANCE

7.1. Amorphous vs. Zeolite Catalysts

Zeolite catalysts are more active for hydrocracking than amorphous catalysts due to the higher acidity of modified Y zeolites, both in strength and in number of acid sites [1]. The higher activity of zeolite-based catalysts is illustrated in Figure 7.1 [2]. It shows that in order to obtain 50% conversion of feedstock, the amorphous catalyst requires a considerably higher reaction temperature than the zeolite-containing catalyst. The deactivation rate of the latter is also lower.

Amorphous and zeolite catalysts also differ in product selectivity. Whereas amorphous hydrocracking catalysts are generally suited to maximize middle-distillate and heating oil yields, zeolite-based catalysts have proven highly effective in maximizing gasoline yields. However, more recently, zeolite-based catalysts have been developed to maximize middle-distillate yields (see Section 7.3).

The effect of amorphous and zeolite catalysts on naphtha and middle-distillate yields is shown in Figure 7.2 [3]. Under similar process conditions, the zeolite catalyst gives better naphtha yields, especially when a high-acidity zeolite is used.

A change in SiO_2/Al_2O_3 ratio (i.e., changing the acidity) and porosity (i.e., changing the diffusion characteristics) of amorphous silica-alumina results in a change in activity and selectivity (see Section 3.2.2). By increasing the acidity and decreasing the porosity of the amorphous catalyst, catalyst activity and naphtha selectivity increase, whereas middle-distillate selectivity decreases [3].

Zeolite catalyst activity increases with zeolite content and zeolite acidity, i.e., with number and strength of acid sites. Changes in zeolite content have a more significant impact on activity and selectivity at low zeolite levels than at high levels (Figure 7.3) [3]. At constant zeolite content, catalyst activity increases with

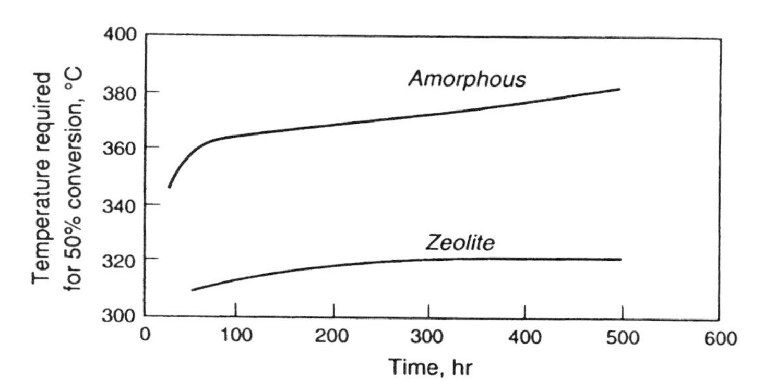

Figure 7.1 Activity of amorphous and zeolite-based catalyst in two-stage hydrocracking of heavy feedstock [2].

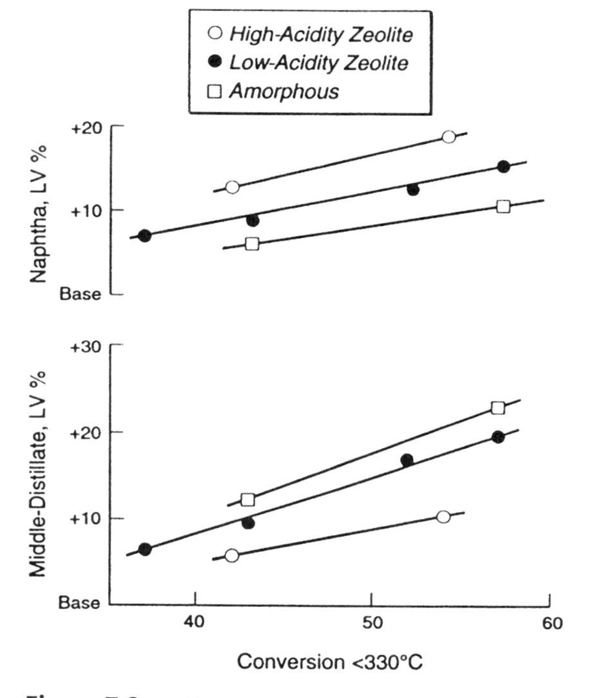

Figure 7.2 Effect of catalyst type on product yields [3]. (Reprinted with permission by Chevron Research and Technology Company, a division of Chevron U.S.A. Inc. Copyright Chevron Research Company 1993.)

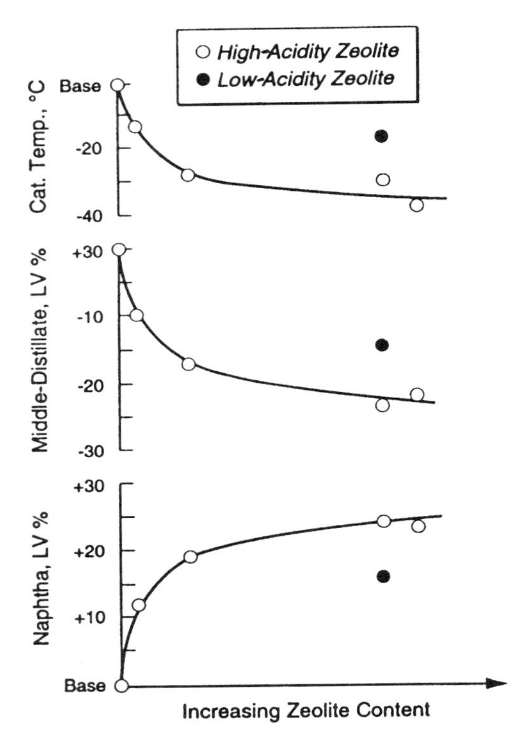

Figure 7.3 Effect of zeolite content in amorphous/zeolite catalyst on activity and selectivity [3]. (Reprinted with permission by Chevron Research and Technology Company, a division of Chevron U.S.A. Inc. Copyright Chevron Research Company 1993.)

zeolite unit cell size due to the corresponding increase in the number of zeolite acid sites. Higher activity usually translates into a shift to lighter products.

Due to their high cracking activity, NH_3 and organic nitrogen resistance, as well as their low coke make, zeolite-based catalysts are being widely used for hydrocracking in an NH_3-containing environment, as well as for conversion of heavy feedstocks. The ability of zeolite-based catalysts to tolerate NH_3 during the hydrocracking process is the result of their higher acidity and activity as compared to amorphous silica-alumina catalysts.

7.2. Gasoline Catalysts

The relative levels of hydrogenation power and acidity for different bifunctional catalysts are shown in Table 3.1. It has already been mentioned that amorphous catalysts have, in general, poor gasoline selectivity. Zeolite-based hydrocracking catalysts, containing Ca,HY, RE,HY, or mildly dealuminated Y zeolites, are strongly acidic because the corresponding zeolites have a substantial number of strong acid sites [1,4]. These catalysts are highly active and gasoline-selective. The strong acidity of the zeolite facilitates the cracking of large hydrocarbon molecules to gasoline range molecules (C_5–C_{11}). An increase in zeolite content and/or zeolite unit cell size (i.e., increase in the number of acid sites) enhances gasoline selectivity, as well as C_4^- yields. It also increases i-C_4/n-C_4, i-C_5/n-C_5, and i-C_6/n-C_6 ratios. At the same time, middle-distillate selectivity decreases [5]. In the presence of H_2S, mixed metal sulfides (Ni/Mo or Ni/W) are the preferred hydrogenation function. In two-stage configurations, where most of the H_2S can be readily removed prior to the second stage, a noble metal (Pd or Pt) or a mixed metal sulfide may be used in the second stage of the process.

A reduction of zeolite crystal size (below 0.5 μm) enhances liquid product yields and reduces light gas yields [6]. Catalysts containing certain zeolite mixtures, such as zeolite β and dealuminated Y, are claimed to have high gasoline selectivity [6,7]. Catalysts with mixtures of zeolites of the same type but of different composition (e.g., stabilized Y zeolites with different degrees of dealumination) have been reported to enhance gasoline selectivity [8]. The production of high-octane gasoline from aromatic feeds, using a catalyst containing MCM-22 zeolite, has also been reported [9].

7.3. Middle-Distillate Catalysts

The less active amorphous catalysts can be used to maximize middle-distillate yields (~C_{11}–C_{24}). Yields up to about 97 vol % have been reported [10]. Furthermore, the middle-distillate yield usually remains constant with time on-stream. However, amorphous catalysts are very sensitive to organonitrogen compounds and to NH_3 derived from them [11].

Zeolite-based catalysts used to maximize middle-distillate yields are more active, less sensitive to NH_3, and give products of better quality. However, the yield of middle distillate obtained with zeolite catalysts is somewhat lower and declines faster with time on stream [11]. The zeolite-based catalysts usually contain lower unit cell size Y zeolites (<24.45 Å) with fewer acid sites, such as dealuminated Y zeolites [12]. Steam-stabilized Y zeolites, prepared by a multiple calcination procedure, have proven especially effective [13]. The middle-distillate catalysts contain less zeolite than gasoline catalysts. Under these conditions, cracking is less severe and the predominant products are in the middle-distillate range.

The general correlation between catalyst composition, required operating temperature, and middle-distillate selectivity is shown in Figure 7.4. The effect of zeolite unit cell size on selectivity for kerosene, a middle distillate, is illustrated in

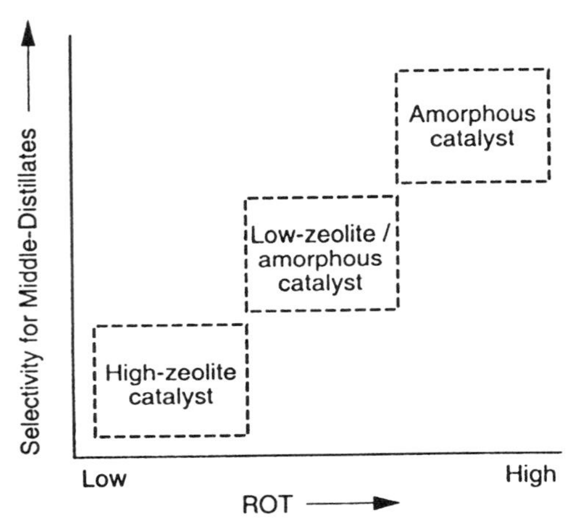

Figure 7.4 Correlation between catalyst composition, required operating temperature, and middle-distillate selectivity at constant conversion.

Figure 7.5. A reduction in zeolite unit cell size enhances kerosene selectivity [2]. Low unit cell size zeolite catalysts also have lower gas selectivity. Their hydrogenation function is provided mostly by nonnoble metal sulfides.

Acid-leached, dealuminated Y zeolites have better activity and middle-distillate selectivity than the corresponding nonleached, dealuminated zeolites (Table 7.1) [14,15]. The data show that the zeolite with a silica-to-alumina ratio of about 8.5 to 11.5 produces a catalyst that is more active and more selective to middle distillates.

The yields and quality of products obtained by hydrocracking a light Arabian vacuum gas oil (VGO) over a zeolite-based catalyst and an amorphous

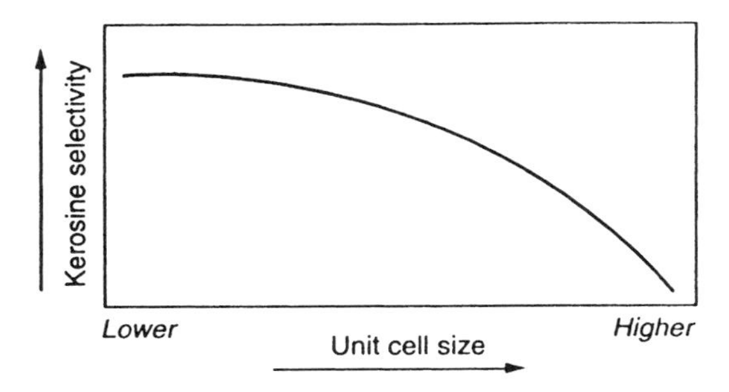

Figure 7.5 Effect of zeolite unit cell size on kerosine selectivity [2].

Table 7.1 Influence of SiO_2/Al_2O_3 Ratio on Catalyst Performance [11,12]

Catalyst	Zeolite SiO_2/Al_2O_3	Activity, temp., °C	Diesel efficiency 150–370°C, %	Jet fuel efficiency 150–290°C, %
1. LZ10	5.2	403	82.9	76.0
2. Acid-leached LZ10	8.8	400	87.0	81.2
3. Acid-leached LZ10	11.3	398	86.7	80.8
4. Acid-leached LZ10	61.3	≫416		

$$\text{Diesel fuel efficiency} = \frac{\text{vol \% (150–370°C) fraction}}{\text{vol \% conversion to 370°C}-}$$

$$\text{Jet fuel efficiency} = \frac{\text{vol \% (150–290°C) fraction}}{\text{vol \% conversion to 290°C}-}$$

catalyst are shown in Table 7.2. Both catalysts are designed to maximize middle distillates. The data show that for the same conversion the less active amorphous catalyst requires a higher reactor temperature. The hydrogen consumption is higher for the zeolite catalyst due to deeper hydrogenation of the product (249 vs. 232 Nm^3/m^3 H_2). This results in a diesel fuel of superior quality: lower aromatics, lower pour point, higher cetane number. However, the amorphous catalyst gives a higher yield of diesel fuel: 76.8 vs. 70.5 vol % of fresh feed. Zeolite-based middle-distillate catalysts have lower deactivation rates and longer cycle lengths.

Table 7.3 shows data obtained with two different zeolite-based catalysts, one designed to maximize naphtha (catalyst A) and the other designed to maximize middle distillates (catalyst B). The two catalysts differ primarily in zeolite content and unit cell size. Good catalytic activity and selectivity for middle distillates has also been reported for zeolite-based catalysts prepared from steamed extrudates [16]. The general correlation between catalyst zeolite content, unit cell size, and selectivity is shown in Figure 7.6.

Zeolite mixtures can be used in the formulation of middle-distillate catalysts. For example, combining two stabilized Y zeolites with different unit cell sizes reportedly results in superior middle-distillate hydrocracking performance, compared to a similar zeolite of average unit cell size [17,18].

Mixtures of different types of zeolite (e.g., zeolite β and Y) [19,20] also show advantages for middle-distillate hydrocracking. The use of blends of amorphous and zeolite-based hydrocracking catalysts for the production of middle distillates has been reported [21]. Good-quality jet fuel was obtained in a two-stage hydrocracking process by using a catalyst containing large-pore MCM-22 zeolite [22].

Due to the considerable flexibility of the hydrocracking process, some zeolite-based catalysts can be used to maximize either naphtha yields or middle-

Table 7.2 Comparison of Zeolite/Amorphous and Amorphous Catalyst Performance in Single-Stage Process (Middle-Distillate Mode) [37], Feedstock: Light Arabian Vacuum Gas Oil

Catalyst	Zeolite/amorphous	Amorphous
LHSV, h^{-1}	Base	Base
Temp., °C	Base	Base + 22
H2 consumption, Nm^3/m^3	249	232
Yields, vol % FF		
C_4	3.4	3.5
C_5–85° (light gasoline)	8.5	7.5
85–150° (heavy naphtha)	19.9	14.3
150–400° (middle-distillate)	80.4	84.5
180–400° (diesel)	70.5	76.8
Unconverted oil	3.0	3.0
Product Properties		
Reactor effluent		
Specific gravity, g/cm^3	0.7958	0.8063
Hydrogen, wt %	14.36	14.17
Diesel (180–400°C)		
Specific gravity, g/cm^3	0.8251	0.8275
Aromatics, wt %	12	16
Pour point, °C	−34	−33
Cetane no.	58.7	57.8

Table 7.3 Comparison of Hydrocracking Yields Obtained with Zeolite-Based Gasoline Catalyst (A) and Zeolite-Based Middle-Distillate Catalyst (B) [37], Once-Through Mode

Liquid yields, vol % feed	Catalyst A (gasoline cat.)	Catalyst B (mid.-dist. cat.)
Butane	5.9	2.9
C_5–85°C (light naphtha)	12.5	7.5
85–150°C (heavy naphtha)	20.0	18.0
150–345°C (jet fuel)	46.3	54.5
345°C+ (unconverted)	30.0	30.0

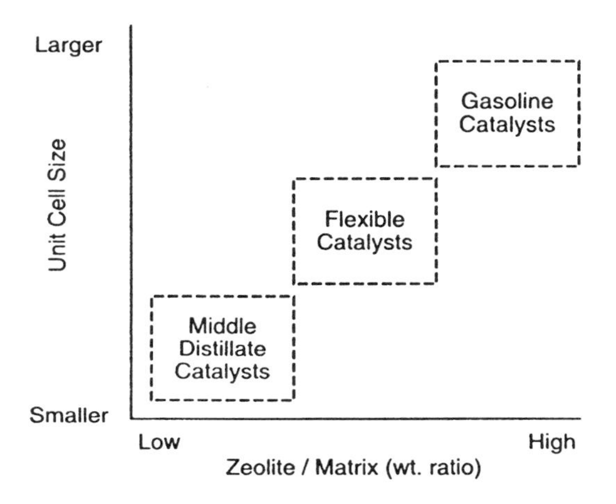

Figure 7.6 Correlation between zeolite/matrix ratio unit cell size and product selectivity.

distillate yields. This is accomplished by adjusting the operating variables, such as reactor temperature and fractionation conditions (see Chapter 11).

7.4. Relation Between Hydrogenation and Cracking Function

The ratio between hydrogenation activity and cracking activity of a hydrocracking catalyst strongly affects product yields and properties. In general, as the hydrogenation activity increases relative to catalyst acidity, total liquid product (C_5^+) yield increases due to the larger volume of hydrogenated product (Figure 7.7) [23]. At the same time, the octane number of light naphtha ($C_5 + C_6$) decreases due to more complete hydrogenation of aromatics and a decrease in the iso-to-normal ratio in product paraffins. An increase in catalyst acidity (i.e., more strong acid sites) vs. hydrogenation activity has the opposite effect. It has already been shown that an increase in catalyst acidity also increases naphtha and C_4^- yields while reducing middle-distillate yields.

Selective poisoning of active sites by nitrogen, sulfur, or polynuclear aromatics will also modify the hydrogenation/cracking ratio and change the product yields and quality. For example, the presence of nitrogen compounds in the feedstock can result in selective poisoning of catalyst acid sites, thus increasing the hydrogenation/cracking ratio of the catalyst. This results in higher liquid product (C_5^+) yield, a decrease in the light naphtha octane number, and an increase in middle-distillate yields [24].

Similarly, by exposing a noble metal catalyst to sulfur-containing feedstock

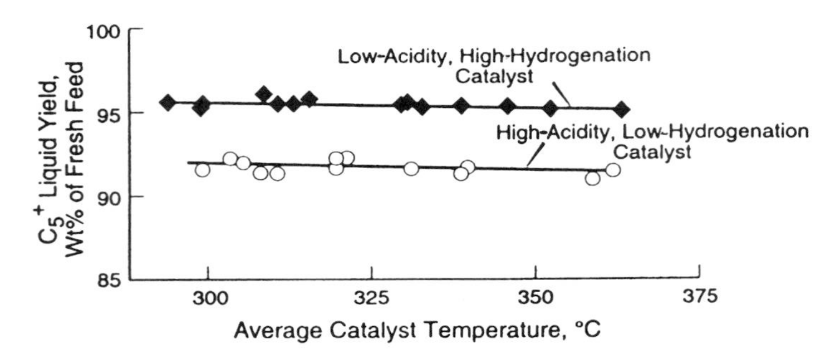

Figure 7.7 Effect of catalyst hydrogenation-to-acidity ratio and of temperature on C_5^+ liquid yield in the hydrocracking of California gas oil [23].

or H_2S, metal hydrogenation sites will be poisoned. This reduces the hydrogenation/cracking ratio, resulting in a decrease in liquid product volume, an increase in light naphtha octane number, and a decrease in middle-distillate yields [24]. The preferential adsorption of polynuclear aromatics on catalytic sites will alter the hydrogenation/cracking activity ratio by blocking some of the acid sites of the support (see also Section 8.3).

Although most catalyst poisons preferentially affect one catalyst component, they may also affect the other components [25]. The relative strengths of adsorption of various species from feedstocks on acid sites and sulfided metal sites are shown in Table 7.4 [26].

The relationship between the two catalytic functions can also be altered by the interaction between support and hydrogenation components. For example, it was shown that nickel reacts with the acid sites of silica-alumina gel forming nickel salts, thus reducing the acidity of the support. It has been suggested that by sulfiding the catalyst H_2S reacts with the nickel salts and regenerates the original

Table 7.4 Relative Adsorption Strength of Different Oil Compounds on Hydrocracking Catalysts [23]

(a) *On acidic sites*
Asphaltenes > resins > N compounds > NH_3 > olefins ⩾ thiophenic compounds > aromatics > H_2O > H_2S ≫ paraffins

(b) *On sulfided active metals*
Asphaltenes > resins > H_2S > S compounds > olefins ⩾ N compounds > aromatics ≫ paraffins

acid sites [25]. A similar reaction is likely to occur during sulfiding of catalysts containing nickel-exchanged zeolites.

The interaction between noble metals and zeolite support has also been investigated. Such an interaction results in electronic deficiency of the reduced noble metal dispersed in the catalyst. Noble metal–zeolite systems have been investigated extensively (see Sections 5.2 and 8.8).

Several methods have been suggested for the quantitative measurement of hydrogenation activity and acidity. For example, noble metal surface area has been related to activity. Adsorption of bases such as pyridine and ammonia has been correlated with acidity. Some key reactions involving pure compounds have also been used to describe catalytic properties. An empirical hydrogenation activity index, based on the aromatic/naphthene ratio in the hydrocracked product, has been used to measure activities under actual hydrocracking conditions.

In summary, to maximize the yield of gasoline and lighter products, strong cracking activity and moderate hydrogenation activity is required. This is obtained by using strong acidic zeolites in combination, for example, with Ni/Mo sulfides. To produce high middle-distillate yields, weaker acidity (i.e., fewer strong acid sites) and strong hydrogenation activity is required to minimize secondary cracking and maximize aromatic saturation. This can be achieved by combining alumina, amorphous silica-alumina, and/or a weak acidic zeolite with a strong hydrogenation component such as Ni/W, Pd, or Pt.

7.5. Relation Between Zeolite Framework Composition and Catalyst Performance

Hydrocracking tests with catalysts containing palladium-loaded, steam-dealuminated Y zeolites having different aluminum content have shown that activity, selectivity, and stability are affected by framework SiO_2/Al_2O_3 ratio (i.e., by the framework aluminum content) of the zeolite [27]. An increase in zeolite framework aluminum content increases hydrocracking activity, but also increases light gas and coke yield, while decreasing heavy liquid (jet and diesel fuel) yield. The stability of the catalyst also decreases with higher aluminum content. The higher activity is due to the increase in zeolite acidity (more acid sites) with the increase in zeolite framework aluminum content. The higher activity translates into more intensive cracking, leading to more conversion of the heavy liquid fraction to lighter liquids and gas as well as to enhanced coke make. Higher coke yields result in faster catalyst deactivation and a shorter cycle length in the hydrocracking unit. According to Bezman, the optimum balance between activity and coke make occurs at ~12 Al/unit cell [27].

Because an increase in zeolite framework SiO_2/Al_2O_3 ratio results in a decrease in zeolite unit cell size, the selectivity of a hydrocracking catalyst can be modified by changing the unit cell size of the zeolite. (Figure 7.8). An increase in

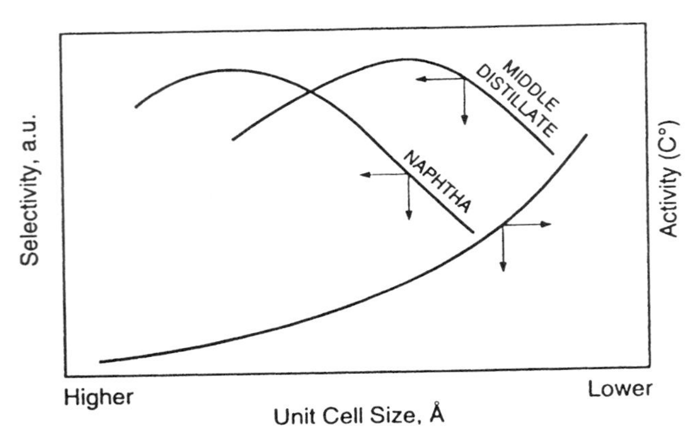

Figure 7.8 Effect of unit cell size of Y zeolite on activity and selectivity. *Note*: activity is expressed as required temperature to obtain a given conversion.

activity and zeolite unit cell size enhances catalyst selectivity for naphtha and lighter products. A decrease in unit cell size reduces catalytic activity but enhances the yield of middle distillates.

Bezman has also observed some inconsistencies in the correlation between hydrocracking activity/selectivity and unit cell size, especially for low-range unit cell sizes [27]. Such inconsistencies were attributed to erroneous data obtained for the number of framework Al atoms from unit cell size measurements. More accurate data were obtained by ^{29}Si-MASNMR or ion exchange measurements.

By increasing the bulk SiO_2/Al_2O_3 mole ratio of the steam-dealuminated zeolite from 6.4 to 11.1 (presumably by leaching some or all of the nonframework aluminum) while maintaining the unit cell size unchanged ($a_0 = 24.35$ Å), Bezman found that the catalytic properties of the zeolite change. That is, in a once-through hydrocracking test, activity and coke yield increase [27]. The higher activity is likely due to better access to the zeolite acid sites following the removal of nonframework aluminum. The higher acidity is also responsible for the higher coke yield. In a recycle test, the high SiO_2/Al_2O_3 ratio zeolite showed not only better activity but superior liquid selectivity and stability. Dealuminated Y zeolites with nonframework aluminum removed have been used in the preparation of hydrocracking catalysts designed to maximize naphtha or middle-distillate yields [14,15].

An increase in palladium content from 0.25 to 0.50 wt % increases the activity and stability of the catalyst but shows little effect on selectivity. The increase in activity and stability are due to higher hydrogenation activity with higher palladium loading.

7.6. Shape Selectivity Effect

Although zeolite catalysts offer major advantages in terms of activity, NH_3 resistance, and coke deactivation, their shape-selective properties, even with large-pore Y zeolites, may result in product selectivity disadvantages, particularly under recycle to extinction conditions.

Haynes et al. [28] showed for a series of polynaphthenic molecules that the relative rate of hydrocracking over an Ni,W/USY zeolite catalyst decreases with increasing number of naphthenic rings. This is due to the decrease in effective diffusivity with increasing molecular kinetic diameter.

The shape selectivity effect is noticed primarily in the hydrocracking of heavy feedstocks, when micropore diffusion limitations for bulky molecules are more pronounced. This is illustrated by comparing the rate constants for hydrocracking over a zeolite-based catalyst (Ni,W/USY) and over an amorphous catalyst (Ni,W/SiO_2·Al_2O_3) as a function of boiling point of the feedstock (Figure 7.9) [29]. The rate constants for cracking increase monotonically for the amorphous

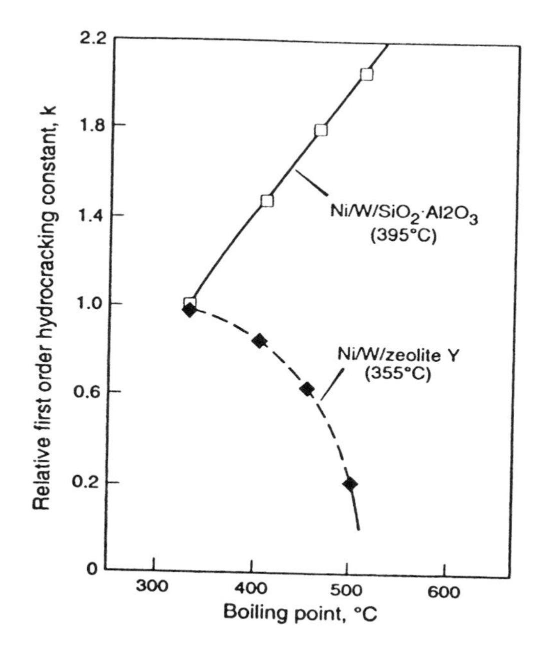

Figure 7.9 Correlation between first-order rate constant and boiling point of fraction for hydrocracking over Ni,W/zeolite and Ni,W/SiO_2-Al_2O_3 catalysts [29].

catalyst with increasing boiling point (increasing molecular weight), whereas the opposite trend is observed for the zeolite catalyst. The decrease in rate constants for the zeolite catalyst is due to an increase in diffusion limitations with rising average molecular weight of the feedstock [29].

The shape selectivity of the zeolite catalyst leads to a gradual buildup of multiring molecules (mostly polynuclear aromatics, PNAs) in the recycle stream. Such an increase in hard-to-crack PNAs requires a gradual increase in reactor temperature in order to maintain constant conversion. The increase in reactor temperature leads to an increase in dry gas and a decrease in heavy naphtha yields. An excessive buildup of PNAs can be avoided by removing ("bleeding") some of the recycle stream from the operation. Other methods can also be used (see Section 11.4).

The shape-selective properties of ZSM-5 zeolite play a key role in its use as a dewaxing catalyst. In catalytic dewaxing, only normal and lightly branched paraffins diffuse into the zeolite pores and are cracked. Isoparaffins remain practically unaffected. This results in products with improved properties (see Chapter 14).

7.7. Effect of Nonzeolitic Component in Zeolite-Containing Supports

In catalysts with maximum zeolite content, the nonzeolitic component is usually a binder (e.g., alumina) that plays only a minor catalytic role. Catalysts with less than maximum zeolite content contain also a nonzeolitic, active component (amorphous or crystalline). The amorphous component can be alumina, silica-alumina, silica-magnesia, silica-titania, etc. (see Section 3.2). Crystalline, nonzeolitic components reported are acid-treated clays, pillared clays, sepiolite, acid aluminum phosphates, and other solid acids.

The nonzeolitic component of the support can have a significant influence on catalyst activity and selectivity, especially when the zeolite content is low. Most of the nonzeolitic components are solid acids with wide variations in acidity. For example, by substituting part of the weakly acidic γ-alumina with strongly acidic, amorphous silica-alumina while maintaining the zeolite type and content unchanged, the catalyst activity and selectivity for middle distillate increases significantly (Table 7.5) [15,30]. Catalyst stability is also improved. Substitution of amorphous silica-magnesia for alumina in catalyst support has a similar effect, although the selectivity for middle distillate is lower [31]. The use of titania as the nonzeolitic component of a zeolite-based catalyst is claimed to improve the deactivation rate of the catalyst [32]. The presence of zinc titanate is also claimed to be effective [33].

It has been reported that zeolite-based catalysts containing pillared clays [34,35], sepiolite (a layered magnesium silicate) [36], or mixtures of pillared clays

Table 7.5 Influence of Amorphous Component on Catalyst Performance [27]

Catalyst support	Activity: reactor temp. for 60% conv. (°C)	Selectivity: conv. to 150–370°C product (vol.%)
LZ10 and γ alumina	396	74.8
LZ10 and SiO_2/Al_2O_3 in γ alumina matrix (25% SiO_2 overall)	395	79.4
LZ10 and SiO_2/Al_2O_3 in γ alumina matrix (40% SiO_2 overall)	390	79.5

and sepiolite show better activity and selectivity than silica-alumina-containing catalysts (Table 7.6) [15,36]. Catalysts containing crosslinked smectite [37], beidelite (a layered clay), or beidelite pillared with Al_2O_3, TiO_2, or ZrO_2 [38], as well as catalysts containing layered silicates [39], have also been described. Catalysts in which the zeolite component has been replaced by rare earth–pillared clays have been reported to be effective in the production of middle distillates [40].

The pore size distribution and pore volume of the nonzeolitic component of the support are of critical importance. The heavy molecules of the feedstock and their conversion products should be able to diffuse through the catalyst pores to and from the zeolite crystals. For that reason, catalysts used for hydrocracking

Table 7.6 Effect of Nonzeolitic Support Component [31]

		Selectivity (vol %)	
Catalyst	Temp. for 60% conversion (°C)	150–288°C (jet fuel)	150–370°C (mid-dist.)
70% Pillared clay 20% Sepiolite 10% LZ10	370	72.2	83.9
90% Pillared clay 10% LZ10	369	69.6	81.4
90% Sepiolite 10% LZ10	383	68.0	79.7
70% Silica-alumina in alumina dispersion 20% alumina binder 10% LZ10	381	70.0	81.5

heavy feedstocks have a specified percentage of large pores (several hundred angstroms in diameter) in order to allow the diffusion of bulky feedstock molecules into the catalyst [41,42].

References

1. K. Tanabe, *Solid Acids and Bases*, Academic Press, New York, 1970, pp. 29, 75.
2. A. Hoek, T. Huizinga, A. A. Esener, I. E. Maxwell, W. Stork, F. J. van de Meerakker, and O. Sy, *Oil Gas J.*, Apr. 22, 1991, p. 77.
3. R. L. Howell, R. F. Sullivan, C-W. Hung, and D. S. Laity, paper presented at the Petroleum Refining Conference, Tokyo, Japan, Oct. 1988.
4. R. Beaumont and D. Barthomeuf, *J. Catal. 30*:288 (1973).
5. P. J. Nat, J. W. F. M. Schoonhoven, and F. L. Plantenga, in *Catalysis in Petroleum Refining, 1989*, Vol. 53 (D. L. Trimm et al., eds.), Elsevier, Amsterdam, p. 399.
6. J. W. Ward, U.S. Patent No. 5,275,720 (1994).
7. D. F. Best, E. M. Flanigen, G. W. Skeels, and J. G. Vassilakis, U.S. Patent No. 5,208,197 (1993).
8. A. L. Hensley, Jr., P. D. Hopkins, S. G. Kukes, and C. L. Marshall, U.S. Patent No. 4,925,546 (1990).
9. G. W. Kirker and S. S. Shih, U.S. Patent No. 4,968,402 (1990).
10. M. E. Reno, B. L. Schaefer, R. T. Penning, and B. M. Wood, NPRA Ann. Mtg., March 1987, San Antonio, TX, AM-87-60.
11. J. W. Ward, in *Applied Industrial Catalysis*, Vol. 3 (B. E. Leach, ed.), Academic Press, 1984, p. 351.
12. K. L. Steigleder, U.S. Patent No. 4,894,142 (1990).
13. R. D. Bezman and J. A. Rabo, U.S. Patent No. 4,401,556 (1983).
14. F. P. Gortsema, R. J. Pellet, A. R. Springer, and J. A. Rabo, U.S. Patent No. 5,047,139 (1991).
15. J. W. Ward, *Fuel Processing Technol. 35*:55 (1993).
16. J. W. Ward, U.S. Patent No. 5,288,396 (1994).
17. K. Z. Steigleder, U.S. Patent No. 4,661,239 (1987).
18. A. L. Hensley, Jr., P. D. Hopkins, and S. G. Kukes, U.S. Patent No. 4,980,328 (1990).
19. S. M. Oleck and R. C. Wilson, U.S. Patent No. 4,568,655 (1986).
20. J. W. Ward, U.S. Patent No. 5,279,726 (1994).
21. J. W. Gosselink, L. R. Groeneveld, and H. Schaper, U.S. Patent No. 5,112,472 (1992).
22. G. W. Kirker, S. Mizrahi, and S. S. Shih, U.S. Patent No. 5,000,839 (1991).
23. R. F. Sullivan and J. W. Scott, in A.C.S. Symposium Series 222, (B. H. Davis and W. P. Hettinger, eds.), Washington, D.C., 1983, p. 293.
24. R. F. Sullivan and J. A. Meyer, *Hydrocracking and Hydrotreating*, A.C.S. Symposium Series 20, (J. W. Ward and S. A. Qader, eds.), Washington, D.C., 1975, p. 28.
25. G. E. Langlois, R. F. Sullivan, and C. J. Egan, *J. Phys. Chem 70*:3666 (1966).
26. J. P. Franck and J. F. LePage, Proc. 7th Int. Congr. Catal., Tokyo, 1981, p. 792.
27. R. Bezman, *Catal. Today 13*:143 (1992).
28. H. W. Haynes, J. F. Parcher, and N. E. Helmer, *Ind. Eng. Chem. Proc. Des. Dev. 22*:401 (1983).

29. I. E. Maxwell, *Catal. Today 1*:385 (1987).
30. J. W. Ward, U.S. Patent No. 4,419,271 (1983).
31. J. W. Ward, U.S. Patent No. 3,838,040 (1974).
32. J. W. Myers and S. L. Parrott, U.S. Patent No. 4,447,555 (1984).
33. L. E. Gardner, U.S. Patent No. 4,394,301 (1983).
34. F. G. Gortsema, J. R. McCauley, R. J. Pellet, J. A. Miller, and J. A. Rabo, Eur. Pat. Publ. No. W.038106614 (1988).
35. J. R. McCauley, U.S. Patent No. 5,202,295 (1993).
36. M. Occelli, U.S. Patent No. 5,076,907 (1991).
37. J. Fijal and J. Shabtai, U.S. Patent No. 4,579,832 (1986).
38. J. S. Holmgren, U.S. Patent No. 5,286,368 (1994).
39. R. F. Socha and M. R. Stapleton, U.S. Patent No. 5,015,360 (1991).
40. F. P. Gortsema, J. R. McCauley, J. G. Miller, P. J. Pellet, and J. A. Rabo, U.S. Patent No. 4,995,964 (1991).
41. A. L. Hensley, G. L. Ott, and L. B. Peck, U.S. Patent No. 4,746,419 (1988).
42. T. Makabe, M. Matsuda, K. Shimokawa, and O. Togari, U.S. Patent No. 4,440,631 (1984).

8

CATALYST DEACTIVATION AND REACTIVATION

8.1. Basic Concepts

Catalyst deactivation is a physical or chemical process that decreases the activity of a given catalyst. A quantitative measurement of deactivation is made by comparing certain data obtained for the deactivated catalyst and the corresponding active (or fresh) catalyst. Such data can be obtained by measuring one of the following [1]:

1. Temperature required for a given conversion
2. Conversion obtained by a set temperature and space velocity
3. Space velocity required for a set conversion at fixed temperatures
4. Reaction rates under differential conversion conditions (e.g., determination of turnover frequencies)
5. Rate constants determined from kinetic studies.

In some instances, the change in catalyst selectivity may be more meaningful than the change in overall activity.

Because various measures of catalyst activity are not equivalent, it is important to maintain consistency with regard to the method used. A common method used in hydrocracking to measure catalyst activity is to determine the temperature required for a given conversion. An increase in required temperature indicates a decrease in activity. In commercial operations, the change in required temperature (Δ) is usually expressed as an average of °C or °F per day for a specific time of operation (100 days, 200 days, etc.).

There are three major categories of catalyst deactivation mechanisms: poisoning, coking (or fouling), and sintering. Other deactivation mechanisms encountered in catalysis are pore blocking via coke or metal deposition in the catalyst pores (or at the pore mouths); volatilization of the catalytically active component; destruction of active sites; and reaction between active component and support with formation of an inactive form. The listed mechanisms may occur singly or in combination. A brief description of these mechanisms follows.

Catalyst poisoning is the result of strong chemisorption of impurities on the catalyst surface, which blocks access of reactants to active sites. For example, the strong chemisorption of basic organic molecules, such as pyridine, at the acid sites of a catalyst blocks the access of other molecules to these sites. The amount of poison required to deactivate a catalyst is usually very small, i.e., of the order of micromoles per gram of catalyst.

Catalyst poisoning can be reversible or irreversible, depending on the degree of affinity between impurity and active site. In fact, various degrees of reversibility can be associated with poisoning. Catalyst poisoning can also be selective or nonselective [1]. In the case of nonselective poisoning, the chemisorption of poison on the catalyst removes active sites in a uniform manner, and there is a linear relationship between catalyst activity and the amount of poison chemisorbed. In selective poisoning, only active sites with certain properties (e.g., strongly acidic sites) are neutralized by the poison, resulting in a nonuniform deactivation of the catalyst. In that case, there is an exponential or hyperbolic relationship between catalyst activity and the amount of poison chemisorbed [1,2].

Coking or fouling of a catalyst is the formation process of hydrogen-deficient carbonaceous residues on the surface of catalysts, usually in reactions involving hydrocarbon processing. Coke formed on the catalyst physically blocks access of reactants to active sites by covering them, leading to catalyst deactivation. In contrast to deactivation by poisoning that is due to strong chemisorption at the active site, deactivation by coke is due to physical blocking of active sites.

Coke is chemically less well defined than catalyst poisons and the mechanism of coke formation is less specific than that of poisoning. While micromoles of impurities can deactivate a catalyst by poisoning, coke can be formed in macroscopic amounts. Some coked catalysts can contain up to 20 wt % coke because coke can be deposited in multiple layers on the catalyst surface. In addition to direct blocking of active sites, the presence of large amounts of coke can lead to pore blocking and can prevent the reactant molecules from reaching the active sites inside the pores.

Sintering describes the loss of active sites due to structure alteration of the catalyst. The process can occur in supported metal catalysts or in unsupported catalytic materials such as zeolites or amorphous silica-alumina. In the former case, sintering is the loss of active sites due to agglomeration of smaller metal aggregates or crystallites to larger ones. Sintering in unsupported catalytic materials is a process that involves the partial or total collapse of the internal pore structure and a corresponding loss of surface area. Sintering is commonly a physical, thermally activated process and leads to materials with a lower surface-to-volume ratio. However, some chemical changes can also occur during the sintering process (e.g., formation of new phases in a collapsed zeolite).

Catalyst deactivation can also occur due to *pore blocking* by coke or by transition metals from the feed (mostly Ni and V) deposited at the pore mouths of the catalyst. The former type of pore blocking occurs, for example, during petroleum cracking processes, whereas the latter type has been observed in some hydrotreatment processes. Such deposits block access of reactant molecules to the active sites in the interior of the catalyst. In commercial processes, many of the deactivation mechanisms take place in parallel. For example, hydrocracking catalysts are commonly deactivated by coke deposition, pore blocking, and metal sintering. In addition, with organic sulfur- or nitrogen-containing feedstocks, catalyst poisoning also occurs.

When operating at high temperatures, in addition to the deactivation mechanisms described, other deactivation mechanisms may take place. *Reaction between metal and support* may occur. For example, at ~1000°C, nickel supported on alumina forms a catalytically inactive nickel aluminate [1]. Some metal-supported catalysts may lose the active component due to formation of *volatile compounds*, such as volatile oxides, during regeneration, e.g., formation of volatile RuO_3-RuO_4 oxides in supported Ru catalysts, loss of Mo from hydroprocessing catalysts at very high oxidative regeneration temperatures (see below).

Catalyst reactivation is a process used to restore a deactivated catalyst to its original, or close to its original, activity. Some authors also call this process *regeneration*, whereas others limit the term regeneration to removal of coke from a catalyst. In this text, the term regeneration will signify only removal of coke from the catalyst.

Thermal treatment is frequently used to reactivate a deactivated catalyst. If deactivation is due primarily to coking, combustion in dilute oxygen under controlled conditions results in oxidation of the carbonaceous deposits to volatile CO_2 and H_2O, with restoration of catalytic activity (regeneration). Organic S and N that may be present in coke are removed during thermal treatment in dilute oxygen as SO_x and NO_x. Thermal treatment may also lead to the removal of certain catalyst poisons that become volatile at treatment conditions (e.g., removal of basic organic molecules chemisorbed on acid sites). Such a process is sometimes called *detoxification*.

Redispersion of metal in supported metal catalysts is called *rejuvenation*. Such redispersion is usually accomplished by oxidation followed by reduction under carefully controlled conditions (see below).

8.2. Coke Formation and Composition

Carbonaceous deposits on catalysts are gradually formed in almost all hydrocarbon conversion reactions. The buildup of coke may become significant in the first few seconds of reaction, as in catalytic cracking processes, or only over a period of months or years, as in hydrocracking and reforming processes. Such carbonaceous deposits ("coke") block the active sites either directly or by pore plugging, causing gradual catalyst deactivation.

The mechanism of coke formation is complex. Aromatics alkylation and olefin oligomerization, followed by dehydrocyclization, aromatization through hydrogen transfer, and condensation lead eventually to coke formation. Appleby [3] suggested a carbenium ion mechanism to explain the formation of higher aromatics (coke precursors) from benzene and naphthalene. Carbenium ions can undergo addition, dehydrogenation, and hydrogen transfer to grow into large polyaromatic molecules, i.e., coke-type products. A similar mechanism was described more recently by Gates et al. [4]. The formation of polyaromatics from benzene through a chain of addition reactions via a carbenium ion mechanism is shown in Figure 8.1. Because of the high stability of the polynuclear aromatic carbenium ion, it can continue to grow on the catalyst surface for a relatively long time before a termination reaction occurs.

The empirical formula of coke varies from $CH_{1.0}$ to $CH_{0.5}$. Fresh coke usually has a higher hydrogen content than aged coke. The nature of coke deposits is complex and varies with conditions under which they are formed. Appleby et al. [3] made a detailed characterization of the coke formed in catalytic cracking. When formed at relatively low temperatures (<400°C), coke deposits consist of ill-defined, large aggregates of polyaromatic molecules with condensed ring structure. Although hydrogen-deficient polyaromatic hydrocarbons are predominant, other species with a higher hydrogen content are also present. When formed at high temperature (e.g., 400–500°C) or after long time on-stream, the carbonaceous deposits tend to be in pseudographitic form, as shown by X-ray diffraction studies. Carbon filaments may also form in the process, especially at catalytic metal sites. Some coke components, especially in fresh, low-temperature coke, are soluble in organic solvents, such as methylene chloride, and can be readily characterized [5]. Insoluble coke components (mostly pseudographitic coke) are usually characterized by physical methods, such as electron microscopy, nuclear magnetic resonance, Raman and infrared spectroscopy, X-ray diffraction, etc. [6–8]. These techniques have proven the heterogeneous nature of coke. The structure and composition of carbonaceous deposits on catalysts as revealed by modern spectroscopic methods was described by Bell [9].

Feedstock type, catalyst composition, reaction temperature, time on-stream, and other process variables affect the yield and composition of coke. For example, aromatic feedstocks form more coke than aliphatic ones under similar processing conditions because polyaromatic structures are more easily formed from an aromatic feedstock [10]. Coking is more pronounced in condensed ring systems (e.g., naphthalene, anthracene) than in the corresponding linked systems (biphenyl, terphenyl). Heterocycles produce less coke than the hydrocarbon analogs. For

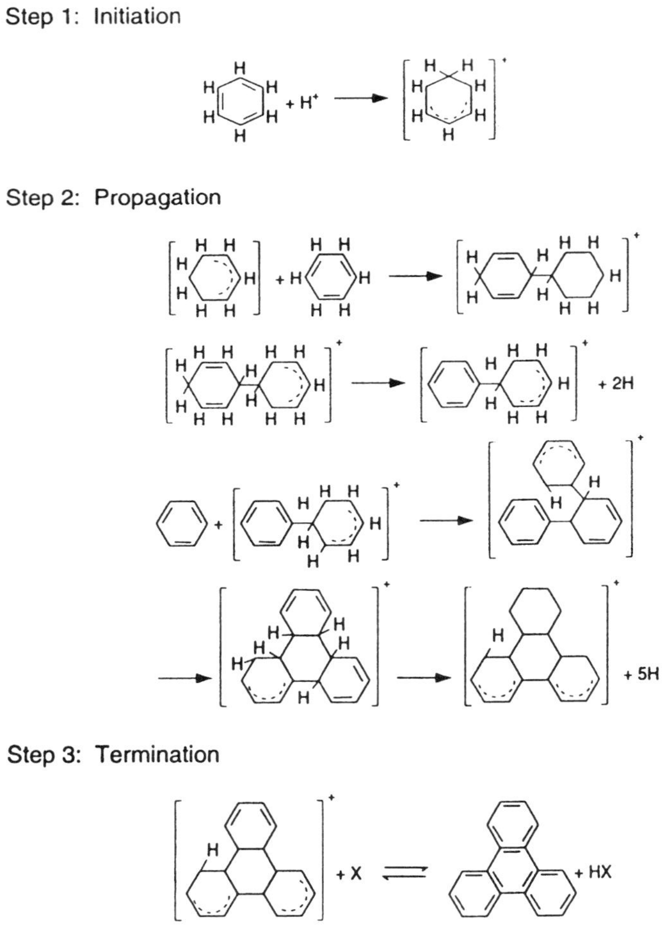

Step 1: Initiation

Step 2: Propagation

Step 3: Termination

Figure 8.1 Formation of polyaromatics from benzene via a carbenium ion mechanism [4]. (Reprinted with permission of McGraw-Hill, Inc., from *Chemistry of Catalytic Processes*, by B. C. Gates, J. R. Katzer, and G. C. A. Schuit.)

alkyl-substituted aromatics, coke deposition increases with the length of the side chains rather than with the number of alkyl substitutions. Olefins, acting as hydrogen acceptors, promote coke formation by removing hydrogen from coke precursors. The coke yield increases with time on-stream [11]. The hydrogen content of coke obtained from aliphatic compounds decreases with increasing reaction temperature and with increasing time on-stream [5]. The polyaromaticity of coke also increases under these conditions.

In the case of zeolite catalysts, zeolite acidity and pore structure affect coke composition and distribution, as well as coking rate [5,12,13]. For example, highly acidic Y zeolites with a three-dimensional, large-pore network and large cavities have a strong tendency to form coke. However, in ZSM-5 zeolites, with very little acidity and narrow pores, the occurrence of reactions leading to coke formation is restricted. The low acidity suppresses carbenium ion formation and hydrogen transfer reactions, whereas the narrow pores impose steric constraints on the transition states that occur during coke formation. Therefore, the coking tendency of ZSM-5 is very low [14]. For ZSM-5, coke formation is a shape-selective reaction (see Section 3.1.3).

Coke formation in Y zeolites involves hydrogen transfer reactions. These are multimolecular reactions and require the proximity of acid sites in order to take place. Suppression of such reactions (e.g., by decreasing acid site density through increased framework Si/Al ratio) reduces coke formation in the zeolite. This explains the lower coke make by dealuminated Y zeolites (see Section 3.1.2). Coke deposition in catalyst pores will reduce the effective diffusivity of reactant and product molecules to and from active sites inside the pores. At high levels of coke deposition, pore blocking may occur [15,16]. The coke deposition on Y zeolite is illustrated in Figure 8.2 [14]. The effect of pore structure on coking and deactivation of zeolites has been reviewed by Guisnet and Magnoux [5].

Walsh and Rollman [17] observed that high hydrogen pressure retards coke deposition. This is due to suppression of dehydrogenation reactions involved in coke formation. The authors also observed that in the conversion of paraffin-aromatics mixtures over Y zeolites, aromatics are the major contributor to coke formation. That contribution increases with increasing aluminum content of the zeolite.

8.3. Coke Deactivation of Bifunctional Catalysts

This topic has been the subject of numerous investigations. Many of these studies were carried out on reforming catalysts having a hydrogenation–dehydrogenation function (Pt or Pt-Re) and an acid function (Al_2O_3 or chlorinated Al_2O_3). Coke formation and deactivation of various Pt-zeolite catalysts [18] as well as of commercial hydroprocessing catalysts [19–21] has also been investigated.

The coking reactions in such systems are very complex and their rate can be defined only by the total amount of coke deposited. Both metal and acid support contribute to coke formation. The metal in a bifunctional catalyst has several roles in the coking process: (a) it acts as a dehydrogenation catalyst, yielding olefins that can polymerize on the acid support; (b) it can destroy coke precursors at high temperature by hydrogeneolysis; and (c) it can promote coke accumulation on the

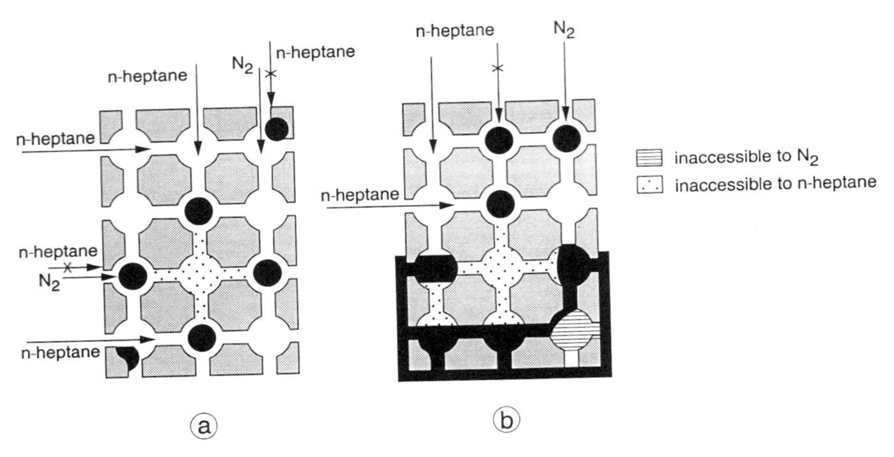

Figure 8.2 Schematic representation of coke distribution in HY zeolite for (a) low coke content and (b) high coke content [14].

support by facilitating the dehydrogenation of polynuclear aromatics; the hydrogen migrates from the support to the metal (inverse spillover effect) [22,39].

Coke can be deposited on the metal and on the acid support. Studies of coke deposition vs. time on Pt/Al_2O_3 catalysts have shown that coke is deposited first on the metallic sites and then on the acid sites [23,24]. Coke can also be deposited on sites that are not active for the main reaction [25]. The higher the metallic dispersion, the lower the coverage by coke of the metallic phase. An increase in the metallic function/acidic function ratio in the catalyst increases the hydrogen content of the coke deposits. On the other hand, the hydrogen content of coke decreases with time on-stream.

An increase in support acidity results in an increase in coke laydown [22]. The location of coke (on metal or support) and the nature of coke (light or graphitic) are more important to catalyst stability than coke content. Coke deposits may also alter the selectivity of bifunctional catalysts. At constant total pressure, coke deposition and catalyst deactivation increases when the H_2/hydrocarbon ratio decreases [26]. The structure of coke is also affected by this ratio [27]. Changes in total pressure will result in changes of coke location and structure [28].

Coke deposition on bifunctional catalysts initially increases rapidly with time on-stream, before a lower rate of coking is reached. In the presence of nitrogen-containing compounds, such as piperidine, these compounds participate in the coke formation process [21]. Organic bases in the feedstock will affect the cracking function of the catalyst by preferential poisoning of acid sites [29,30]. Some of the organic sulfur will also be present in carbonaceous deposits.

Different mechanisms have been proposed for the formation of coke on the metal site and on the acid support of bifunctional catalysts. On metal sites, it is assumed that coke is the result of successive dehydrogenation leading to atomic carbon or partially hydrogenated intermediates that combine to form graphitic coke [31–33]. On acid sites, it is assumed that coke results from polymerization of dehydrogenated intermediates generated at the metal sites [34,35]. Therefore, at least two different types of coke have been identified on bifunctional catalysts: that formed at the metallic sites or in their vicinity, and that formed at the acid sites [36,37]. Coke on metal is more readily removed during regeneration because that coke is more hydrogenated and less polymerized. In the case of noble metal–containing, bifunctional catalysts, an increase in the metal-to-acidity ratio results in a decrease in coke oxidation temperature [36].

8.4. Coke Distribution in Porous Catalysts

Catalyst activity is affected not only by the total amount of coke deposits but by the coke distribution in the catalyst pores. The accumulation of coke within catalyst pores results in a decrease of the effective diameter of the pores. A coke barrier may form that impedes the diffusion of reactant and product molecules in the pore network. The effectiveness of the coke deposit as a barrier depends strongly on its distribution within the pore structure. For example, a given amount of coke concentrated at the pore mouth is a more effective barrier than the same amount of coke uniformly distributed. Many reactions accompanied by coke deposition lead to a nonuniform distribution of coke in the catalyst. Coke may form plugs in the catalyst pores and isolate active, interior regions of the catalyst, making them inaccessible to reactant molecules (see Figure 8.2) [14]. In that case, the loss in activity is much more significant than the amount of coke deposited would indicate. It also results in a nonuniform distribution of intrinsic activity. Plugging by coke has often been observed in zeolite catalysts, where the presence of micropores makes this occurrence more likely. Coke deposits may also interact with metals present in the zeolite. For example, such interaction was observed in the case of PtY zeolites, where coke deposits changed the Pt aggregate morphology [38].

In pellet- or sphere-shaped catalysts, coke distribution is uniform through the catalyst when the catalytic reactions occur unimpeded throughout the catalyst. In some instances, a coke layer (skin) may form near the surface of the pellet or sphere, and create a barrier to reactant molecules. This results in changes in diffusivity and in nonuniform coke distribution in the catalyst.

The coke distribution in a fixed bed of catalyst may vary considerably. With feedstocks entering the top and exiting the bottom of the catalyst bed, heavier coke deposits are usually formed near the top of the bed. Carbonaceous deposits are also formed in the interstices between catalyst particles or on inert material, and result primarily from noncatalytic, thermal reactions. Such deposits increase the pressure drop through the reactor and may cause uneven distribution of reactants through the bed. Catalytic coke affects catalyst activity and selectivity but usually does not affect pressure drop.

8.5. Catalyst Poisoning

Catalyst poisoning is primarily the result of strong chemisorption of impurities on active sites. Poisoning may be reversible or irreversible, depending on the strength of chemisorption of the impurity on the catalyst. Catalyst poisoning may also be selective or nonselective. Selective poisoning is commonly observed on multifunctional catalysts with different types of active sites. In that case, selective poisoning may lead to the poisoning of one type of active site, without affecting the other type or types.

In the present discussion, the catalysts of interest are those having an acid function and a metal function. Since a specific catalyst poison present in the feedstock usually affects only one of the two functions, the poisoning of the two functions will be described separately.

8.5.1. Poisoning of Acid Function

Acidic oxides and mixed oxides play a key role in catalyzing many types of hydrocarbon conversion reactions, and the poisoning of these catalysts by various compounds has been thoroughly investigated. In the early 1950s, Mills et al. studied the interaction of various basic organic compounds such as quinoline and piperidine with silica-alumina catalysts [30]. They found that the severity of poisoning of these catalysts for cracking reactions is a function of the basic strength of the poison. Using quinoline chemisorption, they were able to predict the activity of a cracking catalyst. Later studies, carried out on aluminas with different surface acidities, have shown the importance of acid strength distribution in determining reactivity for a specific reaction and the effect of selective poisoning of such sites [29]. These studies have shown that catalytic activity can arise from a large number of acid sites or from a smaller number of strongly acidic sites. For example, catalytic dehydration of butanol depends on total catalyst acidity, whereas catalytic cracking is much more sensitive to acid strength. In the former case, poisoning of any acid sites reduces catalytic activity (nonselective poisoning), whereas in the latter case poisoning of strong acid sites is sufficient to reduce catalytic activity (selective poisoning).

Numerous studies have been carried out with regard to poisoning of acid zeolites. Many of these studies were carried out by using quinoline or ammonia as poison/probe molecules [40–42]. Titration with quinoline was used to measure acid site density in zeolites [43]. Cumene cracking was often used as a model reaction to investigate the effect of poisons on catalytic cracking. These studies have shown that a single molecule of quinoline per zeolite supercage blocked the cumene from available active sites, regardless of their concentration. Measurements of differential heats of adsorption of NH_3 by microcalorimetric methods have also been used to investigate the poisoning of solid acids [44]. With such measurements it was possible to establish the acid strength distribution in Y zeolites.

8.5.2. Poisoning of Metal Function

Such poisoning is the result of strong interaction of an impurity, usually from the feedstock, and metal, resulting in changes of the surface composition of the metal. This may occur by chemisorption on the metal surface, by reaction with the metal, or by alloy formation. A poison that acts by chemisorption is generally more strongly adsorbed than the reactant. Such chemisorption may be reversible or irreversible, depending primarily on the strength of adsorption. The chemisorbed poison may act by blocking an active site or by altering the adsorptivity of other species by an electronic effect. Several reviews of the subject of metal catalyst poisoning have been published [46,47].

Sulfur is a frequently encountered impurity in commercial catalytic processes, such as hydrogenation, hydrocracking, reforming, methanation, and synthesis. In some commercial hydrocracking units, noble metal catalysts operate in the presence of low concentrations of H_2S. In that case, the noble metal is sulfided and its catalytic activity reduced. Sulfur is strongly adsorbed on a variety of metal catalysts at very low gas phase concentrations and can remain as a very stable adsorbed species under different reaction conditions. Due to the potential impact on such processes, the effect of sulfur on metal catalysts has been studied extensively. Several reviews of this subject have been published [46,48–51]. Many other species, such as water, chlorides, sodium, arsenic, copper, lead, and mercury, may also act as metal poisons in commercial processes [52].

Most of the studies of poisoning by sulfur compounds were carried out on noble metal catalysts, such as platinum and palladium, as well as on nickel catalysts. The nature of the metal–sulfur bond, the structure of the adsorbed sulfur layer, and the influence of this on metal–reactant molecule bonding have been investigated [49]. In the case of platinum poisoning by sulfur compounds, the poisoning effect may be due to both geometrical and electronic factors [43,54]. The strong Pt-S bond modifies the chemical properties of the platinum surface and weakens its interaction with adsorbates [54–56].

The adsorption of sulfur on platinum may be reversible (i.e., sulfur can be readily desorbed at higher temperature under H_2) or irreversible, when sulfur interacts strongly with the metal. Experiments with platinum submitted to variable degrees of presulfiding have shown that the quantity of irreversible sulfur is independent of sulfiding conditions [57].

Sulfur adsorption on metals such as platinum may result in "structure-sensitive" deactivation. In this case, sulfur is adsorbed on particular catalytic sites, significantly reducing the structure-sensitive reactions occurring at such sites. For example, the poisoning of a platinum catalyst by sulfur results in the attenuation of hydrogenolysis (a structure-sensitive reaction) but has a less disturbing effect on hydrogenation-dehydrogenation (a structure insensitive reaction) [58,59].

Small platinum particles in Y zeolites have a high resistance to sulfur poisoning because sulfur adsorption decreases with decreasing particle size. This was attributed to the atomic dispersion of platinum [60], to the electron deficiency of platinum particles in zeolites [62], and to the effect of acidity and electrostatic field of the zeolite on the reactant [63].

Sulfur may be adsorbed not only on the metal but also on the metal catalyst support. For example, sulfur poisoning of alumina-supported platinum catalysts results in sulfur adsorption not only at the metal sites but at the Lewis acid sites of alumina [64].

Only a limited number of studies of bifunctional metal–zeolite catalyst poisoning have been reported. For example, the relative poisoning of metal and acid functions by thiophene and pyridine has been investigated for catalysts Pd/Y; Ni,Pd,W/Y; and Ni,W/Y [45]. Probe reactions were cumene cracking, *o*-xylene isomerization, and *n*-heptane hydrocracking. The data obtained showed that the hydrocracking reactions, having a bifunctional mechanism, are more strongly affected by catalyst poisons than the cracking or isomerization reactions that follow a monofunctional mechanism. Addition of thiophene caused the strongest catalyst deactivation for the hydrocracking reaction due to preferential poisoning of metal sites. Pyridine addition leads to strong deactivation for the cracking and isomerization reactions (due to preferential poisoning of acid sites), with a lesser effect on the hydrocracking reaction.

For Co-Mo-containing hydrocracking catalysts, catalyst presulfiding increases the activity of the metallic function and decreases catalyst deactivation rate. For noble metal hydrocracking catalysts, the presence of sulfur compounds results in lower activity of the noble metal component. In some hydrocracking processes, the noble metal hydrocracking catalysts operate in sulfided form by design in order to reduce hydrogen consumption, to inhibit complete saturation of aromatics, and to maintain product octane rating.

The hydrogenation function of fresh and deactivated commercial hydrocracking catalysts, containing Co-Mo on a zeolite support, has also been investigated by using cyclohexene hydrogenation as a probe reaction [65]. Electron spin resonance (ESR) data have shown that sulfur decreases the oxidation state of molybdenum from Mo^{6+} to Mo^{5+}.

Dautzenberg et al. [66,67] investigated the mechanism of catalyst deactivation during residue hydrocracking. The deactivation occurs primarily through pore mouth plugging and takes place in two stages. The first is coke deposition that reaches an equilibrium level within a relatively short time. The second stage of deactivation corresponds to a gradual accumulation of metal sulfides, resulting in a gradual catalyst activity decline. This activity decline continues until the pore openings of the catalyst are fully blocked. Further deactivation is caused by coke deposition on the catalyst surface. The fact that a totally coke-covered catalyst still has some catalytic activity has been attributed to the autocatalytic effect of the deposited metal sulfides.

8.6. Catalyst Regeneration

A coked catalyst is usually regenerated by combustion in a stream of diluted oxygen or air, although steam or steam–air mixtures have also been used for regeneration. Upon combustion, coke is converted to CO_2 and H_2O. In the absence of excess oxygen, CO also may form. Combustion of coke liberates significant quantities of heat, and this heat may accelerate phase changes or sintering in catalysts [68,69]. To avoid these deleterious effects, the gas stream contains initially small amounts of oxygen (2–5%), and the concentration of oxygen is only gradually increased in order to complete coke burnoff [70].

The bulk of the coke is removed by oxidation at temperatures between 400 and 500°C [71–73]. If the coke contains sulfur, emission of SO_2 starts at a lower temperature than that of CO_2 [71,72]. Sulfur oxides emitted at the lower temperature (about 250°C) originate from oxidation of deposited sulfur and of metal sulfides. Sulfur oxides emitted at a higher temperature (~450–500°C) result from the partial decomposition of sulfates [74].

Upon regeneration, catalyst activity is partially or totally restored. For bifunctional, noble metal catalysts, having an acid and a metallic function, both combustion of coke and redispersion of metal (rejuvenation) is necessary to restore catalyst activity. Depending on the severity of deactivation by coke, periodic shutdowns of the operation are necessary for regeneration of the coked catalyst. The conditions for regeneration of the catalyst depend on the reactivity of the coke. This in turn depends on the chemical nature of the coke deposits (pseudographitic or aromatic), mechanism of formation, composition (H/C ratio, S and N content, type of polynuclear aromatics present), and its distribution in the deactivated catalyst (uniform or nonuniform distribution). If some of the components of the carbonaceous deposit are hydrocarbon polymers, they may decompose and volatilize, and the products burn homogeneously outside the catalyst pores. The type and composition of the deactivated catalyst will also affect the regeneration process.

During the initial stages of catalyst regeneration, the initial oxidation reactions may be diffusion controlled due to the pore blocking by deposited coke and metals. As coke gasification proceeds, the pores become more accessible and mass transfer limitations are reduced. Furthermore, the metal oxides formed during the oxidation process will catalyze coke gasification [71–75].

In catalyst pellets or spheres with a uniform coke distribution, in the absence of diffusion limitations for oxygen, the coke concentration decreases uniformly throughout the pellet or sphere during the combustion reaction. At a sufficiently high temperature the regeneration reaction may become diffusion-limited, and oxygen reacts as fast as it is transported to the coke deposit. In that instance, the reaction moves progressively from the exterior surface to the center, resulting in a progressing shell of coke-free catalyst surrounding the coked catalyst center (Figure 8.3) [76]. This can be visualized by cutting a pellet or sphere in half after partial removal of coke.

The presence of transition metals in the porous catalyst can greatly accelerate the rate of coke oxidation due to the catalytic effect of these metals on the oxidation reaction [36,77]. This effect allows rapid regeneration at lower temperatures and lower oxygen pressures than those required in the absence of these metals.

Hydrogen and sulfur present in coke are removed preferentially in the early stages of the regeneration process. Regeneration of spent, nonnoble metal, bifunctional hydroprocessing catalysts will remove inorganic sulfur from metal sulfides and convert the latter to metal oxides, e.g.,

$$MoS_2/WS_2 + 7/2O_2 \rightarrow MoO_3/WO_3 + 2SO_2$$
$$Ni_3S_2 + 7/2O_2 \rightarrow 3NiO + 2SO_2$$

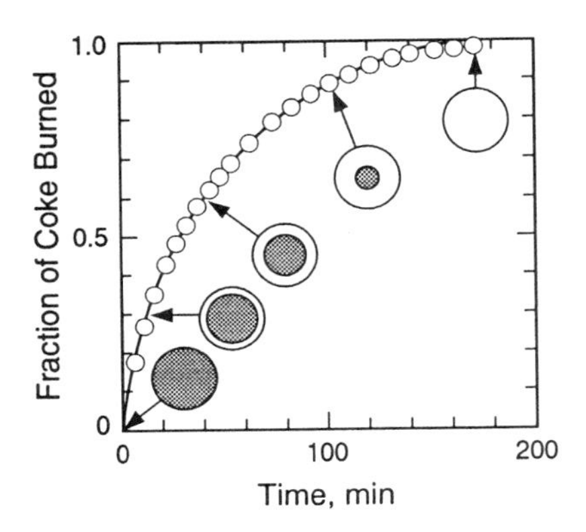

Figure 8.3 Progressive, shell-type regeneration of coked catalyst pellet [76].

Combustion in flowing oxygen will also remove many of the poisons deposited on the catalyst, such as organic sulfur and nitrogen compounds. However, metal poisons such as sodium, copper, or vanadium are not removed by combustion.

During catalyst regeneration, changes may occur in the distribution of catalyst metals on the support. For hydroprocessing catalysts it was shown that regeneration at relatively low temperatures (380°C) leads to migration of both molybdenum and nickel toward the center of the catalyst pellet [77]. Regeneration above 560°C leads to migration of molybdenum and nickel toward the exterior of the pellet, with nickel being more mobile [78]. Regeneration above 700°C leads to volatilization of molybdenum, with the remaining molybdenum concentrated near the surface of the pellet [79].

Deactivated zeolite catalysts are usually regenerated by oxidation in air or in air diluted with nitrogen. This process is often shape-selective, when oxygen circulation in the zeolite pores is restricted by coke deposits. HY zeolite has a high oxidation rate due to free circulation of oxygen through the three-dimensional pore network [80].

In commercial processes, different regeneration methods may be used. In a fixed-bed process, such as hydroprocessing, the reaction step may alternate with a coke burnoff step (cyclical process). These steps are separated by purges, often with an inert gas such as nitrogen, to avoid formation of explosive mixtures.

Temperature, oxygen concentration, and flow rate during regeneration are kept within narrow limits. They should be sufficiently high to allow coke removal at an adequate rate but not too high to prevent unsafe operation or damage to the catalyst. The specifics of each regeneration process may vary significantly.

8.7. Regeneration of Commercial Hydrocracking Catalysts

Hydrocracking catalysts have a relatively long on-stream life. Initial cycles are usually no less than 2 years, and in some instances cycles of 5 years or more have been achieved.

Due to the hydrogen atmosphere in which the catalyst operates, coke deposition is fairly slow. For nonnoble metal catalysts, coke combustion in an oxygen-containing atmosphere generally regenerates catalyst activity. Other methods, such as coke removal using laser irradiation of the catalyst in the presence of an oxidizing gas [81], or "hydrotreatment" of the spent catalyst to solubilize the insoluble carbonaceous deposits [82], have been reported in the patent literature.

The catalyst can be regenerated in situ, i.e., in the hydrocracking unit, or ex situ, i.e., when the catalyst is removed from the unit for regeneration. The degree to which activity is restored depends on the extent of catalyst surface area reduction and zeolite crystallinity loss (if present), which cannot be restored during the regeneration process. The hydrogenation function of the regenerated nonnoble metal catalyst is restored by sulfiding the catalyst.

The ex situ regeneration of spent, commercial hydroprocessing catalysts is usually done by companies specialized in this field. The best known companies are Eurecat and CRI. Both use moving bed regenerators. The details of these processes are of proprietary nature.

The Eurecat process [83,84] uses a continuous rotolouvre, which is a cylindrical drum rotating slowly on a horizontal axis and enclosing a series of overlapping louvres. The spent catalyst passes slowly through the rotolouvre, where it encounters a countercurrent of hot air. The most critical parameter in this process appears to be the regeneration temperature. Preliminary lab tests and data obtained from previous operations are used to set the regeneration temperature. Feed rate, residence time, temperature of combustion air, depth of catalyst bed, and other parameters are adjusted to contain the temperature within desirable limits.

The CRI process uses a porous moving belt as a regenerator [85]. The catalyst is moved with the stainless steel belt through a stationary tunnel furnace vessel where regeneration takes place. The characteristics of moving belt regenerators have been described in the literature [86]. In most commercial processes, the spent catalyst is submitted to de-oiling prior to regeneration.

8.8. Sintering and Redispersion of Metals

In addition to coking and poisoning of acid sites, catalyst deactivation can also result from the deactivation of the hydrogenation function. One cause of such deactivation is poisoning of metal sites, e.g., by sulfur. Another major cause of deactivation is sintering of metals dispersed in the zeolite or located on the nonzeolitic catalyst components. Numerous studies have been carried out on sintering of metals occluded in zeolites as well as on metals deposited on alumina or silica-alumina.

8.8.1. Sintering of Metals

The sintering process of supported metals is affected, in general, by the following factors: nature of the metal, chemical nature and morphology of the support, surface impurities or deposits, nature of gases present, temperature, pressure, and time.

The various stages of a metal sintering process are shown schematically in Figure 8.4 [87]. The sintering mechanism (by coalescence) is discussed in detail in [88]. For a given support, the stability of a metal against sintering generally increases with increasing melting point of the metal. For a given metal dispersed on different supports, the surface free energy of the support affects the stability

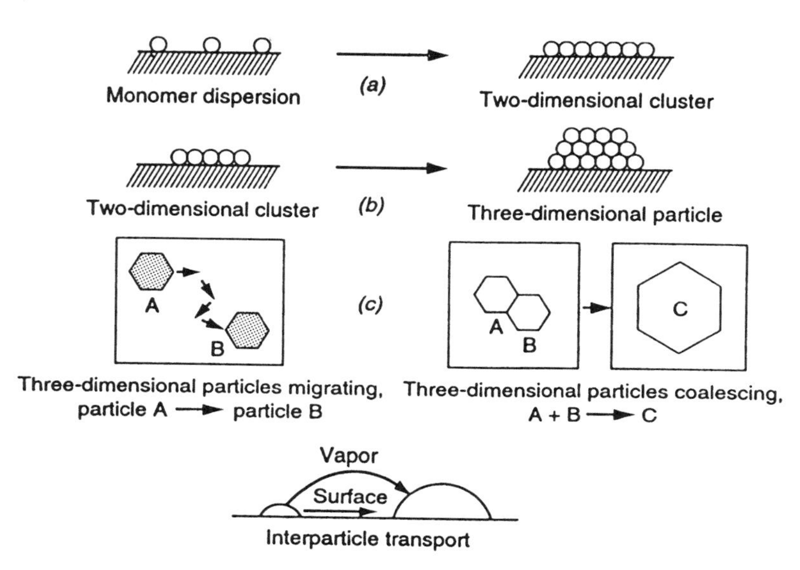

Figure 8.4 Various stages in the formation and growth of metal particles from a monomer dispersion [87]. (Reprinted with permission from Elsevier Science Ltd., Kidlington, U.K.)

against sintering [89]. For example, supported Pt stability against sintering in an oxidizing atmosphere decreases in the order MgO > Al_2O_3 > SiO_2-Al_2O_3 ≃ SiO_2 [92]. Several reviews of this subject have been published [90–93].

The sintering of metals in zeolites is affected by the following factors [94]:

1. Melting point of the metal. The higher the melting point of the metal, the lower the mobility of the atoms. For example, under identical conditions and in the same zeolite, the Pd atoms migrate out of a sodalite cage at 200–300°C, compared to Pt atoms at 300–600°C [95].
2. Interaction of the metal atoms with the support. The metal aggregates interact with the zeolite acid sites or highly charged cations. Such interaction causes an electron transfer or polarization of the metal atoms, resulting in a stronger association of the metal with the zeolite framework.
3. Geometry of the porous network. The zeolite pore geometry can play a major role in metal stability by blocking the migration of metal aggregates (e.g., in Y-type zeolites, with a 7.4-Å cage aperture). In that case, sintering can occur only after an initial disintegration of the aggregate, followed by metal atom migration. Metal migration occurs much easier in straight, parallel channels (e.g., in mordenite or L zeolite). For

example, in Pt-mordenite sintering is observed at 500°C, whereas in Pt-Y zeolites sintering is observed only over 800°C under vacuum [94,96].

4. Nature of the surrounding atmosphere. The atmosphere surrounding the metal zeolite can affect the sintering process by interacting with the surface metal atoms and increasing their mobility. For example, Pd particles in zeolite-based hydrocracking catalysts sinter more easily in the presence of high-pressure steam during catalyst regeneration [97]. Pt clusters encaged in the supercages of Y zeolite are stable up to 800°C under vacuum, but heating at 320°C under H_2 and NH_3 induces migration and sintering into 20- to 30-Å particles [98]. The sintering mechanism of noble metals in zeolites has been investigated in detail and has been described in several reviews [94,99,100].

The noble metal atoms in a zeolite may either diffuse to the external surface of the zeolite and form large particles, or agglomerate inside the zeolite crystal. Such agglomeration inside the zeolite may result in particles larger than zeolite cages. It has been suggested that a local recrystallization of the zeolite lattice takes place, similar to that occurring during steam stabilization, in order to accommodate the larger metal particles [101]. Metal particle growth can also take place in crystal defects.

The morphology of the occluded particles is often determined by the structure of the porous framework. Thus, spherical particles are formed in the isotropic faujasite structure, whereas filament-like particles are formed in the tubular pores of mordenite.

8.8.2. Redispersion of Metals

Redispersion of metals is the opposite process to sintering and is of considerable industrial importance because supported metal catalysts are commonly active when the metal is highly dispersed. Redispersion of metals results in a decrease in particle size accompanied by an increase in specific surface area. Redispersion of sintered metal is usually accomplished by treatment with a strong gaseous oxidant, such as oxygen or chlorine, at elevated temperature. The reagent oxidizes the metal particles, and the oxidized species, such as metal cations, oxides, oxychlorides, or chlorides, migrate over the support. In zeolite catalysts, they migrate to the zeolite cages. The migration may take place by surface diffusion or by gas phase transport. Subsequent decomposition or reduction of the dispersed oxidized species at a relatively low temperature results in dispersed metal. Different models have been proposed to explain the redispersion mechanism of metals [92,102–104].

Numerous studies have been carried out on redispersion of noble metals (mainly Pt and Pd) in zeolites. Several reviews of this subject have been published [94,99,100,105]. Redispersion of noble metals in zeolites is commonly carried out by high-temperature treatment in oxygen, followed by a relative low-temperature reduction in hydrogen. For example, 10-Å Pd aggregates occupying adjacent supercages in Pd-Y zeolites can be oxidized to Pd^{2+} cations at 200°C [95]. These ions migrate to the sodalite cages, leading to metal redispersion. Larger particles occluded in the zeolite crystals or on the external surface are transformed into PdO. Subsequent reduction in hydrogen (at about 150°C) results in dispersed palladium in the zeolite cages [106] (see also Chapter 5). The following reactions take place:

$$Pd_n + n/2O_2 \rightarrow (PdO)_n$$

$$Pd_n + n/2O_2 + 2nH^+ \rightarrow nPd^{2+} + nH_2O$$

$$Pd^{2+} + H_2 \rightarrow Pd + 2H^+$$

Pd aggregates can be readily dispersed by NO. For example, 20-Å Pd aggregates occluded in Pd Y zeolite crystals can be redispersed with NO at room temperature into Pd^{2+} cations [107].

8.8.3. Rejuvenation of Commercial Hydrocracking Catalysts

In addition to regeneration by coke combustion that restores the cracking function of the catalyst, a further treatment is required for noble metal catalysts in order to restore the hydrogenation function. After coke removal by oxidative combustion, the catalyst shows a significant loss in hydrogenation activity due to noble metal agglomeration and redistribution. The hydrogenation activity is restored by redispersion of the noble metals on the catalyst, a process often called catalyst rejuvenation.

Several rejuvenation processes of noble metal hydrocracking catalysts are described in the patent literature. One of the earlier methods used for rejuvenation of palladium containing zeolite catalysts consisted in high-temperature regeneration in a flow of dilute oxygen or air, followed by reduction in hydrogen [108]. Palladium particles of less than 30 Å were obtained by this procedure. A variation of this method consisted in high-temperature oxidative regeneration, followed by a sulfiding step of the noble metal with H_2S, a second high-temperature oxidation step, and finally a reduction in hydrogen [109]. Similar to oxidation, sulfiding the catalyst facilitates the dispersion of the metal particles. The method allowed dispersion to particles having about 10 or 20 Å in diameter. The amount of water present during calcination is closely controlled in order to avoid agglomeration of the metal.

Subsequent methods utilize ammonia in the rejuvenation process. For example, a palladium-containing zeolite catalyst can be rejuvenated by treating the oxidative regenerated catalyst with an aqueous ammonia solution, draining the excess solution, and finally drying and calcining the catalyst [97,110].

Rejuvenation can also be carried out by treatment of the wet catalyst with

Table 8.1 Properties of Regenerated and Rejuvenated Catalyst [113]

Property	After regeneration	After rejuvenation
Surface area (m^2/g)	550	535
Crystallinity (%) (relative to fresh catalyst)	100	100
Relative activity, °C (°F) (temperature requirement)		
Second cycle (compared to fresh catalyst)	+34 (+61)	−5 (−9)

gaseous ammonia until saturation or by treatment with an ammoniacal ammonium salt solution [111]. Urea may also be used for that purpose because it generates ammonia by hydrolysis [112]. It is assumed that the oxidized noble metal is dissolved in the ammonia solution in the zeolite pores and forms soluble tetrammine complex cations, such as $[Pd(NH_3)_4]^{2+}$. The complex ions migrate to the ion exchange sites of the zeolite. Subsequent drying and calcination at a suitable temperature decomposes the complex ammine and yields a highly dispersed noble metal on the catalyst. Calcination also restores the acidity of the zeolite that was temporarily neutralized during the ammonia treatment. The use of an ammoniacal ammonium salt solution for catalyst rejuvenation has the additional advantage of removing sodium ions from the zeolite (initially present in the catalyst or picked up from the feedstock) by ion exchange with ammonium ions, thus increasing catalyst activity.

The properties of a regenerated and rejuvenated noble metal catalyst are compared in Table 8.1 [113]. In this example, the regenerated catalyst is 34°C (61°F) less active than the fresh catalyst. However, after rejuvenation the activity is at least as good as that of the fresh catalyst.

Nonnoble metals do not show significant sintering in sulfided form at hydrocracking temperature. However, metal migration and sintering may occur during oxidative regeneration of the spent catalyst. Careful control of the regeneration temperature and gas composition can prevent sintering [74]. Sulfiding of the regenerated catalyst usually reestablishes the initial metal distribution [71,114, 115]. Catalyst sulfiding will also increase catalyst acidity [116] (see also Section 3.3.5).

References

1. J. B. Butt and E. E. Petersen, *Activation, Deactivation and Poisoning of Catalysts*, Academic Press, New York, 1988.
2. J. B. Butt, *Catalysis: Science and Technology*, Vol. 6 (J. R. Anderson and M. Boudart, eds.), Springer-Verlag, Berlin, 1984, p.1.

3. W. G. Appleby, J. W. Gibson, and G. M. Good, *Ind. Eng. Chem. Proc. Des. Dev. 1*:102 (1962).
4. B. C. Gates, J. R. Katzer, and G. C. A. Schuit, *Chemistry of Catalytic Processes*, McGraw-Hill, New York, 1979, p. 17.
5. M. Guisnet and P. Magnoux, *Appl. Catal. 54*:1 (1989).
6. P. Gallezot, C. Leclercq, J. Barbier, and P. Marécot, *J. Catal. 116*:164 (1989).
7. T. S. Chang, N. M. Rodriguez, and R. T. K. Baker, *J. Catal. 123*:486 (1990).
8. D. A. Best and B. W. Wojiciechowski, *J. Catal. 47*:11 (1977).
9. A. T. Bell, in *Catalyst Deactivation* (E. E. Petersen and A. T. Bell, eds.), Marcel Dekker, New York, 1987, p. 235.
10. H. H. Voge, J. M. Good, and B. J. Greensfelder, *Proc. 3rd World Petr. Congr., 4*: 124 (1951).
11. A. Voorhies, *Ind. Eng. Chem. 37*:318 (1945).
12. L. D. Rollman, *J. Catal. 47*:113 (1977).
13. L. D. Rollman and D. E. Walsh, *J. Catal. 56*:139 (1987).
14. P. Magnoux, P. Cartraud, S. Mignard, and M. Guisnet, *J. Catal. 106*:242 (1987).
15. M. E. Levinter, G. M. Panchekov, and M. A. Tanatarov, *Int. Chem. Eng. 7*:23 (1967).
16. J. B. Butt, S. Delgrado-Diaz, and W. E. Muno, *J. Catal. 37*:158 (1975).
17. D. E. Walsh and L. D. Rollman, *J. Catal. 49*:369 (1977).
18. V. Fouche, P. Magnoux, and M. Guisnet, *Appl. Catal. 58*:189 (1990).
19. R. P. L. Apsil, J. S. Dranoff, and J. B. Butt, *J. Catal. 85*:415 (1984).
20. Keuji Nita et al., in *Catalyst Deactivation 1987* (B. Delmon and G. F. Froment, eds.), Elsevier, Amsterdam, 1987, p. 501.
21. G. C. Hadjiloizou, J. B. Butt, and J. S. Dranoff, *J. Catal. 135*:27 (1992).
22. P. Marécot and J. Barbier, in *Catalytic Naphtha Reforming* (G. J. Antos, A. M. Aitani, and J. M. Parera, eds.), Marcel Dekker, New York, 1994, p. 279.
23. J. N. Beltramini, T. J. Wessel, and R. Datta, in *Catalyst Deactivation* (B. Delmon and G. F. Fromet, eds.), Elsevier, Amsterdam, 1991, p. 119.
24. C. A. Querini, N. S. Figoli, and J. M. Parera, *Appl. Catal. 52*:249 (1989).
25. E. E. Petersen, in *Catalyst Deactivation* (E. E. Petersen and A. T. Bell, eds.), Marcel Dekker, New York, 1987, p. 39.
26. C. G. Myers, W. H. Lang, and P. B. Weisz, *Ind. Eng. Chem. 53*:299 (1961).
27. D. Espinat, H. Dexpert, E. Freund, G. Martino, M. Gouzi, P. Lespade, and F. Cruege, *Appl. Catal. 16*:343 (1985).
28. J. Barbier, in *Catalyst Deactivation 1987* (B. Delmon and G. F. Froment, eds.), Elsevier, Amsterdam, 1987, p. 1.
29. H. Pines and W. O. Haag, *J. Am. Chem. Soc. 82*:2471 (1960).
30. G. A. Mills, E. R. Boedeker, and A. G. Oblad, *J. Am. Chem. Soc. 72*:1554 (1950).
31. J. Barbier, G. Corro, Y. Zhang, J. P. Bournonville, and J. P. Franck, *Appl Catal. 16*:169 (1985).
32. A. Sarkany, H. Lieske, T. Szilagyi, and L. Toth, *Proc. 8th Int. Congr. on Catal.*, Berlin, 1984.
33. D. L. Trimm, *Appl. Catal. 5*:263 (1983).
34. J. Barbier, *Appl. Catal. 23*:225 (1986).

35. Yu. M. Zhorov, G. M. Panchenkov, and Y. N. Kartashev, *Kinet. Katal. 21*:776 (1980).
36. J. M. Parera, N. S. Figoli, and E. M. Traffano, *J. Catal. 79*:484 (1983).
37. C. L. Pieck, E. L. Jablonski, R. J. Verderone, J. M. Grau, and J. M. Parera, *Catal. Today 5*:463 (1989).
38. P. Gallezot, *J. Chim. Phys. 78*:881 (1981).
39. J. M. Parera, R. J. Verderone, and C. A. Querini, in *Catalyst Deactivation 1987* (B. Delmon and G. F. Froment, eds.), Elsevier, Amsterdam, 1987, p. 135.
40. J. Turkevich and Y. Ono, *Adv. Chem. 102*:315 (1971).
41. P. A. Jacobs and C. F. Heylen, *J. Catal. 34*:267 (1974).
42. J. Galuszka, S. Cekiewicz, and A. Baranki, *J. Chem. Soc., Faraday Trans. 75*:1150 (1979).
43. M. S. Goldstein and T. R. Morgan, *J. Catal. 16*:232 (1970).
44. T. Masuda, H. Tahiguchi, K. Tsutsumi, and H. Takahashi, *Bull. Chem. Soc. Jpn. 51*, 1965 (1978).
45. V. Penchev, G. Angelova, and T. Drajev, *React. Kinet. Catal. Lett. 21* (3), 341 (1982).
46. C. H. Bartholomew, P. K. Agrawal, and J. R. Katzer, *Adv. Catal. 31*:135 (1982).
47. J. Barbier, E. Lamy-Patara, P. Marécot, J. P. Bitiaux, J. Cosyns, and F. Verna, *Adv. Catal. 37*:279 (1990).
48. D. W. Goodman and J. E. Houston, *Science 236*, April 24, 1987, p. 403.
49. J. Oudar, in *Metal Support and Metal-Additive Effects in Catalysis* (B. Imelik et al., eds.), Elsevier, Amsterdam, 1982, p. 255.
50. R. J. Madon and H. Shaw, *Catal. Rev. Sci. Eng. 15*:69 (1977).
51. M. R. Mathieu and M. Primet, *C. R. Acad. Sci. Ser. 2*:299, 419 (1984).
52. J. Barbier, A. Morales, P. Marécot, and M. Maurel, *Bull. Soc. Chim. Belg. 88* (7-8), 569 (1979).
53. J. Oudar, S. Pinol, C. M. Pradier, and Y. Berthier, *J. Catal. 107*:445 (1987).
54. J. Biswas, G. M. Bickle, P. G. Gray, D. D. Do, and J. Barbier, *Catal. Rev. Sci. Eng. 30*:161 (1988).
55. H. P. Bonzel and R. Ku, *J. Chem. Phys. 58*:4617 (1973).
56. S. R. Keleman, T. E. Fisher, and J. A. Schwarz, *Surf. Sci. 81*:440 (1979).
57. C. R. Apesteguia, J. Barbier, J. F. Plaza de los Reyes, T. F. Garetto, and J. M. Parera, *Appl. Catal. 1*:159 (1981).
58. C. R. Apestegnia and J. Barbier, *J. Catal. 78*:352 (1982).
59. P. Biloen, J. N. Helle, H. Verbeck, F. M. Dautzenberg, and W.M.H. Sachtler, *J. Catal. 63*:112 (1980).
60. J. A. Rabo, V. Schomaker, and P. E. Pickert, *Proc. 3d Int. Congr. Catal.*, Vol. 2, North Holland, Amsterdam, 1965, p. 1264.
61. R. A. Dalla Beta and M. Boudart, *Proc. 5th Int. Congr. Catal.*, Vol. 2, North Holland, Amsterdam, 1973, p. 1329.
62. P. Gallezot, J. Datka, J. Massardier, M. Primet, and B. Imelik, Proc. *6th Int. Congr. Catal.*, The Chemical Society, London, 1977, p. 696.
63. C. Naccache, N. Kaufherr, M. Dufaux, J. Bandiera, and B. Imelik, in *Molecular Sieves II* (J. R. Katzer, ed.), A.C.S., Washington, D.C., 1977, p. 558.
64. C. R. Apesteguia, S. M. Trevizan, T. F. Garetto, J. F. Plaza de los Reyes, and J. M. Parera, *React. Kinet. Catal. Lett. 20*:1 (1982).

65. S. R. Pookote, J. S. Dranoff, and J. B. Butt, A.C.S. Symposium Series 196, Washington, D.C., 1982, 283.
66. F. M. Dautzenberg, J. Van Klinken, M. A. Pronk, T. S. Sie, and J. B. Wiffles, paper presented at the 5th Int. Symposium on Chemical Reaction Engineering, Houston, TX, March 1978.
67. F. M. Dautzenberg, S. E. George, C. Ouwerkerk, and S. T. Sie, paper presented at the Advances in Catalytic Chemistry II Symposium, Salt Lake City, Utah, May 1982.
68. A. Byrne, R. Hughes, and J. Santamaria-Ramiro, *Chem. Eng. Sci. 40*:1507 (1985).
69. R. V. R. V. Nalitham, A. R. Tarrer, J. A. Guln, and C. W. Curtis, *Ind. Eng. Chem. Proc. Des. Dev. 24*:160 (1985).
70. D. L. Trimm, in *Catalysts in Petroleum Refining 1989* (D. L. Trimm et al., eds.), Elsevier, Amsterdam, 1990, p. 41.
71. Y. Yoshimura, E. Furimsky, T. Sato, H. Shimada, N. Matsubayashi, and A. Nishijama, *Proc. 9th Int. Catal. Congr. 1*:136 (1988).
72. Y. Yoshimura and E. Furimsky, *Appl. Catal. 23*:157 (1986).
73. Y. Yoshimura and E. Furimsky, *Ind. Eng. Chem. Res. 26*:657 (1987).
74. Z. M. George, P. Mohammed, and R. Tower, *Proc. 9th Int. Catal. Congr. 1*:230 (1988).
75. K. Klusacek, H. Davidova, P. Fott, and P. Schneider, *Chem. Eng. Sci 40*:1717 (1985).
76. J. T. Richardson, *Ind. Eng. Chem. Proc. Des. Dev. 11*:8 (1972).
77. P. Weisz and R. D. Goodwin, *J. Catal. 6*:227 (1966).
78. J. M. Bogdanor and H. F. Rase, *Ind. Eng. Chem. Prod. Res. Dev. 25*:220 (1986).
79. A. Stanislaus and K. Al-Dolama, *J. Catal. 101*:536 (1986).
80. P. Magnoux and M. Guisnet, *Appl. Catal. 38*:341 (1988).
81. W. W. Duley, T. M. Steel, and M. F. Wilson, U.S. Patent No. 5,037,785 (1991).
82. M. Inooka, U.S. Patent No. 4,525,267 (1985).
83. J. Wilson, Proc. AIChE Summer Mtg., August 1990, San Diego, CA.
84. G. Berrebi, Proc. Ketjen Catal. Symp., May 1984, Amsterdam.
85. C. J. Barsby and G. B. Theriault, 3d Annual Mtg. Haztech Canada, Nov. 7, 1990, Calgary.
86. R. E. Ellingham and J. Garrett, *Appl. Ind. Catal. 3*:25 (1984).
87. P. Wynblatt and N. A. Gjostein, *Prog. Solid State Chem. 9*:21 (1974).
88. E. Ruckenstein and B. Pulvermacher, *J. Catal. 29*:224 (1973).
89. E. Ruckenstein, in *Metal–Support Interaction in Catalysis, Sintering and Redispersion* (S. A. Stevenson et al., eds.), Van Nostrand Reinhold, New York, 1987.
90. S. E. Wanke and P. C. Flynn, *Catal. Rev. Sci. Eng. 12*:93 (1975).
91. J. W. Geus, in *Sintering and Catalysis* (G. C. Kucynski, ed.), Plenum Press, New York, 1975, p. 29.
92. S. E. Wanke, J. A. Szymura, and T-T. Yu, in *Catalyst Deactivation* (E. E. Petersen and A. T. Bell, eds.), Marcel Dekker, New York, 1987, p. 65.
93. E. Ruckenstein and D. B. Dadyburjor, *Rev. Chem. Eng. 1*:251 (1983).
94. P. Gallezot and G. Bergeret, in *Catalyst Deactivation* (E. E. Petersen and A. T. Bell, eds.), Marcel Dekker, New York, 1987, p. 263.
95. G. Bergeret, P. Gallezot, and B. Imelik, *J. Phys. Chem. 85*:411 (1981).
96. P. Gallezot, A. Alarcon Diaz, J. A. Dalmon, A. J. Renouprez, and B. Imelik, *J. Catal. 39*:334 (1975).

97. J. W. Ward, U.S. Patent No. 4,139,433 (1979).
98. J. Datka, P. Gallezot, J. Massardier, and B. Imelik, in *Proc. 5th Ibero-Amer. Symp. on Catalysis*, Vol. 2 (F. Portela and C. Pulido, eds.), Lisbon, 1979, p. 297., New York, 1987, p. 263.
99. P. Gallezot, in *Metal Clusters* (M. Moskovits, ed.), Wiley, New York, 1986, p. 220.
100. S. T. Homeyer and W.M.H. Sachtler, in *Zeolites: Facts, Figures, Future* (P. A. Jacobs and R. A. van Santen, eds.), Elsevier, Amsterdam, 1989, p. 975.
101. D. Exner, N. I. Jaeger, R. Novak, G. Schulz-Ekloff, and P. Ryder, *Proc. 6th Int. Conf. on Zeolites* (D. H. Olson and A. Bisio, eds.), Butterworths, Guildford, 1984, p. 387.
102. Y. F. Chu and E. Ruckenstein, *J. Catal. 55*:287 (1978).
103. I. Sushumna and E. Ruckenstein, *J. Catal. 108*:77 (1987).
104. E. Ruckenstein, in *Catalyst Deactivation 1991* (C. H. Bartholomew and J. B. Butt, eds.), Elsevier, Amsterdam, 1991, p. 585.
105. P. A. Jacobs, in *Metal Microstructures in Zeolites* (P. A. Jacobs et al., eds.), Elsevier, Amsterdam, 1982, p. 71.
106. G. Bergeret and P. Gallezot, *J. Phys. Chem. 87*:1160 (1983).
107. M. Che, J. F. Dutel, P. Gallezot, and M. Primet, *J. Phys. Chem. 80*:2371 (1976).
108. C. G. Wight and R. H. Haas, U.S. Patent No. 3,197,397 (1965).
109. R. C. Hansford and R. H. Haas, U.S. Patent No. 3,287,257 (1966).
110. J. W. Ward, U.S. Patent No. 4,190,553 (1980).
111. D. E. Clark, U.S. Patent No. 5,206,194 (1993).
112. D. F. Best, U.S. Patent No. 4,645,750 (1987).
113. J. W. Ward, in *Applied Industrial Catalysis*, Vol. 3 (B. E. Leach, ed.), Academic Press, New York, 359 (1984).
114. J. S. Jepsen and H. F. Rase, *Ind. Eng. Chem. Prod. Des. Dev. 20*:467 (1981).
115. T. C. Ho, *Catal. Rev. Sci. Eng. 30*:117 (1988).
116. G. E. Langlois, R. F. Sullivan, and C. J. Egan, *J. Phys. Chem. 70*:3666 (1966).

9

CATALYST CHARACTERIZATION AND TESTING

To ensure proper performance in the unit, the hydrocracking catalyst is submitted to a series of analyses and tests. Characterization and testing of hydrocracking catalysts is of crucial importance for both the catalyst manufacturer and the refiner. The catalyst manufacturer uses the evaluation to ensure that the manufactured catalysts are in compliance with established quality specifications. He also uses it in the development of new catalysts, designed to meet changing market demands. The refiner uses the evaluation to monitor the properties of fresh and spent catalysts from his refinery. Furthermore, he uses it to compare the quality of catalysts obtained from different suppliers.

In addition to characterizing and testing the finished catalyst, the catalyst manufacturer characterizes the different components used in the catalyst manufacturing process. The quality of these components is critical to the catalytic and physical properties of the catalyst. When a zeolite is used in the catalyst formulation, its chemical composition, structural characteristics [structure type, unit cell size, magic angle spinning nuclear magnetic resonance (MASNMR) spectra], and a series of physicochemical properties (acidity, surface area, crystallite size, etc.) are determined. The amorphous silica-alumina component is usually characterized by determining the silica-to-alumina ratio, surface area, pore size distribution, and pore volume. In some instances, additional measurements are used. The data obtained from these measurements should meet the established specifications in order to make these materials acceptable as catalyst components.

A number of techniques are being used for the elemental analyses of hydrocracking catalysts. Chemical analysis is the oldest method in use. Several physical methods, such as atomic absorption spectroscopy (AAS), X-ray fluorescence spectroscopy (XRF), and inductive coupled plasma (ICP), are used for the elemental analysis of the bulk catalyst. Electron probe microanalysis (EPMA) can be used to measure the composition across the catalyst pellet or extrudate. Other physical methods can be used to obtain information about the surface composition of catalyst particles (see below).

The physicochemical characterization of hydrocracking catalysts combines a variety of methods. Several methods can be used to measure catalyst acidity. Some methods are designed to measure the bulk properties of the catalyst, such as compact bulk density (CBD), crush strength, abrasion loss, crystalline phase characteristics (type and amount of crystalline phases present, zeolite unit cell size), surface area, pore size distribution and pore volume, average length and diameter of extrudates, and mercury density. Other methods are used to characterize the catalyst surface (e.g., scanning electron microscopy, transmission electron microscopy, scanning transmission electron microscopy, Auger, X-ray photoelectron spectroscopy) or to describe aluminum coordination and distribution (e.g., magic angle spinning NMR, fast atom bombardment–mass spectrometry). Many of these methods are used to characterize catalysts as well as catalyst components. Furthermore, some methods can be used to characterize the changes that occur in the catalyst during the catalytic process. These changes are then related to changes in catalyst activity and selectivity.

The names and corresponding abbreviations of some instrumental methods used to characterize solids and surfaces are listed in Table 9.1 [1]. The characteristics and applications of selected methods used to describe catalytic materials are summarized in Table 9.2. Some of the methods listed are fairly complex and are

Table 9.1 Selected Methods Used for Characterization of Solids and Surfaces [1]

Abbreviation	Name of method
AAS	Atomic absorption spectroscopy
AEM	Analytical electron microscopy
AES	Auger electron spectroscopy
AFM	Atomic force microscopy
BET	BET method
Chem.	Chemisorption
DRIFTS	Diffuse reflectance Fourier transform spectroscopy
DSC	Differential scanning calorimetry
DTA	Differential thermal analysis
EELS	Electron energy loss spectroscopy
EMA	Electron microprobe analysis
EPMA	Electron probe microanalysis
EPR	Electron paramagnetic resonance
ESCA	Electron spectroscopy for chemical analysis
EXAFS	Extended X-ray absorption fine structure
FAB-MS	Fast atom bombardment mass spectroscopy

Table 9.1 Continued

Abbreviation	Name of method
FEM	Field emission microscopy
FIM	Field ion microscopy
FTIR	Fourier transform infrared spectroscopy
Hg-Por.	Mercury porosimetry
HREELS	High-resolution electron energy loss spectroscopy
ICP	Inductive coupled plasma
IMP	Ion microprobe
IR	Infrared spectroscopy
ISS	Ion scattering spectrometry
LEED	Low-energy electron diffraction
LMMS	Laser microprobe mass spectrometry
Magn.	Magnetic susceptibility measurements
MASNMR	Magic angle spinning NMR
MBS	Molecular beam scattering
Micr.	Optical microscopy
Micr. Cal.	Microcalorimetry
Moss.	Mossbauer spectroscopy
NMR	Nuclear magnetic resonance
NS	Neutron scattering
PAS	Photoacoustic spectroscopy
Physisorp.	Physisorption
PIXE	Proton-induced X-ray emission
Raman	Raman spectroscopy
RED	Radial electron distribution
SEM	Scanning electron microscopy
SEXAFS	Surface-sensitive extended X-ray absorption fine structure
SIMS	Secondary ion mass spectroscopy
STEM	Scanning transmission electron microscopy
STM	Scanning tunneling microscopy
TEM	Transmission electron microscopy
TGA	Thermogravimetric analysis
TP	Transport properties
TPD	Temperature-programmed desorption
TPO	Temperature-programmed oxidation
TPR	Temperature-programmed reduction
UPS	UV photoelectron spectroscopy
UV-vis	UV-visible spectroscopy
XES	X-ray emission spectroscopy
XLBA	X-ray line broadening analysis
XRF	X-ray fluorescence spectroscopy
XPS	X-ray photoelectron spectroscopy
XRD	X-ray diffraction

Table 9.2 Selected Instrumental Methods Used for the Characterization of Solid Catalysts and Surfaces

Category of analysis	Method	Characteristics of method	Applications
Diffraction analysis	XRD	• Gives a distinctive X-ray pattern showing the X-ray scattering intensity at different diffraction angles (2θ)	• Identification of structure, composition, degree of crystallinity, crystallite size, unit cell size
	Neutron diffraction	• Scattering of neutron beam by light elements	• Location of protons and light elements, structural details
	EXAFS	• Synchroton X-ray radiation used to obtain radial structure functions	• Local coordination and interatomic distances, mostly in deposited or impregnated metals
	XLBA	• Uses relation between width of X-ray diffraction lines and size of small crystals	• Measurement of particle size of metal dispersed on support
Pore structure analysis and adsorption	Adsorption of N_2, Hg, H_2O, He, Ar, or organic matter	• Measures pore structure and size from adsorption of gases or vapors • Determines interaction between surface and adsorbent	• Determination of pore volume, pore size, and surface area from adsorption isotherms N_2: for pores < 60 nm Hg: for pores > 20 nm • He: for skeletal density determination • Org: estimate pore opening in new zeolites • Ar: estimate pore types and size distribution (applied to v. small pores: <2 nm); identify zeolites in mixture
	Chemisorption	• Measures specific chemisorption of H_2, CO, or O_2 on metal particles dispersed on a support	• Measurement of metal catalyst (Pt, Pd) dispersion on a support

Table 9.2 Continued

Category of analysis	Method	Characteristics of method	Applications
Thermal analysis	TPD	• Measures rates of adsorption/desorption using mass spectromety	• Acid site characterization (e.g., NH_3 ads) • Evaluate activation energy of desorption • Study dehydroxylation of solids
	TGA	• Measures rates of adsorption/desorption using microbalance	
	DSC	• Measures rates of phase changes and sintering	• Zeolite stability; phase transition, desorption, and phase transformation kinetics
	TPR/TPO	• Measures rates of reduction/oxidation	• Characterization of redox properties of metals on catalysts
	Microcalorimetry	• Measures heats of adsorption	• Acid (base) sites characterization from heats of ads. of base (acid) molecules
Spectroscopic analysis	IR	• Measures OH stretching frequencies directly	• Measurement of number, strength, and type of acid sites
		• Measures interaction of solid acids with probe molecules (e.g., Py, NH_3)	• In zeolites, measures absorbance of SiO_2-Al_2O_3 framework at different frequencies
	UV-visible	• Use of Hammet indicators • Use of benzene adsorption (UV)	• Characterization of acidity of solids
	MASNMR	• Measures chemical shift of Si, Al, H^+, P	• Determination of type and concentration of structural units (SiO_4, AlO_4) in zeolites
	Xe-NMR	• Measures NMR spectrum of adsorbed ^{129}Xe	• Characterization of pore size and shape
	Mossbauer	• Measures Mossbauer spectrum	• Fe in catalyst (location, valency)

Table 9.2 Continued

Category of analysis	Method	Characteristics of method	Applications
	Raman	• Measures vibrational characteristics of material	• Characterization of SiO_2-Al_2O_3 framework in zeolites • Identification of zeolite structure and that of template
Surface analysis	SEM	• Bombardment of surface with e^- beam and detection of backscattered e^-s	• Examination of particle morphology • Identification and location of impurities • Observe distribution of components in catalyst • Identification of mesopores in zeolites
	TEM	• Electrons transmitted through thin section of sample	• Imaging of zeolite lattice structures • Observe intergrowth and faulting • Measurement of particle size distribution
	STEM	• Surface analysis through bombardment with e^-beam and detection of transmitted e^-beam	• Observe structure and topography of catalytic material • Observe pore opening in zeolites
	XPS (ESCA)	• Bombardment of surface with X-ray beam and observation of ejected photoelectrons	• Location and oxidation state of metals on zeolite • Dispersion of supported catalyst
	Auger	• Bombardment with e^-s to ionize core e^-s and observation of following relaxation process	• Elemental analysis, oxidation state
	SIMS	• Bombardment of surface with positive ionic beam (ion sputtering technique)	• Elemental analysis
	FAB-MS	• Bombardment of surface with atomic beam (atom sputtering technique)	• Identification of Al gradient in dealuminated zeolites

used mostly for fundamental research. A more detailed description of instrumental methods used to characterize solid catalysts can be found in several reviews and chapters in various books [1–5].

In the following sections, methods used to characterize catalyst acidity and metal dispersion are discussed in more detail due to the important role these properties play in hydrocracking. Subsequently, several key instrumental methods used for catalyst characterization are described.

9.1. Characterization of Catalyst Acidity

Since the acidity of the catalyst support is reflected in the cracking activity of the hydrocracking catalyst, support acidity plays a key role in catalyst performance. The complete description of acidic properties of solid surfaces requires the determination of type, number, strength, and location of acid sites. A variety of methods have been developed that can be used to characterize the acidity of solid acids.

Two types of acid sites can be found on solid acids: Broensted sites, characterized as proton donors; and Lewis sites, characterized as electron pair acceptors. Dehydration, e.g., by heating, converts a Broensted acid to a Lewis acid. Conversely, hydration converts a Lewis acid to a Broensted acid.

Figure 9.1 shows the effect of temperature on the number of Broensted and

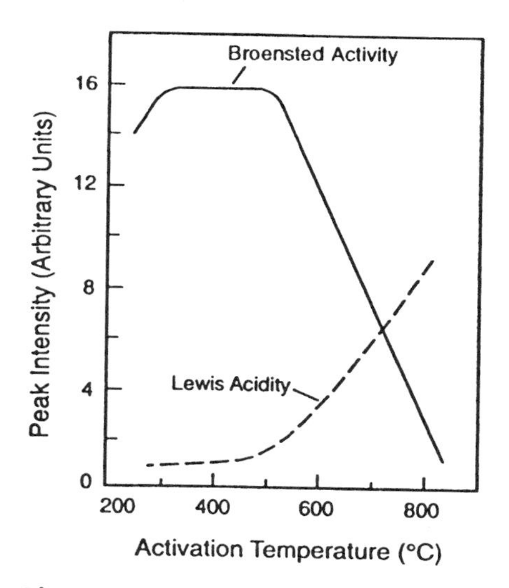

Figure 9.1 Effect of calcination temperature on concentration of Broensted and Lewis acid sites in Y zeolite [6].

Lewis acid sites in a H^+-Y zeolite, as measured by pyridine adsorption [6]. The number of Broensted acid sites converted to Lewis sites increases with temperature. For every two Broensted sites lost, one Lewis site is formed. Superacid sites have also been identified in some solid acids, such as acid salts and zeolites. Superacids have very high proton donor strength [7,8].

The acid strength of a solid indicates how readily the solid is able to donate protons or to accept electron pairs. Acid strength can be measured by a variety of methods. Some of these methods can be used to determine both the strength and number of acid sites. A number of reviews of this subject have been published [9–12]. A brief description of these methods follows.

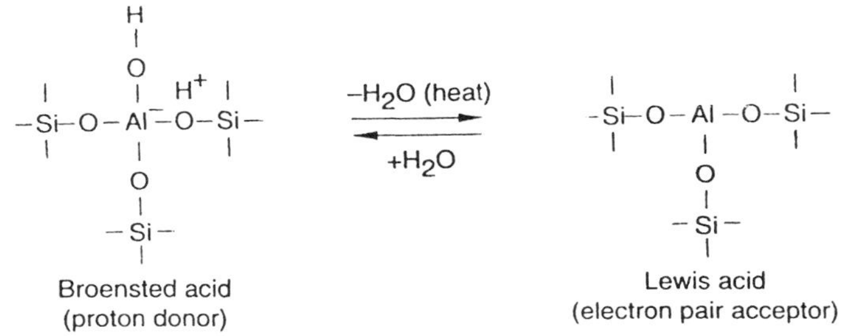

$-H_2O$ (heat)
$+H_2O$

Broensted acid
(proton donor)

Lewis acid
(electron pair acceptor)

9.1.1. Indicator Methods

The earliest methods used to measure acid strength were the indicator methods, based on the adsorption of suitable indicators on the solid acid [9–12]. The indicator is usually a neutral organic base, which upon adsorption on the solid is changed to its conjugated acid form. The neutral base form of the indicator has a different color than the conjugated acid form.

The acid strength is often expressed by the Hammett acidity function, H_0:

$$H_0 = \mathrm{p}K_a + \log \frac{[\mathrm{B}]}{[\mathrm{BH}^+]}$$

where K_a is the equilibrium constant of dissociation of the acid form of the indicator BH^+, and $\mathrm{p}K_a = -\log K_a$. [B] and $[BH^+]$ are the concentration of the neutral base and of the conjugated acid, respectively. The lower the value of H_0, the more acidic the surface.

To measure the acid strength, samples of the pulverized solid suspended in a nonaqueous, inert solvent (benzene, decalin, cyclohexane) are tested with a battery of neutral basic indicators (Hammett indicators). Different indicators will change color at different $\mathrm{p}K_a$ values (see Table 9.3) [10]. This procedure gives an approximate value of the acid strength, which typically can be measured over the

Table 9.3 Basic Indicators Used for the Measurement of Acid Strength [10]

Indicators	Color		pK_a	$[H_2SO_4]$ (%)[a]
	Base form	Acid form		
Neutral red	Yellow	Red	+6.8	8×10^{-8}
Methyl red	Yellow	Red	+4.8	—
Phenylazonaphthylamine	Yellow	Red	+4.0	5×10^{-5}
p-Dimethylaminoazobenzene (dimethyl yellow or butter yellow)	Yellow	Red	+3.3	3×10^{-4}
2-Amino-5-azotoluene	Yellow	Red	+2.0	5×10^{-3}
Benzenazodiphenylamine	Yellow	Purple	+1.5	2×10^{-2}
4-Dimethylaminoazo-1-naphthalene	Yellow	Red	+1.2	3×10^{-2}
Crystal violet	Blue	Yellow	+0.8	0.1
p-Nitrobenzeneazo-(*p*′-nitro) diphenylamine	Orange	Purple	+0.43	—
Dicinnamalacetone	Yellow	Red	−3.0	48
Benzalacetophenone	Colorless	Yellow	−5.6	71
Anthraquinone	Colorless	Yellow	−8.2	90

[a]wt % of H_2SO_4 in sulfuric acid solution which has the acid strength corresponding to the respective pK_a.

pK_a range of about +6.8 (very weak) to about −8 (very strong). Other types of indicators that react with acids to form carbenium ions, such as aryl alcohols (H_R indicators), may also be used.

The number of acid sites in a solid can also be measured by titration with a base, such as *n*-butylamine ($pK_a \sim +10$) in the presence of different color indicators [12]. However, amine titration is of limited use for measuring zeolite acidity because the indicator molecules are not likely to enter the zeolite pores. The method is also difficult to apply to colored samples.

9.1.2. Gas Phase Sorption Methods

Acid strength and the number of acid sites can be measured by gaseous base adsorption methods. The amount of gaseous base that a solid acid can adsorb chemically from the gaseous phase is a measure of the number of acid sites on its surface. Upon heating, the adsorbed base molecules are removed first from weak acid sites. As the temperature increases, the base molecules are progressively removed from stronger acid sites. Hence, the amount of adsorbed base evacuated at various temperatures can give a measure of acid strength. The bases commonly used in such measurements are ammonia, pyridine, and *n*-butylamine. More recently, adsorption of isopropylamine has been used to measure the number of acid sites [13]. These methods can be applied to the whole catalyst as well as to its acidic components. Typically, the base compound is adsorbed on the solid in a thermogravimetric balance, evacuated, and heated gradually. The decrease in weight with temperature is measured. The amount of thermally desorbed base can also be determined by mass spectroscopy. This is the temperature-programmed desorption (TPD) method [12,14]. The higher the temperature required for desorption, the stronger the acidity of the corresponding acid sites. This technique has the advantage of allowing the study of a catalyst under conditions similar to those of reaction and can be applied to zeolites. It can also be applied to colored samples (see also Section 9.3.5).

The indicator methods and gaseous base adsorption methods do not distinguish between Broensted and Lewis acid sites because both types of acid sites react with basic molecules. Therefore, these methods give the sum of Broensted and Lewis acid sites.

9.1.3. Infrared Spectroscopic Methods

Infrared (IR) spectroscopy is widely used to characterize the acidity of zeolites and solid catalysts. Zeolites show strong absorption bands in the 3500–3750 cm^{-1} region of the IR spectrum due to stretching vibrations of OH groups. The band near 3750 cm^{-1} is assigned to nonacidic silanol groups. Strongly acidic OH groups are responsible for a band in the 3630–3650 cm^{-1} region. Weaker acidity is associated with a band near 3550 cm^{-1} [15,16].

Recent studies by Klinowski et al. [17] have shown that the IR frequency of the strongly acidic band depends on the Si/Al ratio in the zeolite framework and shifts to lower frequencies as the ratio increases (Figure 9.2). Furthermore, it was shown that the acidic band in the 3630–3650 region is actually composed of several bands of different acid strength. These bands were associated with different Si(nAl) structural groups identified by ^{29}Si-MASNMR spectroscopy (see Section 9.3.3).

In dealuminated zeolites, the intensity of acidic OH bands is significantly reduced as compared to those in the parent zeolite due to lower Broensted acidity. However, the intensity of the silanol band often increases with dealumination. Furthermore, a new acidic band, associated with nonframework species, appears in the OH stretching region at about 3610 cm^{-1} [18]. A band in the same region has been associated with strong acidity in ZSM-5 and other high-silica zeolites [19]. Hydroxyl nests, left in the framework vacancies upon dealumination, generate a broad IR absorption band in the OH stretching region [20].

The 300–1250 cm^{-1} (mid-IR) region of the faujasite spectrum is usually designated as the framework region. The bands in this region have been assigned to different vibrations in the framework (Figure 9.3) [21]. Upon dealumination, some of these bands shift to higher frequencies [18,22]. By measuring the fre-

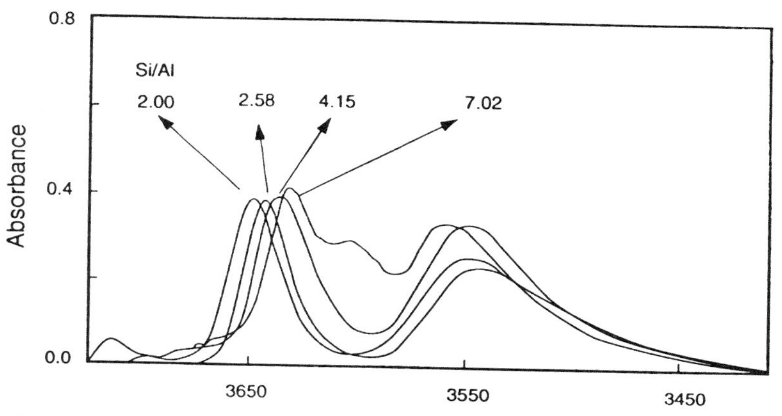

Figure 9.2 Infrared spectra of OH stretching region for Y zeolites with different Si/Al ratios [17].

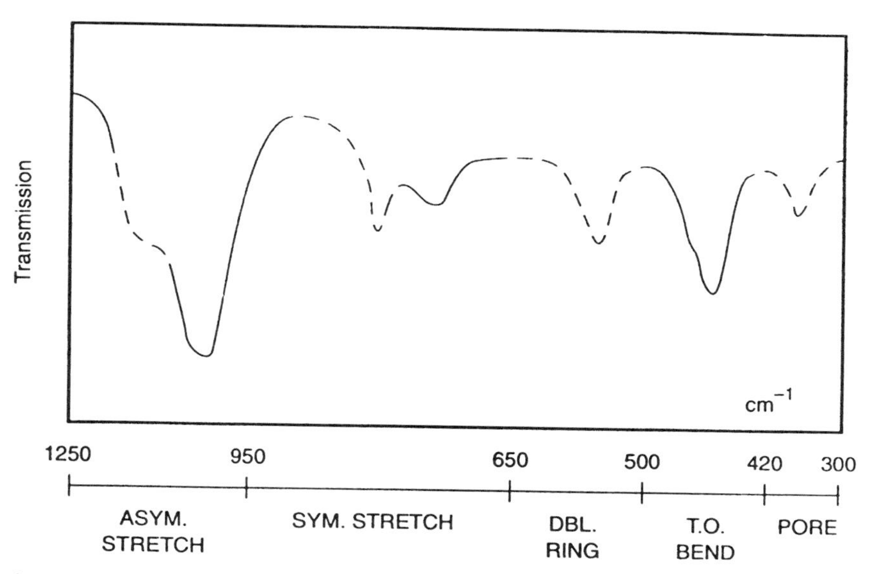

Figure 9.3 Infrared spectrum of framework region for Y zeolites, showing the assignment of different bands. Dashed line indicates bands affected by framework dealumination [21,24].

quency shift of the 800 cm^{-1} and 1050 cm^{-1} IR bands, it is possible to estimate the aluminum content of the framework [23,24].

Infrared spectroscopic studies of ammonia and pyridine adsorbed on acid sites make it possible to distinguish between Broensted and Lewis acid sites, and to assess the number of these sites independently [25–27]. NH_3 molecules react with Broensted acid sites and form NH_4^+ ions, or can be bonded to Lewis acid sites as coordinatively bound NH_3. Similarly, pyridine reacts with Broensted acid sites forming pyridinium ions and with Lewis acid sites forming coordinatively bound pyridine. The species formed at the Broensted acid sites absorb in a different region of the IR spectrum than the species formed at the Lewis acid sites [18]. By measuring the area of the corresponding absorption bands in the IR spectra, one can determine quantitatively the number of Broensted and Lewis acid sites of a solid. This number is commonly expressed in mmol/g or μmol/g. Infrared spectroscopy using pyridine as an adsorbate is extensively used to measure the Broensted and Lewis acidity of solids.

9.1.4. Catalytic Methods

The catalytic activity of a solid acid can be used as a measure of its acidity and its acid strength. Several model reactions have been suggested.

A frequently used test to measure the intrinsic acid activity of a solid catalyst is the α test [28]. The test uses *n*-hexane as the probe molecule. The "α value" of a catalyst is defined as the ratio of the first-order rate constant for *n*-hexane cracking over the sample to that obtained over an arbitrary standard, measured at 538°C. Measurements made at other temperatures can be extrapolated to 538°C using a temperature correction factor. For example, using the α test, an essentially linear relationship between catalytic activity and the number of protonic sites in H-ZSM-5 was established over a range of SiO_2/Al_2O_3 ratios from 35 to 60,000 (~15 ppm Al) (Figure 9.4) [28].

Another catalytic reaction used to compare the acidity of zeolites is the cracking of *n*-butane [29]. A plot of the first-order rate constants, K_a, for *n*-butane cracking in this test (2 mol % butane in He, 500°C) vs. zeolite aluminum content is shown for a variety of zeolites in Figure 9.5. The results show the presence of significant catalytic acidity in high-silica zeolites such as ZSM-5 and mordenite.

Some of the other catalytic reactions used to examine zeolite acidity are isomerization/hydrocracking of *n*-decane [12], isomerization of xylene [30], skeletal isomerization of olefins [31], disproportionation of ethylbenzene [32], cumene cracking [33], *n*-heptane cracking [34], and so on.

The rate of reaction over a solid acid depends on both the number and strength of acid sites, and different acid-catalyzed reactions require catalysts with different acid strength. A list of reactions requiring a minimum H_R value for the acid catalyst is shown in Table 9.4 [12].

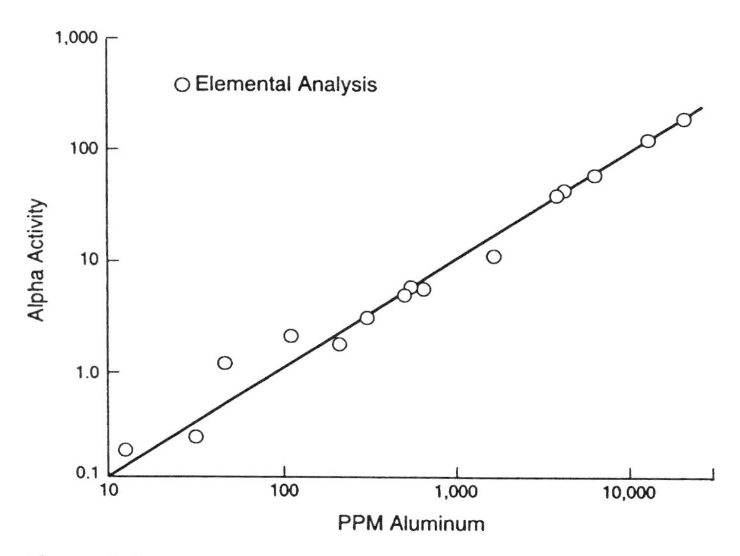

Figure 9.4 Correlation between acid activity and aluminum content in HZSM-5 zeolite [28].

The catalytic properties of a solid can be changed significantly by changing the number and strength of its acid sites. For example, partial dealumination of zeolites reduces the number of acid sites and, in the case of dealuminated Y zeolites, increases the strength of the remaining acid sites [35] (see Section 3.1.2).

9.1.5. Other Methods

Microcalorimetric measurements of the adsorption heats of bases on solid acids [36,37], conductometric titration [11], UV-visible spectroscopy [38], aromatics adsorption [39], NMR [40], luminescence [41], and electron spin resonance [42], spectroscopic methods have been used to measure the acidity of solids. This subject, as well as the correlation between surface acidity and catalytic activity, has been described in several reviews [9,11,12] and in chapters of different treatises [16,38,43–46].

9.2. Measurement Methods of Metal Dispersion

Optimum dispersion is of particular economic importance in the case of supported metal catalysts containing expensive active materials such as noble metals. By improving metal dispersion one can improve catalyst activity for a given amount of metal. Similarly, improved metal dispersion can result in a catalyst having a

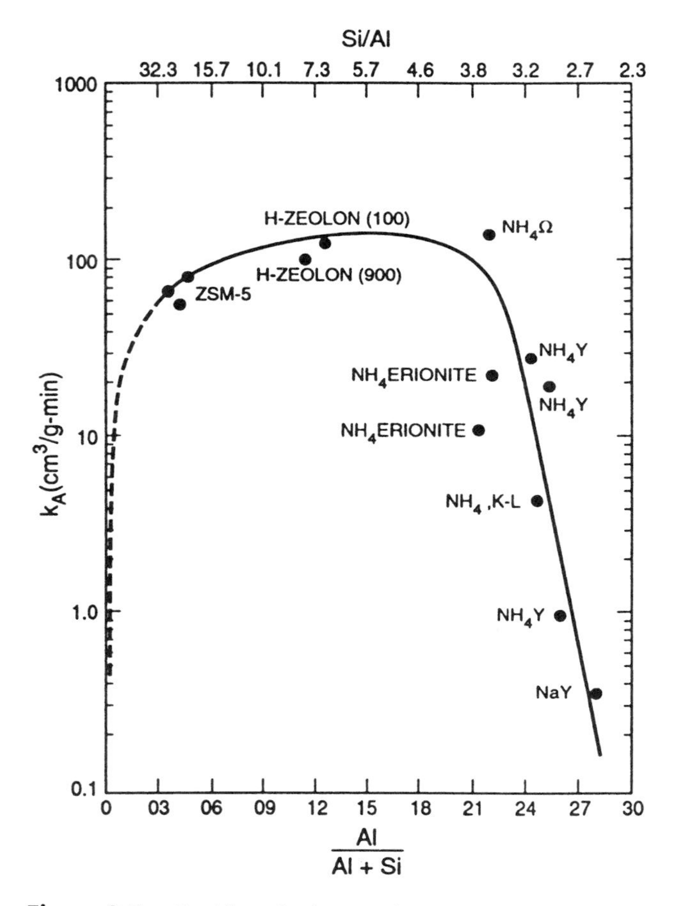

Figure 9.5 Cracking of *n*-butane: the pseudo-first-order rate constant k_A vs. aluminum content of zeolite [29].

Table 9.4 Minimum Broensted Acid Strength Required for Different Types of Acid-Catalyzed Reactions [12]

Acid strength	H_R value	Reaction type
Low	$< +4$	Dehydration of alcohols
↓	$< +0.82$	Cis-trans isomerization
	< -6.63	Double-bond migration
	< -11.5	Cracking of alkylaromatics
	< -11.6	Skeletal isomerization
High	< -16.0	Cracking of paraffins

reduced amount of active metal, but with the same activity as one with a larger amount of less dispersed metal. Dispersion measurements also allow the determination of the number of exposed active sites of supported metals and thus make possible the study of reaction mechanisms and kinetics.
The most common methods used to measure metal dispersion are chemisorption, X-ray diffraction, TEMS and XPS peak intensity measurements. Other methods, such as extended X-ray absorption fine structure spectroscopy and ^{129}Xe-NMR, are also now being used to measure metal dispersion.

9.2.1. Chemisorption

Chemisorption is the most widely used method for measuring metal catalyst dispersion [47]. The gases used in these measurements are mostly hydrogen and carbon monoxide. A typical example is the selective chemisorption of hydrogen on supported Pt, which occurs instantly at room temperature and readily reaches a complete monolayer coverage on the exposed Pt surface.

The specific Pt surface can be easily measured by H_2 chemisorption provided the following conditions are fulfilled: (a) The H/Pt surface stoichiometry is ~1 and the effective area per H atom on the surface is known; (b) the physical adsorption is negligible, both on metal and on support; and (c) no secondary phenomena occur during H_2 chemisorption, such as dissolution of H_2 in the metal, strong metal–support interaction, or spillover of hydrogen from metal to support. However, these conditions are not always fulfilled [47,48]. Furthermore, the method is not applicable to multimetal catalysts, in which the metals have similar chemisorption properties (no selective chemisorption). Still, for monometallic supported catalysts, the method is widely used. By knowing the amount of H_2 chemisorbed, one can readily obtain the dispersion of supported Pt, as well as the metal surface area. The measurement is carried out in an adsorption apparatus, and the amount of H_2 adsorbed is determined by a volumetric or gravimetric method. Other techniques to measure adsorbed H_2, such as the continuous flow method or the pulse technique, may also be used.

A variation of the traditional chemisorption method is the H_2-O_2 titration method proposed by Benson and Boudart [49] to measure Pt dispersion on a supported catalyst. The method consists in adsorption of H_2 on Pt, followed by "titration" with O_2. Water is the final reaction product. Conversely, O_2 can be adsorbed on Pt and then "titrated" with H_2. From the amount of gas used for titration one can calculate Pt dispersion.

Chemisorption of H_2 on Pd is complicated by the tendency of Pd to dissolve H_2 and form hydrides, to undergo sintering during evacuation at high temperature, and to weakly chemisorb excess hydrogen [50]. Chemisorption of CO can be used as an alternative. However, CO can adsorb on Pd in linear or bridged form: Pd=C=O or $^{Pd}_{Pd}$ > C=O. The stoichiometry for CO/Pd is 1 for the linear form and

½ for the bridged form. This makes it difficult to use CO chemisorption to calculate Pd dispersion. The H_2-O_2 titration method has been used more recently to measure Pd dispersion on a support [51].

Chemisorption methods have also been used to measure the dispersion of supported, nonmetallic components, such as MoO_3 and MoS_2. Low-temperature O_2 chemisorption has been used to measure the surface area of supported MoO_3 [52,53]. Oxygen chemisorption was used to measure the active surface area of MoS_2 in hydrodesulfurization (HDS) catalysts [54]. However, these methods need further development.

9.2.2. X-ray Diffraction

The X-ray diffraction method estimates the size of small crystallites, using the relationship between crystallite size and the width of the diffraction lines. X-ray line broadening analysis (XLBA) is one of the oldest methods used to measure crystallite sizes, described by Scherrer in 1918 [55]. More recently, it has been used to determine the size distribution of supported metals [56]. The correlation between line broadening and crystallite size is used to measure crystallite sizes in the 0.01- to 1-μm size range [57].

The X-ray diffraction method has some limitations. XLBA provides information on the dispersion of a supported catalyst only if it is in the form of a separate crystalline phase (i.e., it does not apply at high degrees of dispersion, such as atomic dispersion). Furthermore, the result may be affected by the presence of other crystallites in the system.

An alternative X-ray method for particle size determination consists of the analysis of small-angle X-ray scattering (SAXS). The theory of this method has been developed by Guinier and applied to heterogeneous catalysts by Whyte [58].

9.2.3. Transmission Electron Microscopy

Transmission electron microscopy (TEM) is another method used to determine metal dispersion on a support. The method allows for the determination of the size and shape of a discrete number of particles. By using these data, one can estimate various average parameters of the population of particles, such as particle size distribution or mean particle size.

To obtain a reliable particle size distribution from electron images, the following assumptions apply: (a) the measured size is equal to the true size of the particle; (b) all particles have the same probability of being observed, regardless of size; (c) contrast arising from support cannot be confused with contrast arising from particles.

The resolving power of modern electron microscopes is currently down to about 3 Å [59]. This allows the identification of isolated, heavy atoms. However, in the case of supported catalysts, the visibility of the smallest particles will be hindered by the contrast arising from the support. Therefore, particle size distribution measurements involving small particles below a limiting size are subject to error. For example, for Pt supported on γ-Al_2O_3 the limit is set at 25 Å [60]. For amorphous supports the limit is lower ($<$20 Å). One should note that a particle size of about 10 Å corresponds generally to 100% dispersion because all atoms are exposed on the surface.

9.2.4. XPS Peak Intensity Measurement

The XPS method can be used to assess the dispersion of metals on supports. The method can be applied to any catalytic system, without the limitations inherent to the other methods described. Models have been developed that show the relationship between XPS peak intensities and dispersion [61,62]. So far, however, only the qualitative interpretations of XPS data have been found reliable.

9.2.5. EXAFS Method

Extended X-ray absorption fine structure spectroscopy (EXAFS) can be used to obtain structural information about supported metal catalysts. It is a method used for determining metal atom coordination numbers, interatomic distances and the nearest-neighbor metal atoms, as well as the vibrational motions of the metal atoms. The average coordination number is indicative of metal dispersion. EXAFS has been used to measure bond distances and the dispersion of Pt and other noble metals on different supports [63,64]. The application of EXAFS in catalysis has been described in several reviews [65–67].

Due to the limitations of the different methods described, measurements are often carried out using several methods. The data obtained are compared and attempts made to interpret observed discrepancies. Examples of such interpretations can be found in the literature [47,63].

9.3. Selected Instrumental Methods

The instrumental methods used for catalyst characterization can be classified, somewhat arbitrarily, in the following categories [68]: (a) Diffraction methods; (b) Adsorption methods; (c) spectroscopic methods; (d) surface analytical methods; and (e) thermal analytical methods (Table 9.2).

Diffraction methods are primarily used to identify the type, structure, and amounts of crystalline materials, such as zeolites and aluminas. Adsorption methods are used to determine surface area, pore volume, and pore size distribution. Spectroscopic methods are used to characterize the acid sites of solid acids and to describe aluminum coordination and distribution in zeolites. Surface analytical methods are used to characterize the morphology of solids, to provide elemental mapping, and to evaluate the oxidation state of elements present.

Thermal methods are used to characterize the number and strength of acid sites, to establish thermal and hydrothermal stability, and to evaluate stability to oxidation/reduction.

Some of these methods, used for characterization of solid acids and for measurement of metal dispersion, have already been described. The description of selected instrumental methods follows.

9.3.1. X-ray Diffraction Methods

X-ray diffraction (XRD) involves the diffraction of X rays by the lattice of crystalline materials. In addition to the measurement of small crystallite sizes already described, X-ray diffraction can be used to identify crystalline phases, to measure the amount of those phases in a mixture, to determine crystal structures, and to determine the unit cell size of the crystalline material. In the case of zeolites, XRD methods can also be used to estimate the aluminum content of the zeolite framework by measuring unit cell size.

Each crystalline material has a characteristic XRD pattern. This makes it possible to identify different crystalline phases in a mixture. In the case of hydrocracking catalysts, the XRD pattern may show peaks of some or all of its components, such as zeolites, crystalline alumina, and transition metals or metal oxides. Collections of zeolite XRD patterns are available [46,69,70]. A collection of XRD patterns for aluminas has been published by Alcoa [71]. Available programs for analysis of powder diffraction data have been reviewed [72].

a. Quantitative analysis and percent crystallinity. The peak intensities in an XRD scan can be used to estimate the amount of a specific crystalline phase present in a catalyst. This is accomplished by comparing the peak intensities to some standard defined as having 100% crystallinity. Peak areas are often used instead of peak intensities for better accuracy. In the case of zeolites, a standardized test (ASTM D-3906) is used to measure the percent crystallinity of a faujasite-containing material [73].

b. Zeolite unit cell size measurement. X-ray diffraction is often used to determine the unit cell size of a zeolite. This, in turn, provides information about the zeolite framework composition and its potential catalytic performance (see Section 7.5).

The simplest case is that of zeolites crystallized with a cubic unit cell, such as X or Y zeolites. In this case, all sides of the unit cell are equal in size ($a = b = c = a_0$) and all angles are 90°. The measurement of unit cell size by XRD has been standardized as ASTM D-3942 [73].

Unit cell size measurements are often used to evaluate the framework aluminum content of faujasite-type zeolites. Because the Si-O-Al bond is longer than the Si-O-Si bond, the unit cell increases slightly with aluminum content.

Several correlations between unit cell size and framework aluminum content have been proposed. An early correlation was developed by Breck and Flanigen [74] for low Si/Al ratio (1.5–2.5) sodium faujasite. More recently, correlations have been proposed by Fichtner-Schmittler et al. [75] and by Sohn et al. [76]. These latter correlations have been derived for dealuminated and decationized zeolites. The number of aluminum atoms per unit cell, N_{Al}, is given by the following correlations (Figure 9.6):

$N_{Al} = 115.2\ (a_0 - 24.191)$ (Breck and Flanigen)

$N_{Al} = 112.4\ (a_0 - 24.233)$ (Fichtner-Schmittler et al.)

$N_{Al} = 107.1\ (a_0 - 24.238)$ (Sohn et al.)

where a_0 is the unit cell size in Å. Knowing N_{Al}, one can calculate the framework Si/Al ratio, R: $N_{Al} = 192/(R + 1)$. The total aluminum content of the zeolite, including both framework and nonframework aluminum, can be obtained by chemical analysis.

In addition to aluminum content, the unit cell size of a faujasite-type zeolite is affected by the type and amount of cations present, as well as by the degree of hydration of the zeolite [44,77]. Cation exchange, formation of cationic aluminum upon steam-dealumination, and removal of nonframework aluminum will affect

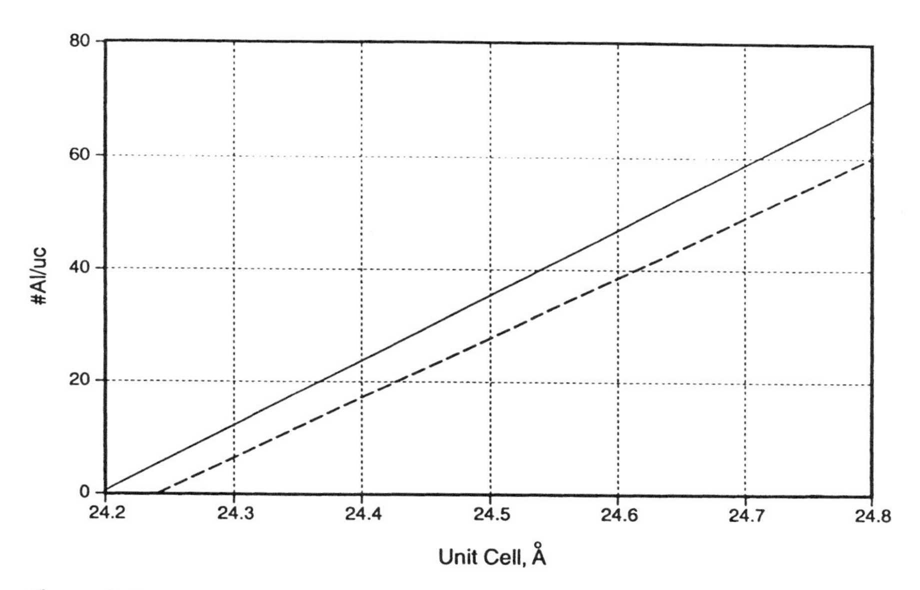

Figure 9.6 Correlation between unit cell size and framework aluminum content. Number of (Si+Al) atoms for unit cells is 192 [76–76]; —— Breck; – – – Sohn.

unit cell size. These effects may explain reported unit cell sizes lower than the zero aluminum limit of the Fichtner-Schmittler and Sohn correlations [78–90].

Other diffraction methods, such as *neutron diffraction* [81], *EXAFS* (extended X-ray absorption fine structure spectroscopy) [82], and *SAXS* (small angle X-ray scattering) [83,84], have found a more limited application in the study and characterization of catalytic materials.

9.3.2. Adsorption Methods

Adsorption methods are used to characterize the dispersed metals as well as the support of bifunctional catalysts. The use of chemisorption to measure metal dispersion is described in Section 9.2.1. Adsorption methods used to measure catalyst acidity are described in Section 9.1.2. Nitrogen adsorption methods are commonly used to determine the surface area, pore volume, and pore size distribution of catalysts and catalyst supports. Water adsorption is used to determine the total pore volume of a material. Helium adsorption (helium picnometry) is used to measure the skeletal density of porous solids. Furthermore, adsorption data can provide structural information about porous materials. For example, adsorption of probe molecules of different sizes, such as *n*-hexane, *n*-butane, O_2, N_2, Ar, He, H_2O, and CO_2, has been used to estimate the size of pore openings in new zeolites.

Several books provide comprehensive reviews of gas adsorption by porous and nonporous solids, as well as the use of adsorption for surface area and pore size distribution measurements [85–89].

Nitrogen adsorption is commonly used to determine the total surface area of a catalyst or catalyst component, such as zeolites, alumina, or amorphous silica-alumina. Nitrogen adsorption can be used to estimate the volume of pores ranging in size up to ~600 Å. For materials with larger pores (macropores), the mercury intrusion method is used.

By plotting the volume of nitrogen adsorbed by the catalytic material at different pressures at liquid nitrogen temperature (−195°C, −319°F), one can obtain the adsorption isotherm. Nearly all adsorption isotherms may be grouped into six types shown in Figure 9.7 [85,90]. Upon desorption, some isotherms show hysteresis loops. The type of isotherm is related to the structure of the solid sample. Microporous materials, such as conventional Y or ZSM-5 zeolites, have a type I isotherm. Mesoporous materials have a type IV isotherm.

Among the physical properties of hydrocracking catalysts (or of their components) most frequently measured by adsorption methods are surface area and pore size distribution. A brief discussion of the measurement of these properties follows.

a. Surface area. The surface area of zeolite-based catalysts is the result of the combined contribution of zeolite and matrix component. Whereas the surface area of fresh Y zeolites is high (usually over 800 m^2/g for conventional Y

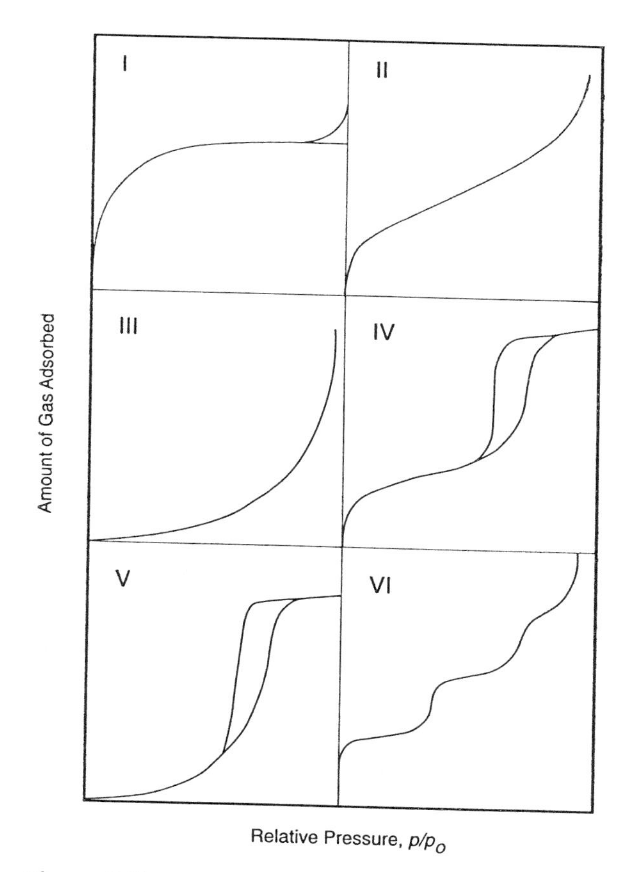

Figure 9.7 Different types of sorption isotherms [85].

and over 600 m^2/g for USY zeolites), the surface area of the matrix can vary, depending on composition and preparation method. In general, the surface area of amorphous silica-alumina is higher than that of alumina. An increase in the surface area of an active matrix increases the activity of the catalyst. This applies especially to amorphous catalysts or to those with a low zeolite content. The surface area of the amorphous component decreases by exposure to high temperatures or steam. Spent catalysts have lower surface areas than the corresponding fresh catalysts.

The principal method of measuring the total surface area of a zeolite consists of determining the adsorption isotherm of nitrogen at liquid nitrogen temperature. From the adsorption isotherm one can calculate the surface area of the zeolite using the BET (Brunauer, Emmett, Teller) equation. Some commercial laboratories calculate surface areas from the volume of gas absorbed at low p/p_0 values (0.02–0.03).

The surface area of a catalyst is also determined by the BET method, from the corresponding nitrogen adsorption isotherm obtained at liquid nitrogen temperature (ASTM D-3663).

b. Pore size distribution and pore volume. The pore size distribution and pore volume of the catalyst are determined by composition, preparation method, and thermal treatment. According to IUPAC, pores can be classified in the following groups, based on their average width: micropores ($\phi < 20$ Å), mesopores ($\phi = 20$–500 Å), and macropores ($\phi > 500$ Å). Y zeolites have only micropores ($\phi \simeq 7.4$ Å), whereas steam-dealuminated Y zeolites have both micropores and mesopores (see Section 3.1.2). The amorphous components of most hydrocracking catalysts contain primarily mesopores and macropores.

The pore size distribution in the catalyst plays a key role in its performance. A predominance of small pores ($\phi < 50$ Å) in the matrix can cause pore plugging by coke and can lead to diffusion problems, especially when processing heavy feedstocks. Furthermore, in an amorphous matrix such a pore system has poor thermal and hydrothermal stability. A predominance of macropores is usually associated with a lower surface area, which diminishes the catalytic role of the matrix. This role is especially important in amorphous catalysts and in low-zeolite catalysts. To be effective, such catalysts normally have a somewhat narrow pore size distribution [91]. Calcination at high temperature or in the presence of steam increases the average pore radius of the amorphous component and broadens the distribution of pore radii. The pore volume is also reduced under these conditions.

The volume of pores in the range of 20–600 Å is usually determined from the nitrogen sorption isotherm (ASTM D-4222 and D-4641), whereas that of pores in the range of 600–20,000 Å is determined by mercury intrusion porosimetry (ASTM D-4284). The pore volume can also be measured by adsorption of water, oxygen, *n*-hexane, CO_2, Ar, or *n*-butane.

The sorption properties of zeolites have been thoroughly investigated using a variety of probe molecules [87–89,92]. Data obtained from these studies have often provided substantial structural information.

This can be illustrated in the case of conventional and dealuminated Y zeolites. The nitrogen adsorption isotherms, measured at −195°C, show that while adsorption on NaY rapidly reaches saturation at low pressure ($p/p_0 = 0.05$), the adsorption on steam-dealuminated Y (USY) and acid-leached, steam-dealuminated Y (USY_l) increases with increasing p/p_0 values (Figure 9.8) [93]. NaY yields a completely reversible type I isotherm, characteristic of micropore filling common in many zeolites. However, USY and USY_l yield isotherms with hysteresis loops, close to type IV. The differences in the shape of isotherms have

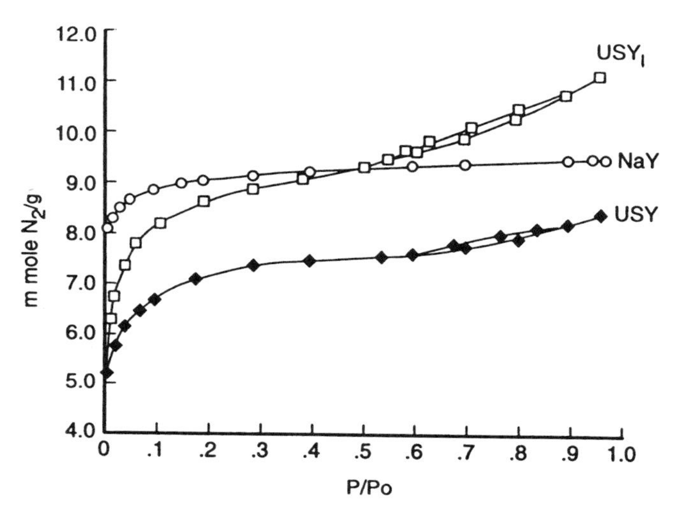

Figure 9.8 Nitrogen adsorption isotherms for NaY, ultrastable Y (USY), and leached ultrastable Y (USY_l) (SiO_2/Al_2O_3 = 68) [93].

been attributed to the formation of secondary pores during framework dealumination. The hysteresis loops observed were attributed to capillary condensation in the secondary pores [92]. The adsorption isotherms obtained for Y zeolites dealuminated with $SiCl_4$ are similar to those obtained for NaY zeolite, suggesting the predominance of micropores [94].

When the porous material contains micropores and mesopores, the pore volume due to mesopores can be calculated by using the so-called t plot [95,96]. To obtain such a plot, the volume of nitrogen adsorbed is plotted against t rather than p/p_0, where t represents the thickness of the adsorbed layer. The resulting plot is linear. Extrapolation of the linear t plot to $t = 0$ yields the micropore volume. By subtracting this value from the total pore volume one can obtain the mesopore volume. In the absence of micropores, the t plot can be extrapolated to the origin (Figure 9.9) [96]. The t-plot method has been used to estimate the amount of zeolite in a catalyst by measuring the micropore and mesopore volume (ASTM D-4365). The method gives accurate results for catalysts with zeolites having only micropores. The t-plot method can also be used to estimate the average particle size of zeolites.

Water adsorption is frequently carried out to measure the total pore volume of a material. By this method, water is added gradually to a weighed amount of outgassed sample until incipient wetness is reached, i.e., until the particles begin to stick together and form a cake. The amount of water added corresponds roughly to the pore volume of the sample.

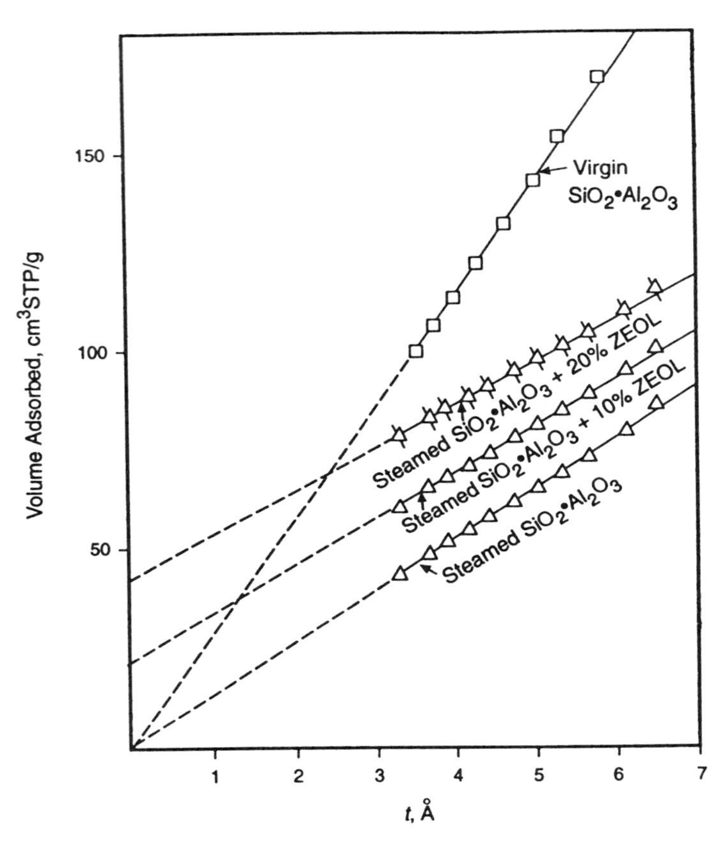

Figure 9.9 Nitrogen adsorption as a function of *t* [96].

9.3.3. NMR Methods

^{13}C-NMR and ^{1}H-NMR methods have been used for decades by organic and physical chemists for identification and structural characterization of organic compounds. The more recent solid state catalytic applications use ^{27}Al-, ^{29}Si-, ^{1}H-, ^{129}Xe-, ^{15}N-, and ^{31}P-NMR techniques to characterize the structure of a variety of catalytic materials. Magic angle spinning nuclear magnetic resonance (MASNMR) techniques are used to characterize the chemical bonds involving a particular element. The values of the chemical shifts observed in the spectra are used to characterize the local environment of that element (Si, Al, P, et al.) in the material under investigation.

The most frequently used methods to obtain structural information about zeolites and other silicon and aluminum containing materials are ^{29}Si-MASNMR and ^{27}Al-MASNMR. The pioneering work in this field was done by Lippmaa et al. [97] and by Engelhard et al. [98]. The application of these methods to aluminosilicates is described in detail in the literature [99–102].

^{29}Si-MASNMR. The structure of most zeolites consists of silicon and aluminum tetrahedra linked through oxygen. The tetrahedral aluminum has a formal negative charge, compensated by a cation. Two adjacent aluminum tetrahedra will not link together due to repulsion of the two negative charges (Loewenstein's rule) [103]. Therefore, in aluminosilicate frameworks, an aluminum will always be surrounded by four silicon tetrahedra. The silicon, however, may be linked to 0, 1, 2, 3, or 4 aluminum atoms in the aluminosilicate structure. Consequently, silicon may exist in five different structural units in the zeolite framework, designated Si(0Al), Si(1Al), Si(2Al), Si(3Al), and Si(4Al). For each of these structural units there is a corresponding silicon chemical shift in the ^{29}Si-MASNMR spectrum, as shown in Figure 9.10A [104]. (The peak corresponding to Si(4Al) in the spectra of these materials is too weak to be observed.) Such spectra provide information about the ordering of silicon and aluminum atoms in zeolites. Furthermore, by measuring the peak areas of different peaks in the spectrum, one can calculate the number of aluminum atoms per unit cell [68].

Upon progressive dealumination, the maximum peak intensities shift to Si(1Al) and ultimately to the Si(0Al) peak, with total or near-total disappearance of the Si(4Al) and Si(3Al) peaks (Figure 9.12A). This allows the use of ^{29}Si-MASNMR spectra to monitor the dealumination of the zeolite framework.

^{27}Al-MASNMR. This method is used to characterize the different types of aluminum structures in zeolites and in other silica-alumina materials, both crystalline and amorphous. Tetrahedral aluminum as it occurs in the faujasite framework has a chemical shift of about 60 ppm relative to the octahedrally coordinated aqueous aluminum cation used as a standard [105]. This chemical shift may have a slightly different value in other zeolites. Alumina, amorphous silica-alumina, and hydrothermally dealuminated zeolites may contain octahedral aluminum, characterized by a peak with a chemical shift of about 0 ppm [106]. Pentacoordinated aluminum, observed in hydrothermally treated faujasite and amorphous silica-alumina, generates a chemical shift at about 30 ppm [107].

The ^{27}Al-MASNMR spectra of thermally dealuminated Y zeolites are shown in Figure 9.10B. An increase in severity of hydrothermal treatment results in a decrease in the chemical shift assigned to tetrahedral aluminum and an increase in the one corresponding to octahedral aluminum. Pentacoordinated aluminum is also formed upon hydrothermal dealumination.

Because framework and nonframework aluminum give rise to different signals, leaching of nonframework aluminum from hydrothermally dealuminated faujasite can be monitored by observing the changes occurring in the corresponding ^{27}Al-MASNMR spectra.

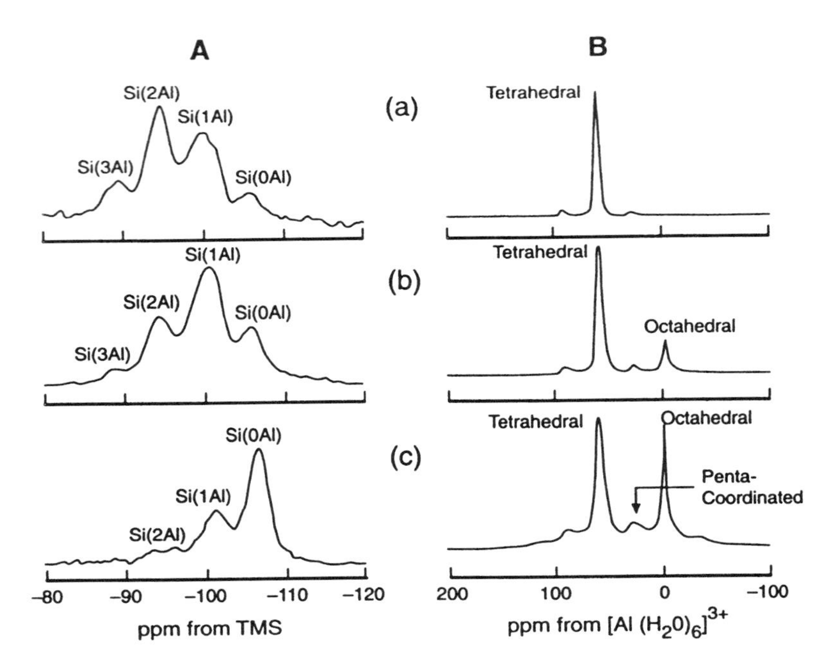

Figure 9.10 ^{29}Si (at 79.80 MHz) (A) and ^{27}Al (at 104.22 MHz) (B) MASNMR spectra of conventional and dealuminated Y zeolites: (a) parent zeolite NH_4, NaY; (b) after calcination in air at 400°C/1 hr; (c) after heating at 700°C/1 hr in the presence of steam [104].

^{31}P-MASNMR has been used to characterize zeolites containing phosphorus in the framework (ALPO- and SAPO-type molecular sieves).

^{129}Xe-MASNMR has been used to characterize the pore structure of zeolites [108,109] and catalysts [110]. ^{129}Xe-MASNMR has also been used to estimate the average number of Pt atoms per cluster for Pt supported on NaY zeolite [111,112].

9.3.4. Surface Characterization Methods

A variety of instrumental methods have been developed to characterize the morphology and surface of solid materials. Electron microscope methods and several spectroscopic methods are commonly used for that purpose. Electron microscopes are often used with attachments for elemental analysis. The thickness of the surface layer analyzed is usually about 1 μm.

Electron microscope methods (TEM, SEM, STEM). These methods involve bombardment of the surface with electrons under vacuum and detection of the backscattered electrons. The electron microscope is typically used to observe the surface structure and topography of catalytic materials at high resolution. It can be used to observe the occurrence and location of zeolites embedded in the matrix of a catalyst particle. It is also used to obtain information about the size of supported metal particles and changes in their size, shape, and location during catalyst use. Furthermore, electron microscopy can be used to observe the pore openings in zeolites.

Scanning electron microscopy (SEM) typically operates at a magnification of 20,000 or more and can resolve structures of 0.1 μm or less. SEM is commonly used to identify the morphology of zeolite crystals. The method is also used to identify impurities that may have crystallized with the main component, using differences in morphology. Special procedures have been developed to observe dynamic events in a controlled environment, such as the growth of carbon filament on metal surfaces in a reducing environment [113].

Transmission electron microscopy (TEM) and scanning transmission electron microscopy (STEM) operate at a higher magnification, i.e., about 100,000 or more. Using high-resolution transmission electron microscopy (HREM), it is possible to resolve features as small as 2 Å, about the size of an atom. Microdiffraction patterns for regions as small as 10 Å in diameter can be obtained. These methods are often used to observe structures at the molecular level in both zeolites and catalysts (Figure 9.11) [114]. TEM micrographs have been used to identify intergrowth and faulting in zeolites [115], as well as the presence of mesopores [116]. TEM is often used to measure metal dispersion on a support (see Section 9.2.3). TEM and STEM can also be used to observe the growth and size of metal crystallites.

Both STEM and SEM frequently have attachments (e.g., an electron microprobe) that allow them to carry out elemental analysis. Such high-resolution elemental analysis was used to identify the existence of an aluminum gradient in some dealuminated zeolites [117].

Electron microprobe is a method often used to establish the local chemical composition of different structures observed by using the electron microscope. It is often applied in conjunction with SEM or STEM. The method consists of bombardment of the surface (typically 1 μm) with an electron beam to produce a characteristic X-ray emission spectrum, which is used to identify the elements present in the bombarded spot. The method has been used to establish the location of different heavy metals in catalyst particles or the distribution of impurities in spent catalysts by cutting the catalyst particle into thin slices and analyzing each slice of material.

X-ray or ultraviolet photoelectron spectroscopy (XPS or UPS). XPS, sometimes identified as ESCA (electron spectroscopy for chemical analysis), is another method used for surface characterization. It involves bombardment of a sample with X rays in ultrahigh vacuum and measurement of the energies of the emitted photoelectrons. Depending on the energy of the incident radiation, emission of the

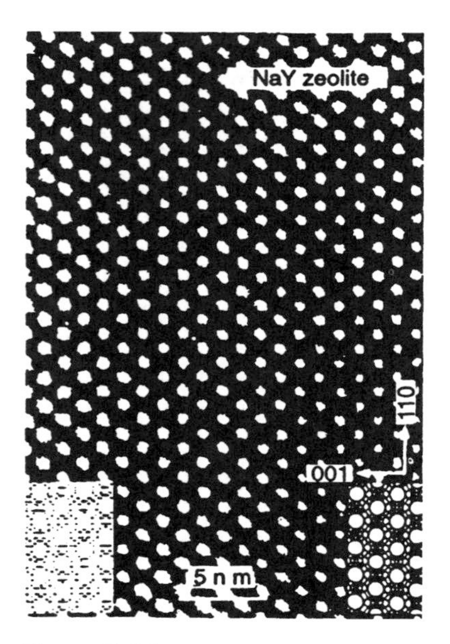

Figure 9.11 High-resolution electron micrograph of NaY zeolite in the [110] direction. The lower inset is the computer-simulated image [114].

photoelectrons will be from the valence band (UPS) or from both the valence and core levels (XPS). Each element has a characteristic spectrum of photoelectron emission energies. These energies also depend on the valency of the element in the sample. This allows the use of XPS to identify the major elements in a sample and to estimate their oxidation state [118]. XPS has been used to identify aluminum gradients in dealuminated zeolites [119,120].

Auger electron spectroscopy (AES). The method is similar to XPS and involves bombardment of a sample with high-energy electrons and measurement of the energy of electrons emitted from an outer shell. AES is being used for elemental analysis of solid surfaces and for oxidation state analysis. The method can also be used for depth profiling. This is accomplished by sputtering successive layers of the material by bombardment with a beam of ions and recording the surface concentration of the elements of interest.

Fast atom bombardment–mass spectroscopy (FAB-MS) and secondary ion mass spectroscopy (SIMS) have been used to study surface composition and depth profiles of materials such as catalysts in order to assess their compositional uniformity. Both are ion-sputtering techniques. These techniques have been used to identify aluminum gradients in zeolites dealuminated by different methods [121,122].

9.3.5. Thermal Analysis Methods

Thermal analysis methods are commonly used to estimate the number and strength of acid sites, as well as thermal or hydrothermal stability and stability to oxidation–reduction.

Microcalorimetric methods measure heats of adsorption resulting from the interaction between an adsorbate and a solid acid. Since the strength of the interaction is measured by the heat of adsorption, microcalorimetric measurements can be used to measure the strength of acid sites on a solid acid.

Thermal analysis methods involving temperature programming provide kinetic information with regard to rates of adsorption–desorption (TPD, TGA), rates of oxidation–reduction (TPO, TPR), and rates of phase change and sintering (DSC).

Temperature-programmed desorption (TPD) can be used to characterize different components of hydrocracking catalysts. The method is used primarily to characterize the acidity of solids, using basic molecules (ammonia, pyridine) as adsorbates (see Section 9.1.2). Combined with infrared spectroscopy, pyridine and ammonia TPD experiments have been used to characterize Broensted and Lewis acidity in zeolites (see Section 9.1.3).

Temperature-programmed reduction (TPR) is often used to characterize the reducibility of supported metal catalysts. In TPR experiments, the sample is heated in a reducing environment and the rate of disappearance of the reducing agent (e.g., H_2 in an H_2/N_2 gas mixture) is observed by measuring the change in thermal conductivity of the gas mixture. The rate of change in hydrogen concentration is proportional to the rate of catalyst reduction. Because different ionic species are reduced at different temperatures and rates, the TPR method can be used to identify and characterize different Pt species present in a catalyst. These findings have been correlated with different preparation and activation methods. The TPR method has also been applied to the study of metal oxide reduction in supported and unsupported catalyst systems.

To gain a better understanding of the nature of the catalyst under investigation, a combination of several instrumental methods is often applied. Furthermore, in situ methods have been developed to observe catalysts under reaction conditions, i.e., at high temperatures in a strongly reducing environment under pressure. These methods may involve, for example, combining thermal and spectroscopic methods in a single experiment. Results obtained from such experiments are more representative of the properties of catalysts under operating conditions.

9.4. Catalytic Evaluation

To assess the catalytic performance of hydrocracking catalysts, different laboratory and pilot plant testing methods have been developed. The data obtained from such tests are being used to compare the activities, selectivities, and stabilities of different commercial or experimental catalysts and to predict their performance in commercial units. Testing hydrocracking catalysts in a pilot plant in order to predict their commercial performance is fairly complex. The difficulty arises primarily from trying to simulate an adiabatic operation in a relatively small system.

The process configurations used for testing hydrocracking catalysts are similar to those used in commercial units. Single-stage or two-stage configurations can be used (see Chapter 10). The test can be carried out in a once-through or recycle mode. Each catalyst manufacturer has developed his own testing methods and equipment. The following is an outline of various testing schemes.

9.4.1. Once-Through Test

The performance of a catalyst may be tested in the relatively simple, inexpensive once-through configuration. In this operation, only 40–80 vol % conversion of feedstock to middle distillates and lighter products is achieved. The products are separated by fractionation and analyzed. Both product yields and quality are measured. A simplified flow diagram of a once-through, single-stage testing configuration is shown in Figure 9.12.

When hydrotreated feed is used for the test, a hydrotreating step is not necessary. In this case, the feed is pumped directly into the hydrocracking reactor. However, the sulfur and nitrogen content of the feed may be adjusted prior to hydrocracking by addition of calculated amounts of sulfur- and nitrogen-containing organic compounds, such as thiophene and *tert*-butylamine. The addition of these compounds to the feed ("doping") is designed to simulate commercial operating conditions, when first-stage hydrocracking takes place in the presence of sulfur and nitrogen compounds.

The major components of the testing unit are reactor, scrubber, high-pressure separator (HPS), and low-pressure separator (LPS). A metered amount of "raw" feed or "doped" feed is mixed with high-pressure hydrogen and pumped to the top of the reactor filled with a specified volume of catalyst. The catalyst extrudates or spheres are usually diluted with glass beads prior to loading. The reactor is usually a stainless steel pipe. The reactor is heated to the temperature required to obtain a specified single-pass conversion (e.g., 50 vol %) to middle distillates and lighter products. The reactor effluent enters a scrubber where H_2S and NH_3 formed in the reactor are removed with a scrubbing solution, such as dilute caustic. Some of the dry gas is also separated in the scrubber, sampled, and metered. The scrubber effluent enters a high-pressure separator, where more dry gas is separated from the liquid product and spent scrubbing solution is removed. From the high-pressure separator the effluent enters a low-pressure separator, where flash gas is separated from liquid product. The test is usually run for several

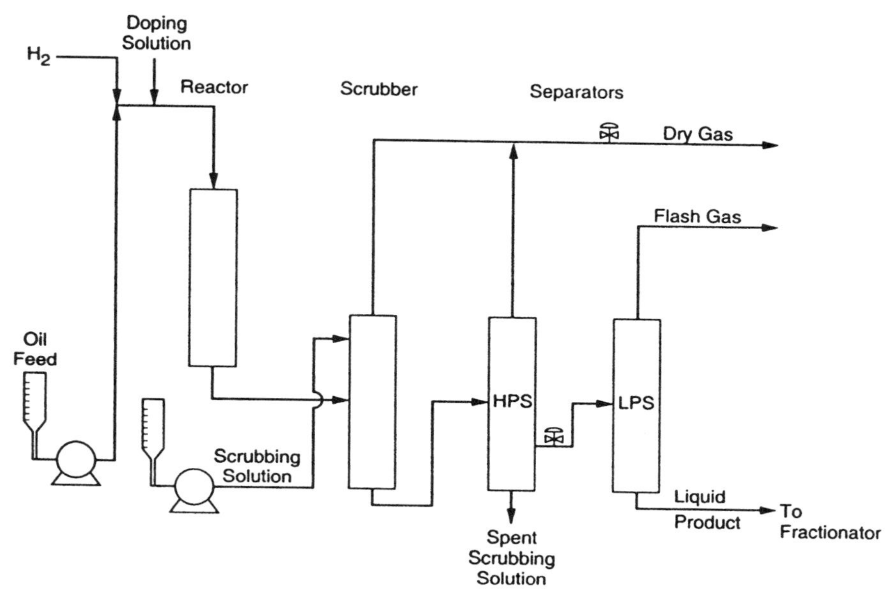

Figure 9.12 Simplified flow diagram of once-through, single-stage testing unit.

days. The gases and liquid product are submitted to analyses. The liquid product is either fractionated or analyzed by simulated distillation. The yield and quality of different fractions is determined. The fractions usually obtained from the liquid product are light gasoline, heavy naphtha, jet fuel, diesel fuel, and unconverted oil.

9.4.2. Pilot Plant Recycle Test

Commercial units often operate with recycle of unconverted feed in order to obtain total conversion. This may result in the gradual buildup of polynuclear aromatics in recycle feed, which might adversely affect unit performance (see Section 11.4). Therefore, the correct assessment of catalyst performance in a commercial recycle operation requires testing under recycle conditions. Because the composition of recycle oil is different from that of the initial feed, the yields, product quality, and catalyst deactivation rate will be different in a recycle operation compared to once-through.

The testing is usually done in integrated pilot plant recycle units using a single-stage or two-stage process configuration. Pilot plant testing may be carried out under isothermal or adiabatic conditions. Running the test under isothermal conditions allows the simulation of the commercial process under strictly controlled conditions, while requiring only small amounts of feedstock. Such a test also has considerable flexibility and allows the quantitative evaluation of the effects of process variables. Testing under isothermal conditions is commonly used commercially. Testing under adiabatic conditions provides a better simulation of the commercial process and gives a more accurate assessment of catalyst stability under commercial process conditions. It also allows the measurement of heat rises and the determination of kinetic data.

The flow diagram of a single-stage Unicracking pilot plant unit with recycle is shown in Figure 9.13. The fresh feed is mixed with high-pressure hydrogen and pumped to the top of the heated hydrotreating reactor filled with hydrotreating catalyst. In this reactor hydrodesulfurization, hydrodenitrogenation, and partial hydrogenation of unsaturated compounds takes place. The reactor effluent is combined with recycle oil and recycle gas, and passed through the heated hydrocracking reactor filled with hydrocracking catalyst. Most of the hydrocracking takes place in this reactor. The effluent from the reactor is cooled and passed first through an HPS and then through an LPS. The lighter components are separated and form the recycle gas. Flash gas released from depressurized liquid in the LPS is analyzed or combined with gases removed in the stabilizing column. Scrub

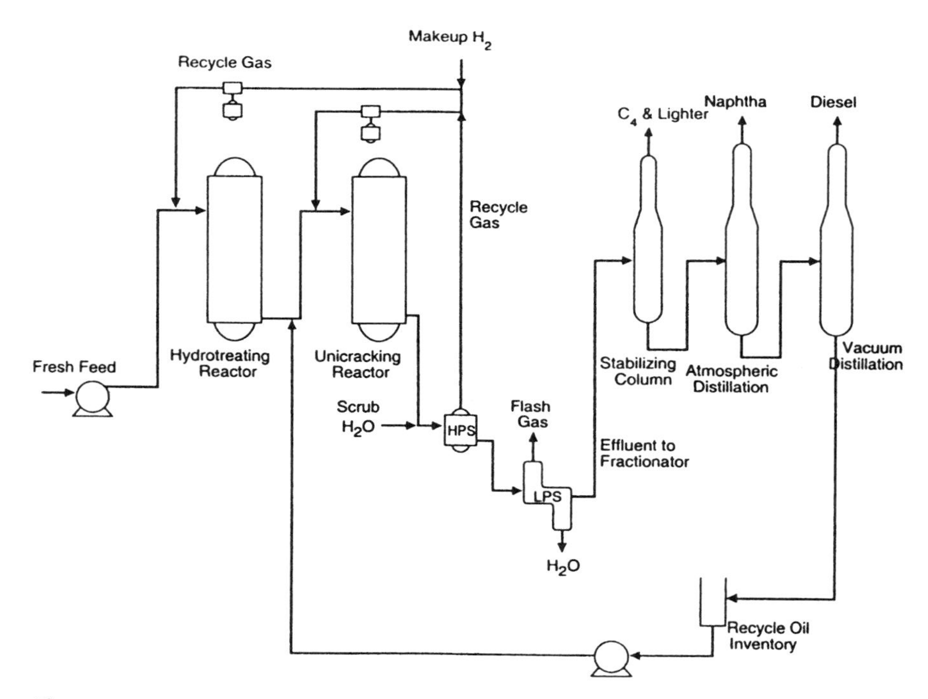

Figure 9.13 Single-stage Unicracking pilot plant unit with recycle.

water is used to dissolve and eliminate ammonia generated during reaction. The hydrocarbon effluent from the LPS is submitted to fractionation. This is accomplished first in a stabilizing column, where gaseous components are removed; then in an atmospheric distillation column, where light liquid fractions are separated; and finally in a vacuum column, where the diesel fraction is separated. The unconverted oil left at the bottom of the vacuum distillation column is recycled.

The testing in a two-stage hydrocracking unit is similar to the one described for the single-stage unit, except for using an additional hydrocracking reactor and additional HPS. The feedstock for this reactor is the recycle oil that results from the fractionation of the effluent obtained from the first hydrocracking reactor.

In order to achieve steady-state operation and to be able to assess catalyst stability, pilot plant tests are usually run for several weeks. Because the catalyst gradually deactivates during operation, reactor temperature is gradually increased in order to maintain constant conversion. This temperature increase, which is a measure of the catalyst deactivation rate, is sometimes called temperature increase requirement (TIR) and is expressed in °C or °F/day.

The operating conditions of the unit have to be closely monitored. Control and adjustment of reactor temperature profiles, uniform distribution of gas and liquid phases across the catalyst bed, and the ability to obtain 100% mass balance all have to be achieved in order to obtain a valid simulation of a commercial process.

In small, bench-scale pilot plants, the catalyst can be tested without dilution. Larger pilot plants usually require dilution of the catalyst with inert material of varying sizes and packing to allow a better temperature control. Catalyst loading in the reactor should be such as to minimize flow maldistribution. For kinetic modeling, a powdered catalyst can be used to minimize mass transfer and heat transfer limitations. However, to simulate a commercial operation the testing of extrudates or pellets is recommended.

Catalyst activation by sulfiding can be carried out in a separate pilot plant before loading in the testing unit. Activation is typically done in an H_2S environment. Alternatively, catalyst sulfiding can be carried out in situ in the testing unit as part of the startup process that simulates commercial operation.

In many commercial laboratories, catalyst testing is carried out in the following sequence: (a) once-through testing to establish catalyst activity; (b) once-through testing with product analysis to establish product quality, (c) recycle testing under specific conditions; and (d) recycle testing under different conditions with different feedstocks.

Modern pilot plant designs incorporate on-line analysis and data acquisition capability. This minimizes manpower requirements and enhances data quality. The major operating variables are monitored at least every 2 hr. The steps are carried out over a fixed time period in which all input and output streams are accurately measured.

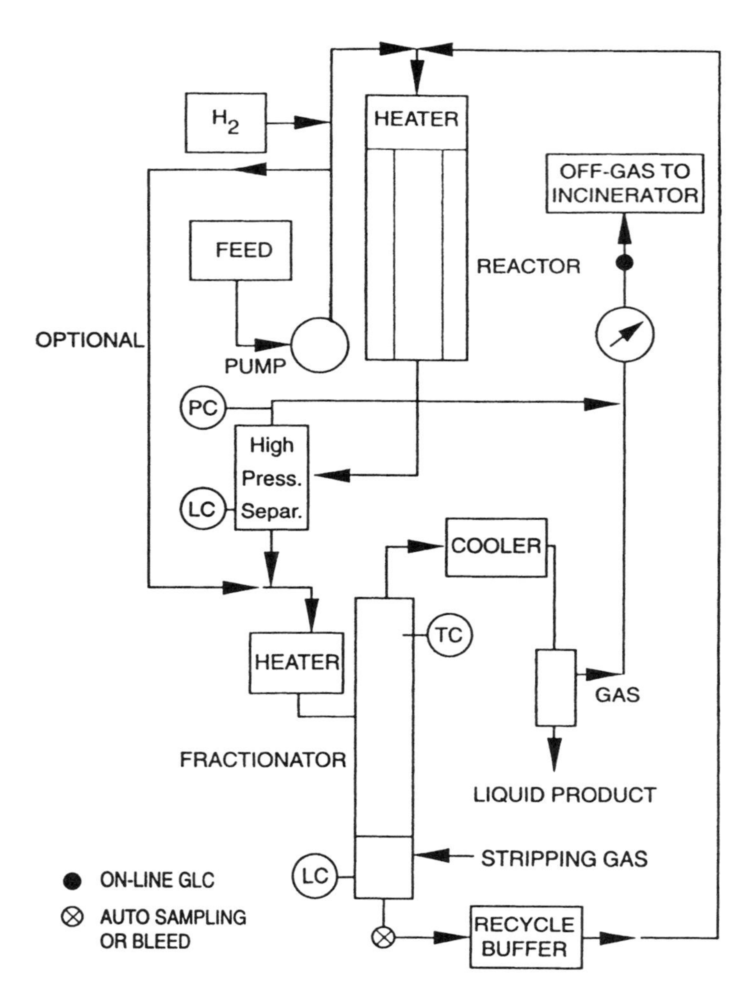

Figure 9.14 Small-scale recycle unit [123].

9.4.3. Small-Scale Recycle Test

A small-scale recycle unit has been designed by a group from Shell. The flow scheme of such a unit is shown in Figure 9.14 [123]. While other tests may require several hundred milliliters of catalyst, the small-scale test can be carried out with as little as 10 ml of catalyst. The volume of catalyst used in this test can vary from 10 ml to 100 ml. The unit consists of a gas and liquid supply system, a single reactor, a high-pressure separator, and a product workup section. Fresh feed is mixed with fresh hydrogen, heated, and fed to the reactor. The reactor is loaded with extrudates, which are diluted with inert particles. The reactor effluent is cooled and fed to an HPS. The off-gas from the HPS is depressurized, passed through a gas meter, and analyzed by gas chromatography. The liquid leaving the HPS is preheated and fed to the heated stripper/fractionator column. High cut points (up to 400–410°C) and sharp separation can be achieved with this stripper/fractionator. The distillate top fraction is condensed and collected. Noncondensable gases are analyzed on-line, whereas liquid product is analyzed off-line. The fractionator bottoms are collected in a buffer vessel and the liquid is then recycled to the reactor. Residence time in the unit can be as low as a few hours, approaching residence times of commercial units. The unit is fully automated and can be operated in different modes (e.g., the fresh feedstock can be fed directly to the reactor or to the stripper/fractionator column). Results obtained from the analyses of the off-gas, liquid product and recycle streams are plugged into calculation models, which allow the assessment of catalyst performance in a commercial unit.

References

1. F. Delannay (ed.), *Characterization of Heterogeneous Catalysts*, Marcel Dekker, New York, 1984.
2. J. M. Walls (ed.), *Methods of Surface Analysis*, Cambridge Univ. Press, New York, 1989.
3. J. L. G. Fierro (ed.), *Spectroscopic Characterization of Heterogeneous Catalysts. Parts A and B*, Elsevier, Amsterdam, 1990.
4. M. L. Deviney and J. L. Gland (eds.), *Catalyst Characterization Science-Surface and Solid State Chemistry* A.C.S. Symposium Series 288, Washington, D.C., 1985.
5. M. J. Kelley, in *Successful Design of Catalysts: Future Requirements and Development*, Elsevier, New York, 1989, p. 61.
6. J. W. Ward, *J. Catal. 9*:225 (1967).
7. C. Mirodatos and D. Barthomeuf, *J. Chem. Soc., Chem. Commun. 336* (1985).
8. J. H. Lunsford, *Fluid Catalytic Cracking II* (M. L. Occelli, ed.), A.C.S. Symposium Series 452, 1991, p. 1.
9. H. A. Benesi and B. H. C. Winquist, *Adv. Catal. 27*:97 (1978).
10. K. Tanabe, *Solid Acids and Bases*, Academic Press, New York, 1970.
11. L. Forni, *Catal. Rev. 8*:65 (1978).
12. P. A. Jacobs, in *Characterization of Heterogeneous Catalysts* (F. Delannay, ed.), Marcel Dekker, New York, 1984, p. 367.
13. A. I. Biaglow, C. Gittleman, R. J. Gorte, and R. J. Madon, *J. Catal. 129*:88 (1991).
14. H. Itoh, T. Hattori, and Y. Murakami, *Chem. Lett.* 1147 (1981).
15. J. B. Uytterhoeven, R. Schoonheydt, B. V. Liengme, and W. K. Hall, *J. Catal. 13*:425 (1969).
16. J. W. Ward, in *Zeolite Chemistry and Catalysis* (J. A. Rabo, ed.), A.C.S. Monograph 171, Washington, D.C., 1976, p. 118.
17. B. Gil, E. Broclawik, J. Datka, and J. Klinowski, *J. Phys. Chem. 98*:930 (1994).

18. J. Scherzer and J. L. Bass, *J. Catal. 28*:101 (1973).
19. S. G. Hedge, R. Kumar, R. N. Bhat, and P. Ratnasamy, *Zeolites 9*:231 (1989).
20. G. W. Skeels and D. W. Breck, Proc. 6th Intl. Zeolite Conf., Reno, USA, July 1983, p. 87, Butterworths.
21. E. M. Flanigen, H. Khatami, and H. A. Szymanski, *Adv. Chem. Ser. 101*:201 (1971).
22. J. Scherzer, *Catalytic Materials: Relationship between Structure and Reactivity* (T. E. Whyte et al., eds.), A.C.S. Symposium Series 248, 1984, p. 157.
23. P. Pichat, R. Beaumont, and D. Barthomeuf, *J. Chem. Soc., Faraday Trans. 1, 70*:1402 (1975).
24. E. M. Flanigen, *Zeolite Chemistry and Catalysis* (J. A. Rabo, ed.), A.C.S. Monograph 171, Washington, D.C., 1976, p. 80.
25. A. Macedo, F. Raatz, R. Boulet, A. Janin, and J. C. Lavalley, *Innovation in Zeolite Materials Science* (P. Grobet et al., eds.), Elsevier, Amsterdam, 1988, p. 375.
26. F. Lonyi, J. Engelhard, and D. Kallo, *Zeolites 11*:169 (1991).
27. J. Datka and M. Boczar, *Zeolites 11*:397 (1991).
28. D. H. Olson, R. M. Lago, and W. O. Haag, *J. Catal. 61*:390 (1980).
29. H. Rastelli, Jr., B. M. Lok, J. A. Duisman, D. E. Earls, and J. T. Mullhaupt, *Can. J. Chem. Eng. 60*:44 (1982).
30. V. R. Choudhary and D. P. Akolekar, *J. Catal. 125*:143 (1990).
31. J. P. Damon, B. Delmon, and J. M. Bonnier, *J. Chem. Soc., Faraday Trans. 1, 73*:372 (1977).
32. H. G. Karge, K. Hatada, Y. Zhang, and R. Fiedorow, *Zeolites 3*:13 (1983).
33. J. Turkevich, F. Nazaki, and D. N. Stamires, Proc. 3rd Intl. Congress on Catal., Amsterdam, 1964, p. 586.
34. J. Arribas, A. Corma, V. Fornes, and F. Melo, *J. Catal. 108*:135 (1987).
35. R. Beaumont and D. Barthomeuf, *J. Catal. 30*:288 (1973).
36. G. I. Kapustin, T. R. Brueva, A. L. Klyachko, S. Beran, and B. Wichterlova, *Appl. Catal. 42*:239 (1988).
37. A. Macedo, A. Auroux, F. Raatz, E. Jacquinot, and R. Boulet, A.C.S. Symposium Series 368, Washington, D.C. 1988, p. 98.
38. P. A. Jacobs, *Carboniogenic Activity of Zeolites*, Elsevier, Amsterdam, 1977, p. 41.
39. C. Naccache, C. F. Ren, and G. Coudurier, *Zeolites: Facts, Figures, Future* (P. A. Jacobs et al., eds.), Elsevier, Amsterdam, 1989, p. 661.
40. A. Thangaraj, P. R. Rajamohanan, P. M. Suryavanshi, and S. Ganapathy, *J. Chem. Soc., Chem. Commun. 493* (1991).
41. P. Szedlacsek, S. L. Suib, M. Deeba, and G. S. Koermer, *J. Chem. Soc., Chem. Commun. 1531* (1990).
42. Y. Ben Taarit, C. Naccache, and B. Imelik, *J. Chim. Phys. 70*:728 (1973).
43. S. Bhatia, *Zeolite Catalysis: Principles and Applications*, CRC Press, Florida, 1989.
44. B. W. Wojciechowski and A. Corma, *Catalytic Cracking: Catalysts, Chemistry and Kinetics*, M. Dekker, New York, 1986.
45. *Catalysis by Acids and Bases* (B. Imelik et al., eds.), Elsevier, Amsterdam (1985).
46. R. Szostak, *Molecular Sieves: Principles of Synthesis and Identification*, Van Nostrand Reinhold, New York, 1989.
47. J. L. Lamaitre, P. G. Menon, and F. Delannay, *Characterization of Heterogeneous Catalysts* (F. Delannay, ed.), M. Dekker, 1984, p. 299.

48. S. E. Wanke, J. A. Szymura, and T-T. Yu, *Catalyst Deactivation* (E. E. Petersen and A. T. Bell, eds.), Marcel Dekker, 1987, p. 65.
49. J. E. Benson and M. Boudart, *J. Catal. 4*:704 (1965).
50. J. J. F. Scholten, *Preparation of Catalysts II* (B. Delmon et al., eds.), Elsevier, Amsterdam, 1979, p. 685.
51. J. E. Benson, H. S. Hwang, and M. Boudart, *J. Catal. 30*:146 (1973).
52. B. S. Parekh and S. W. Weller, *J. Catal. 47*:100 (1977).
53. N. R. Ramakrishnan and S. W. Weller, *J. Catal. 67*:237 (1981).
54. S. J. Tauster, T. A. Pecoraro, and R. Chianelli, *J. Catal. 63*:515 (1980).
55. P. Scherrer, *Goett. Nachr. 2*:98 (1918).
56. P. Ganesan, H. Kud, A. Saavedra, and R. J. de Angelis, *J. Catal. 52*:310, 320 (1978).
57. R. D. Bonetto, H. R. Viturro, and A. G. Alverez, *J. Appl. Cryst. 23*:136 (1990).
58. T. E. Whyte, Jr., P. W. Kirklin, R. W. Gould, and H. Heinemann, *J. Catal. 25*:407 (1972).
59. F. Delannay, *Characterization of Heterogeneous Catalysts* (F. Delannay, ed.), M. Dekker, New York, 1984, p.71.
60. P. C. Flynn, S. E. Wanke, and P. S. Turner, *J. Catal. 33*:233 (1974).
61. C. Defossé, P. Canesson, P. G. Rouxhet, and B. Delmon, *J. Catal. 51*:269 (1978).
62. F. P. J. M. Kerkhof and J. A. Moulin, *J. Phys. Chem. 83*:1612 (1979).
63. G. H. Via, G. Meitzner, F. W. Lytle, and J. H. Sinfelt, *J. Chem. Phys. 79*:527 (1982).
64. F. W. Lytle et al., *Catalyst Characterization Science* (M. L. Deveney and J. L. Gland, eds.), A.C.S. Symposium Series 288, Washington, D.C., 1985, p. 280.
65. F. W. Lytle, R. B. Greegor, and E. C. Marques, *Proceedings of the 9th Intern. Congr. Catalysis* (M. J. Phillips and M. Ternan, eds.), Chemical Inst. of Canada, Ottawa, 1988, v. 5, p. 54.
66. S. M. Heald and J. M. Tranquada, *Phys. Methods Chem.*, Wiley, New York, 2nd ed., 1990, v. 5.
67. J. H. Sinfelt, G. H. Via, and F. W. Lytle, *Catal. Rev. Sci. Eng. 26*:81 (1984).
68. A. W. Peters, *Fluid Catalytic Cracking: Science and Technology* (J. S. Magee and M. M. Mitchell, Jr., eds.), Elsevier, New York, 1993, p. 183.
69. W. M. Meier and D. H. Olson, *Atlas of Zeolite Structure Types* 2nd ed., Butterworth, 1987.
70. R. von Ballmoos and J. B. Higgins, *Zeolites 10*:313-520 (1990).
71. K. Wefers and C. Misra, "Oxides and Hydroxides of Aluminum," Alcoa Technical Paper No. 19, Revised, Aluminum Company of America, 1987.
72. D. K. Smith and S. Gorter, *J. Appl. Cryst. 24*:369 (1991).
73. ASTM Standards on Catalysts, 3rd edition, 1988.
74. D. W. Breck and E. M. Flanigen, *Molecular Sieves*, Soc. Chem. Ind., London, 1968, p. 47.
75. H. Fichtner-Schmittler, U. Lohse, G. Engelhardt, and V. Patzelova, *Cryst. Res. Tech. 19*, K1 (1984).
76. J. R. Sohn, S. J. DeCanio, J. H. Lunsford, and D. J. O'Donnell, *Zeolites 6*:225 (1986).
77. P. O. Fritz, J. H. Lunsford, and C. M. Fu, *Zeolites 8*:205 (1988).
78. V. Bosachek, V. Patzelova, Z. Tvaruzkova, D. Freude, U. Lohse, W. Schirmer, H. Stach, and H. Thamm, *J. Catal. 61*:435 (1980).
79. G. H. Kuhl, *J. Phys. Chem. Solids 38*:1259 (1977).

80. A. Corma, V. Fornes, and F. Rey, *Appl. Catal. 59*:267 (1990).
81. J. M. Newsam, *Science 231*:1093 (1986).
82. *EXAFS Spectroscopy: Techniques and Applications* (B. K. Teo and D. C. Joy, eds.), Plenum Press, New York, 1981.
83. G. Porod, *Small-Angle X-ray Scattering* (O. Glatter and O. Kratsky, eds.), Acad. Press, New York, 1982, Part II, Ch. 2.
84. G. Cocco, L. Schiffini, G. Strukul, and G. Garturan, *J. Catal. 65*:348 (1980).
85. S. J. Gregg and K.S.W. Sing, *Adsorption, Surface Area, and Porosity*, Academic Press, New York, 1982 (2nd ed.).
86. S. Lowell and J. E. Shields, *Powder Surface Area and Porosity*, Chapman and Hall, New York, 1984 (2nd ed.).
87. *Characterization of Porous Solids* (K. K. Unger, J. Rouquerol, and K. S. W. Sing, eds.), Elsevier, New York, 1988.
88. *Characterization of Porous Solids, II* (F. Rodrigies-Reinoso, J. Rouquerol, K. S. W. Sing, and K. K. Unger, eds.), Elsevier, New York, 1988.
89. R. M. Barrer, *Zeolites and Clay Minerals as Sorbents and Molecular Sieves*, Acad. Press, New York, 1978.
90. S. Brunauer, L. S. Deming, W. S. Deming, and E. Teller, *J. Amer. Chem. Soc. 62*:1723 (1940).
91. A. G. Bridge, D. R. Cash, and J. F. Mayer, NPRA Ann. Mtg., March, 1993, S. Antonio, TX, AM-93-60.
92. U. Lohse, H. Stach, H. Tamm, W. Schirmer, A. A. Isirikjan, N. I. Regent, and M. M. Dubinin, *Z. anorg. allg. Chem. 460*:179 (1980).
93. J. Scherzer, *Catalytic Materials: Relationship Between Structure and Reactivity* (T. E. White, Jr. et al., eds.), A.C.S. Symposium Series 248, Washington, D.C., 1984, p. 157.
94. H. K. Beyer and I. Belenykaia, *Catalysis by Zeolites* (B. Imelik et al., eds.), Elsevier, Amsterdam, 1980, p. 203.
95. B. C. Lippens and J. H. de Boer, *J. Catal. 4*:319 (1965).
96. M. F. L. Johnson, *J. Catal. 52*:425 (1978).
97. E. Lippmaa, M. Mägi, A. Samoson, G. Engelhardt, and A. R. Grimmer, *J. Amer. Chem. Soc. 102*:4889 (1980).
98. G. Engelhardt, U. Lohse, E. Lippmaa, M. Tarmak, and M. Mägi, *Z. anorg. allg. Chem. 482*:49 (1981).
99. J. M. Thomas and J. Klinowski, *Advances in Catalysis*, Acad. Press, New York, 1985, v. *33*, 199.
100. C. A. Fyfe, J. M. Thomas, J. Klinowski, and G. C. Gobbi, *Angew. Chem. Int. Ed. Engl. 22*:259 (1983).
101. J. Klinowski, *Progr. NMR Spectr. 16*:237 (1984).
102. G. Engelhardt and D. Michel, *High-Resolution Solid-State NMR of Silicates and Zeolites*, J. Wiley, New York, 1987.
103. W. Loewenstein, *Am. Mineral 39*:92 (1954).
104. J. Klinowski, J. M. Thomas, C. A. Fyfe, and G. C. Gobbi, *Nature (London) 296*:533 (1982).
105. J. B. Nagy, Z. Gabelica, G. Debras, E. G. Derouane, J. P. Gilson, and P. A. Jacobs, *Zeolites 2*:59 (1982).

106. C. A. Fyfe, C. G. Gobbi, J. S. Hartman, J. Klinowski, and J. M. Thomas, *J. Phys. Chem. 86*:1247 (1982).
107. J. P. Gilson, G.C.E. Edwards, A. W. Peters, K. Rajagopalan, R. F. Wormsbecker, T. G. Roberie, and M. P. Shatlock, *J. Chem. Soc., Chem. Commun.* No. 2, 91 (1987).
108. T. Ito and J. P. Fraissard, *J. Chem. Soc., Faraday Trans.* (1) *83*, 451 (1987).
109. T. T. P. Cheung and C. M. Fu, *J. Phys. Chem. 93*:3740 (1989).
110. T. T. P. Cheung, *J. Catal. 124*:511 (1990).
111. R. A. Dalla Betta and M. Boudart, in *Proceedings of 5th Intern. Congr. Catalysis* (J. W. Hightower, ed.), North-Holland, Amsterdam, 1973, v. 2, p. 1329.
112. M. Boudart, R. Ryoo, G. P. Valenca, and R. Van Grieken, *Catal. Lett. 17*:273 (1993).
113. R. T. K. Baker, *Catal. Rev. Sci. Eng. 19*:161 (1979).
114. G. W. Qiao, J. Lu, J. Zou, and K. H. Kuo, *J. Catal. 103*:170 (1987).
115. D. E. W. Vaughan, *Fluid Catalytic Cracking: Science and Technology* (J. S. Magee and M. M. Mitchell, Jr., eds.), Elsevier, Amsterdam, 1993, p. 83.
116. F. Maugé et al., *Catalysis by Acids and Bases* (B. Imelik et al., eds.), Elsevier, New York, 1985, p. 94.
117. Q. L. Wang et al., *Zeolites 10*:703 (1990).
118. S. Lars, T. Andersson, S. T. Lundin, S. Jaras, and J. E. Otterstedt, *Appl. Catal. 9*:317 (1984).
119. M. B. Ward and J. B. Lunsford, *J. Catal. 87*:524 (1984).
120. J. N. Ness, D. J. Joyner and A. P. Chapple, *Zeolites 9*:250 (1989).
121. J. Dwyer, F. R. Fitch, G. Qin, and J. C. Vickerman, *J. Phys. Chem 86*:4574 (1982).
122. J. Dwyer, F. R. Fitch, F. Machado, G. Olin, S. M. Smyth, and J. C. Vickerman, *J. Chem. Soc., Chem. Commun. 422* (1981).
123. A. van Dijk, A. F. de Vries, J. A. R. van Veen, W. H. J. Stork, and P. M. M. Blauwhoff, *Catalysis Today 11*:129 (1991).

10

HYDROCRACKING PROCESSES

The hydrocracking processes described in the literature are designed to upgrade a variety of petroleum feedstocks, such as vacuum gas oils, straight run gas oils, coker gas oils, deasphalted oils, FCC cycle oils, thermally cracked stocks, straight run and cracked naphthas, by adding hydrogen and cracking to a desired boiling range. The particular products are liquefied petroleum gas (LPG), light naphtha, heavy naphtha, jet fuel, diesel fuel, heating oil, petrochemical feedstocks, FCC feedstock, ethylene cracker feedstock, and lube oil base stock. Hydrogen addition is accomplished by a metal function on the hydrocracking catalyst. The catalyst adds hydrogen to organic sulfur, nitrogen, and oxygen compounds to form H_2S, NH_3, and H_2O, respectively. In addition, the catalyst saturates olefins and multi-ring aromatics. Part of the hydrogen addition function of the hydrocracking process is sometimes done in a separate reactor with hydrotreating catalyst. Cracking is accomplished by an acidic catalyst function. In general, the cracking catalyst function selectively converts high-boiling hydrocarbons to products in the desired boiling range of naphtha, jet fuel, or diesel fuel. The product boiling range can be controlled by a combination of process variables and catalyst modifications. The following parameters determine product distribution:

1. Process configuration (i.e., single-stage, two-stage, once-through, etc.)
2. Feed/recycle ratio
3. Fractionation cut point
4. Conversion level
5. Catalyst type

In general, the hydrocracking process operating conditions are: 300–450°C (570–840°F) catalyst bed temperature, 85–200 bar (1250–2915 psi) pressure, 0.5–2.5 hr^{-1} liquid hourly space velocity, 505–1685 nm^3/m^3 (3000–10,000 scf/b) hydrogen-to-oil ratio, 200–590 nm^3/m^3 (1200–3500 scf/b) hydrogen consumption. Due to high hydrogen partial pressures and the use of dual function catalysts, the rate of catalyst coking and deactivation is very low, resulting in on-stream cycle lengths of several years.

The typical hydrocracking process utilizes reactors with fixed catalyst beds. Due to hydrogen consumption and the overall exothermic nature of the hydrocracking process, multiple beds are usually used and cooling is achieved by introducing hydrogen recycle gas between the catalyst beds (see Section 10.5).

The feedstocks used in hydrocracking processes contain sulfur, nitrogen, and, in the case of resid feedstocks, metals such as nickel and vanadium. Because such compounds have a deleterious effect on hydrocracking catalysts, the feedstock typically requires hydrotreatment prior to contact with the hydrocracking catalyst. For that reason, most of the hydrocracking processes involve both hydrotreating and hydrocracking steps [1].

Different companies that license hydrocracking technology (Table 1.3) have trade names for their processes, such as UOP's Unicracking processes [2] or Chevron's Isocracking processes [3]. A joint venture by Mobil R&D, Akzo Chemicals, and Kellogg is also offering hydrocracking technology (MAK Hydrocracking) [4]. IFP, Shell, Exxon, Texaco, and BP have developed their own hydrocracking technologies. Several companies have developed residue hydrocracking processes (see Chapter 14).

10.1. Single-Stage Recycle Hydrocracking

The processes used in hydrocracking can be divided into two major categories: single-stage and two-stage. The simplest configuration is that of a single-stage process, utilizing a single catalyst in a single reactor or two reactors in series. A simplified flow diagram for such a process is shown schematically in Figure 10.1. Many of the units designed to maximize diesel product utilize this configuration and employ amorphous catalyst. The fresh feed combined with unconverted oil is passed downward through the catalyst bed in the presence of hydrogen. The effluent from the reactor is passed through high- and low-pressure separators, where hydrogen is recovered. The separated hydrogen, together with fresh make-up hydrogen, is recycled to the reactor. The liquid product is sent to fractionation, where the final products are separated from unconverted oil.

Fractionation yields light ends (C_4^-) and liquid products. Of the light ends, C_3 and C_4 can be used as feedstock for the alkylation unit, recovered as LPG, hydrogen plant feed stock, or fuel. Iso-C_4 is used in the manufacture of oxygenated gasoline additives (MTBE, ETBE).

The following liquid products are obtained by fractionation: light naphtha (C_5–80°C) (C_5–175°F) used as a gasoline pool component; heavy naphtha (80–150°C) (175–300°F) used as a reformer feedstock to be converted into high-octane gasoline or aromatics for petrochemicals; jet fuel/kerosene (150–290°C) (300–550°F) used as fuel for turbine engines; and diesel fuel (290–370°C) (550–700°F). The fractionator bottoms containing the unconverted feed (370°C+)

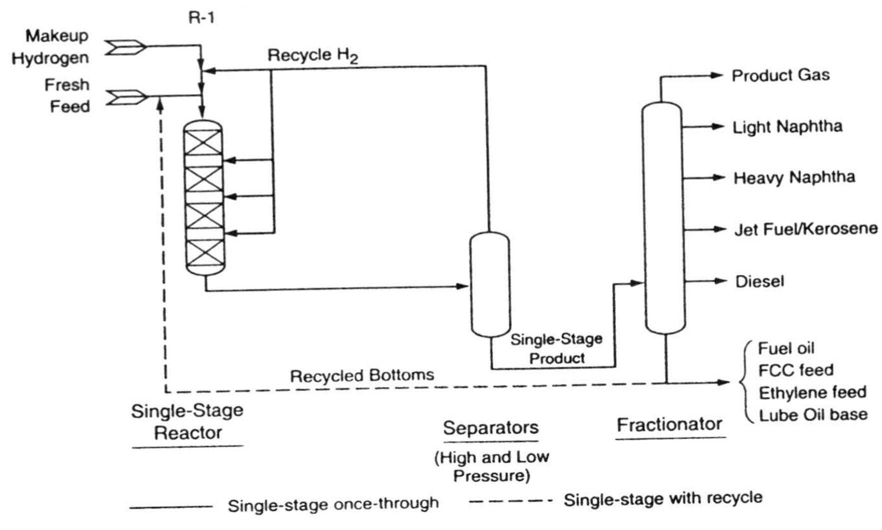

Figure 10.1 Simplified flow diagram of single-stage, single-catalyst hydrocracking process, with and without ("once-through") recycle.

(700°F+) material is recycled to the reactors so it can be converted to (370°C−) (700°F−) products.

In addition to boiling range, these products are characterized by the following properties: light naphtha (mostly $C_5 + C_6$) is characterized by octane rating; heavy naphtha is characterized by octane rating, aromatic content, and specific gravity; jet fuel/kerosene is characterized by smoke point, flash point, freeze point, and specific gravity; diesel fuel is characterized by pour point, cloud point, flash point, cetane number, and specific gravity. These liquid products may also be characterized by hydrocarbon type (paraffins, isoparaffins, naphthenes, aromatics), sulfur and nitrogen content, viscosity, and volatility (see Chapter 1).

To be of superior quality, the liquid products obtained in the hydrocracking process should have the following properties: light naphtha with a high isoparaffin content; heavy naphtha with a high content in single-ring hydrocarbons; jet fuel with low freeze point and high smoke point; diesel fuel with low pour point, low cloud point, and high cetane number; heavy products (370°C+) (700°F+) that are hydrogen-rich.

The above description of fractionation applies to all of the hydrocracking flow schemes.

In commercial hydrocrackers, 40–80 vol % of the feed is converted in one pass. The unconverted oil (fractionator bottoms) is recycled back to the reactor.

Another configuration of a single-stage process employs two catalysts either in the same reactor ("stacked beds") or in different reactors linked in series (Figure 10.2). In this case, the first catalyst (a hydrotreating catalyst) converts organic sulfur and nitrogen from hetero compounds in the feedstock to hydrogen sulfide and ammonia, respectively. The deleterious effect of H_2S and NH_3 on hydrocracking catalysts is considerably less than that of the corresponding organic hetero compounds. The hydrotreating catalyst also facilitates the hydrogenation of aromatics. A more detailed flow diagram of a single-stage, dual-catalyst Unicracking process is shown in Figure 10.3.

In the single-stage, two-reactor configuration, the products from the first reactor (R-1) are passed over a hydrocracking catalyst in the second reactor (R-2), where most hydrocracking takes place. The conversion occurs in the presence of NH_3, H_2S, and small amounts of unconverted organic hetero compounds. After separation and fractionation of the effluent, the unconverted oil is recycled to the first or second reactor for further processing. By recycling the unconverted feedstock, 100% conversion can be achieved (recycle-to-extinction hydrocracking).

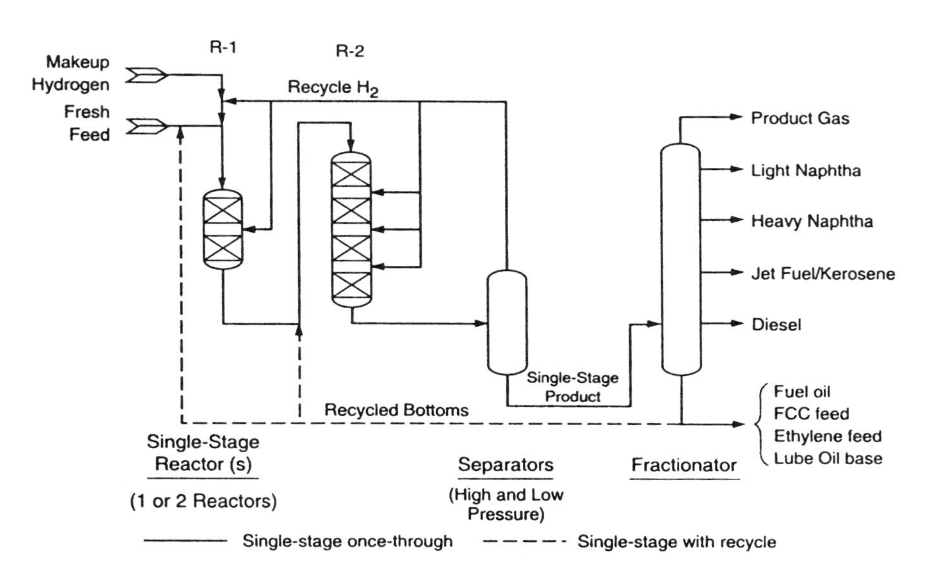

Figure 10.2 Simplified flow diagram of single-stage, dual-catalyst hydrocracking process, with and without recycle.

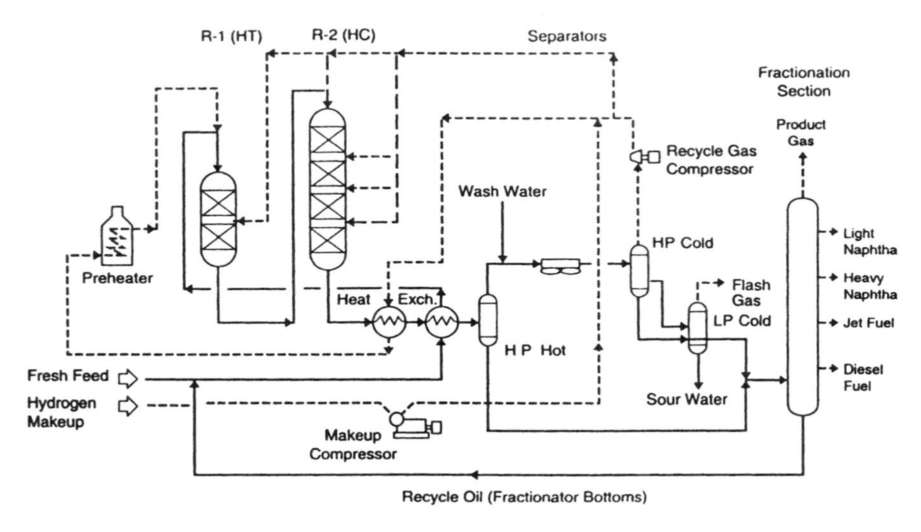

Figure 10.3 Detailed flow diagram of single-stage, dual-catalyst Unicracking process, with recycle.

The hydrotreating catalyst in the first reactor is designed to convert the hetero compounds in the feedstock. Typically, such catalysts comprise sulfided molybdenum or tungsten and nickel or cobalt on an alumina support. The reactor operates at temperatures varying from 300 to 450°C (570–840°F), and hydrogen pressures between 85 and 200 bar (1250–2915 psig). Under these conditions, in addition to hetero atom elimination, significant hydrogenation occurs and some cracking also takes place.

The hydrocracking catalyst in the second reactor is designed to optimize the yields and quality of the desired products. Its design is also dependent on the feedstock processed. The correlation between catalyst design, feedstock composition and products desired is discussed in Chapters 7 and 11.

Stacked bed configurations with hydrocracking catalysts of different composition or different particle size (graded catalyst beds) have also been described [5–8]. Such configurations can be used to fine-tune the selectivity for specific products. Furthermore, the process configuration may involve either a single train of reactors or multiple trains running in parallel.

A key characteristic of the single-stage process is that the effluent from the first reactor is not subjected to intermediate vapor/liquid separation and fractionation prior to being passed into the second reactor. In single-stage processing, hydrocracking takes place in the presence of small amounts of unconverted sulfur- and nitrogen-containing organic hetero compounds, and/or in the presence of hydrogen sulfide and ammonia. In such a process, mostly zeolite-based catalysts are being used because they have sufficient cracking activity in the presence of ammonia or small amounts of nitrogen-containing compounds.

For feeds with relatively low endpoints, a single-stage recycle configuration is commonly used to achieve total conversion. However, such a configuration can be less selective for liquid product than a two-stage configuration because all conversion must be accomplished in a single stage.

10.2. Once-Through Hydrocracking

A variation of the single-stage process is the "once-through" process. In contrast to the single-stage process with recycle, in the once-through process the unconverted oil resulting from the first pass is not recycled (Figure 10.2). The fractionator bottoms are used as steam cracker feed, FCC feed, or lube oil base. The once-through process can achieve up to 90% conversion of fresh feedstock to high-value, lighter products. However, middle distillates made by this process are generally higher in aromatics and are therefore of poorer burning quality than those produced by recycle hydrocracking. By selecting the proper hydrocracking catalyst, the yield of desired products can be optimized.

Since conversion of the heaviest molecules in the feedstock is not required, once-through hydrocracking can be carried out at lower temperatures and in many cases at lower hydrogen partial pressures than recycle hydrocracking, where total conversion of the feedstock is usually the objective. Catalyst deactivation is reduced by the elimination of the recycle stream, although sometimes the once-through conversion of the unit is so great compared to that achieved in recycle hydrocracking that the gain achieved by not recycling the unconverted recycle is negated by the high once-through conversion. The reduction in plant operating pressure and elimination of recycle stream processing results in lower capital cost.

10.3. Two-Stage Recycle Hydrocracking

A widely used hydrocracking process configuration is the two-stage process, which allows a larger throughput (higher feed rates) than the single-stage process. In this case, the hydrotreating and some of the hydrocracking takes place in the first stage. The effluent from the first stage is separated and fractionated, with the unconverted oil hydrocracked in the second stage. A common and versatile configuration of the two-stage hydrocracking process employs three reactors. A simplified flow diagram of the two-stage Unicracking process is shown in Figure 10.4. Hydrotreating is carried out in the first reactor, R-l, over hydrotreating

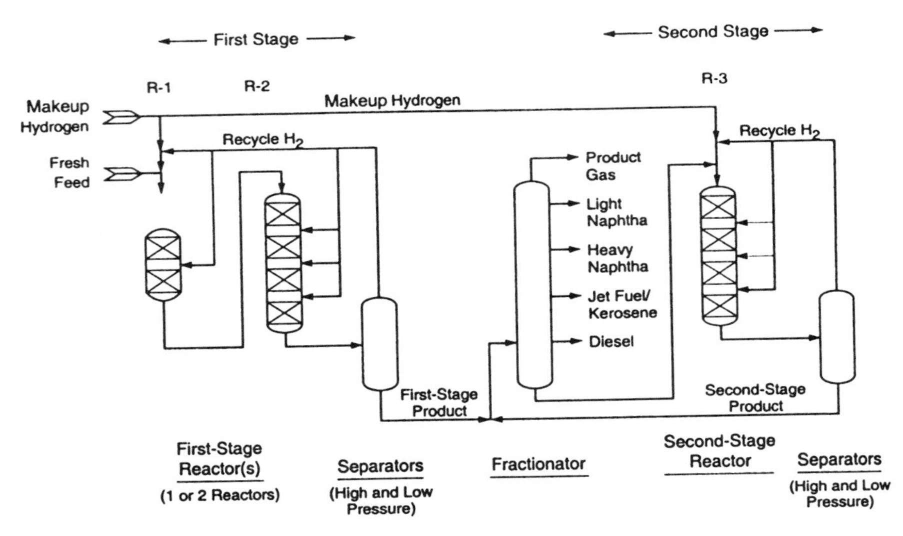

Figure 10.4 Simplified flow diagram of two-stage Unicracking process, with recycle.

catalyst, and partial hydrocracking is done in the second reactor, R-2, over hydrocracking catalyst. The effluent from the second reactor is separated and fractionated. The unconverted oil (fractionator bottoms) is passed into the third reactor, R-3, containing hydrocracking catalyst. The effluent from the third reactor is separated and fractionated, with the fractionator bottoms recycled to the third reactor. The conversion in R-1 and R-2 represents the first stage of the process, whereas the conversion in R-3 represents the second stage of the process. A separation and fractionation step separates the two stages. The recycle gases are washed, or stripped, of ammonia and, depending on the process, of most or part of hydrogen sulfide.

The first and second reactor in the two-stage configuration use the same catalysts as in single-stage configuration. The hydrocracking catalyst in the third reactor is operated in the near absence of ammonia and, depending on the process, in the absence or presence of hydrogen sulfide. The near absence of ammonia and hydrogen sulfide in the third reactor allows the use of either noble metal or base metal sulfide hydrocracking catalysts. The base metal sulfide catalysts are most frequently used. Only in cases where a very hydrogenated product is desired are sulfur-free systems used. The catalyst selected also depends on the desired product slate (see below). Stacked catalyst beds may be used in both first and second stage [9].

The hydrocracking in the second stage can be carried out at lower temperatures than in the first stage due to the absence of ammonia in the reaction environment. Thus, first-stage hydrocracking temperatures are in the range of 300–450°C (570–840°F), whereas second-stage hydrocracking temperatures are in the range of 270–370°C (520–700°F). Three-stage hydrocracking, using different catalysts at each stage, has also been described [10].

Chevron's two-stage Isocracker configuration, with no separate hydrotreating reactor, is shown schematically in Figure 10.5 [3]. The effluent from the first-stage reactor is sent to separation and fractionation. The fractionator bottoms are sent to second-stage hydrocracking in the second reactor. The second-stage product is separated and fractionated, with the unconverted fraction recycled to the second-stage reactor. The hydrocracking catalyst in the first-stage reactor has a high hydrogenation/acidity ratio, causing feed hydrogenation but only mild cracking, since conversion occurs in the presence of sulfur, nitrogen, and, in some instances, traces of metals. The hydrogenation/acidity ratio in the catalyst used in the second-stage reactor depends on product requirements. Catalysts with low hydrogenation-to-acidity ratio are desired for the production of naphtha [11]. These catalysts will maximize aromatics and branched paraffins in the products. When middle-distillate (kerosene and diesel) production is desired, particularly from high-endpoint feedstocks, the second-stage catalyst will require a high hydrogenation-to-acidity ratio [11]. The high ratio is also required to produce a good jet fuel smoke point and high-cetane diesel. Both amorphous and zeolite-based hydrocracking catalysts are used in Chevron's Isocracking process.

10.4. Separate Hydrotreat Hydrocracking

Some commercial units have a two-reactor, two-stage configuration, in which the first reactor contains a hydrotreating catalyst and the second reactor a hydrocracking catalyst. The process configuration is similar to the one shown in Figure 10.5. The effluent from the first reactor is separated and fractionated, with the fractionator bottoms fed to the second reactor. The products from the second reactor are separated and fractionated, with the unconverted oil recycled to the second reactor. In this configuration, hydrotreating is separated from hydrocracking. This process uses the same catalysts as those used in single- and two-stage, three-reactor configurations.

The key characteristics of different hydrocracking processes are summarized in Figure 10.6. The two-stage configuration offers more flexibility than the single-stage configuration. It is better suited for processing heavier feedstocks, rich in aromatics and nitrogen. Separation of ammonia and hydrogen sulfide between the first and second stage offers more flexibility in the selection of a hydrocracking catalyst for the second stage and results in a deeper hydroconversion of the feedstock.

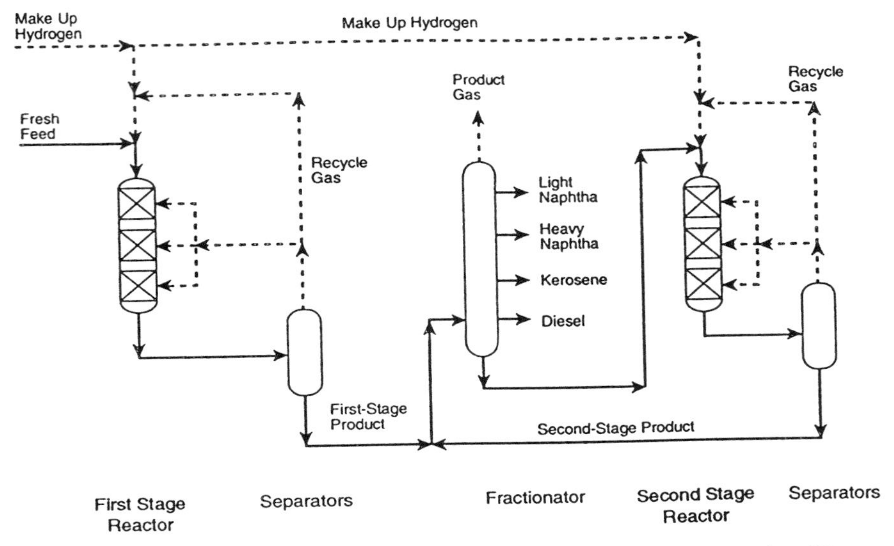

Figure 10.5 Flow diagram of Chevron's two-stage Isocracker configuration [3].

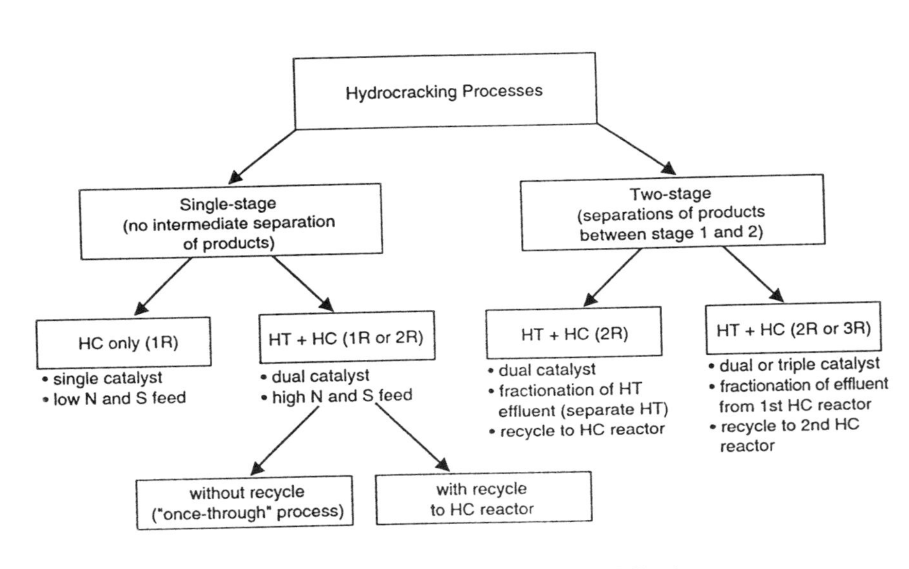

Figure 10.6 Classification of hydrocracking processes.

10.5. Hydrocracking Reactors: Design and Operation

The detailed design of different hydrocracking reactors is proprietary information and is described almost exclusively in the patent literature. Only general characteristics of such reactors are described in this section.

The reactors used in hydrocracking processes are downflow, fixed-bed catalytic reactors. They are usually cylindrical vessels fabricated from 2¼ Cr–1 Mo material with austenitic stainless steel weld overlay or liner, for added corrosion protection, closed at the upper and lower ends by curved heads provided with flanged connections (Figure 10.7). The opening at the top of the reactor is used to introduce the feed and to fill the vessel with catalyst. The opening at the bottom of the reactor is used as an outlet for the liquid product. A grating above this outlet

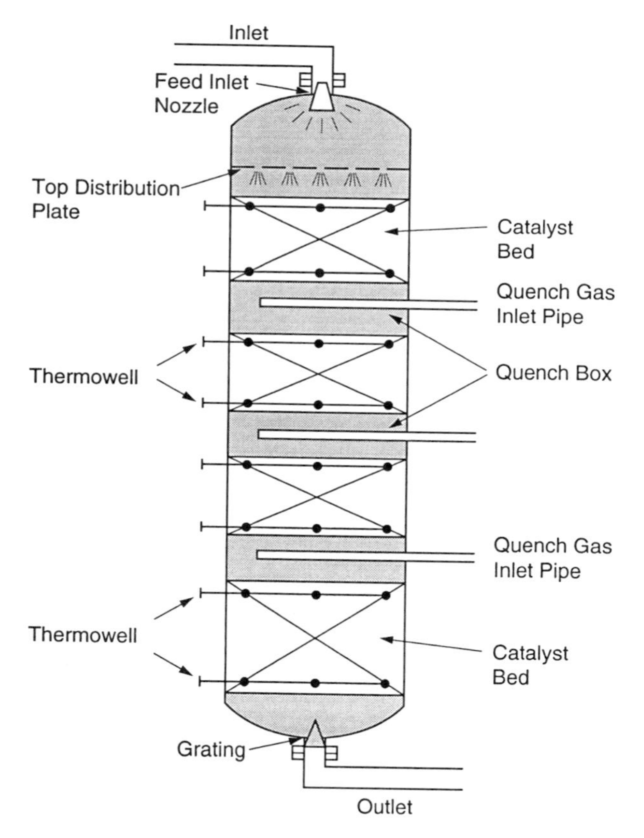

Figure 10.7 Schematic drawing of hydrocracking reactor.

prevents the escape of catalyst particles. At the opening on top of the reactor is an inlet nozzle that disperses the feed entering the reactor [12]. Below the inlet nozzle is a vapor/liquid distribution tray that facilitates the uniform spreading of fluid over the top of the first catalyst bed. The reactor may contain up to five or six catalyst beds separated by quench zones. Such a multibed arrangement of the catalyst allows cooling and uniform redistribution of fluids exiting one catalyst bed prior to entering the next catalyst bed.

Each bed of catalyst rests on a perforated plate or screen. Such a support permits fluids to exit the bottom of the catalyst bed and to enter the next bed after passing through a quench zone. Rashig rings or ceramic balls are placed on top of the bed to keep the catalyst in place. Layers of inert granular material are also placed at the bottom of each catalyst bed. The catalyst beds may be of the same or of different sizes. The catalyst particles may be in extruded, pilled, or spherical form.

The quench zones or quench boxes separating successive catalyst beds have the following functions: (a) to cool the partially reacted fluids with hydrogen quench gas; (b) to assure a uniform temperature distribution in the fluids entering the next catalyst bed; and (c) to mix efficiently and disperse evenly the fluids over the top of the next catalyst bed. Since hydrocracking is an exothermic process, the fluids exiting one catalyst bed have to be cooled prior to entering the next catalyst bed, in order to avoid overheating and to provide a safe and stable operation. This is accomplished by thorough mixing with cool hydrogen. Furthermore, the temperature distribution in the cooled fluids entering the next catalyst bed has to be uniform in order to minimize the radial temperature gradients in successive catalyst beds. Unbalanced temperatures in a catalyst bed may result in different reaction rates in the same bed. This can lead to temperature excursions and to different deactivation rates of the catalyst. In addition to a uniform temperature distribution, it is also important to achieve a good mass flow distribution. The effective vapor/liquid mixing and uniform distribution of fluids over the top of the catalyst bed, accomplished in the quench zone, reestablishes an even mass flow distribution through the bed.

Different quench box designs are described in the patent literature [13–15]. One quench box design is shown schematically in Figure 10.8. The key components of a quench box are (a) a quench conduit with a sparger, which feeds cold quench gas into the box; (b) a quench tray or collection tray for receiving vapors and liquid exiting from the catalyst bed above the quench box and cooled by the quench gas; (c) a mixing chamber for effective mixing of fluids supplied by the quench tray; (d) a perforated first distribution tray or splash tray for dispersion of fluids exiting the mixing chamber over the cross-sectional area of the reactor; and (e) some quench boxes have a second distribution tray to uniformly distribute the reactants into the next catalyst bed.

Different designs have been proposed for the sparger attached to the quench

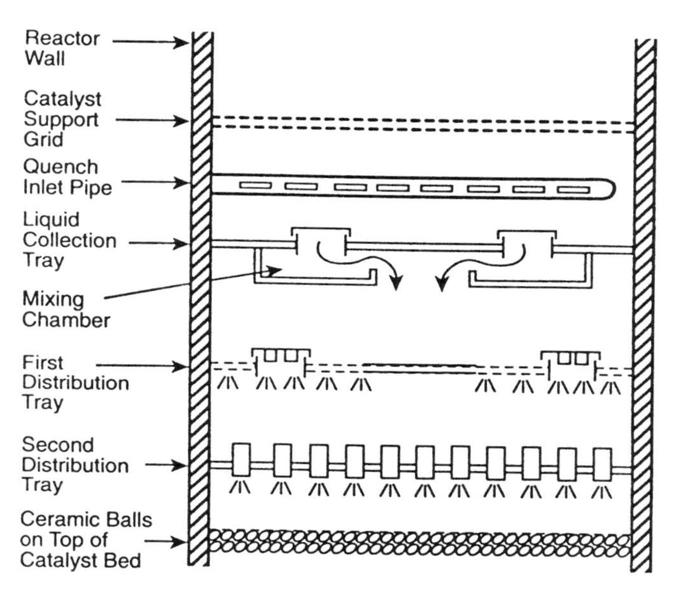

Figure 10.8 Schematic drawing of quench box.

conduit that feeds cool hydrogen into the quench box (Figure 10.9) [12–14,16]. A well-designed sparger allows the dispersion of quench gas in the form of jets, in a pattern usually normal to the downward flow of fluids exiting the catalyst bed above the quench box. This results in an intimate mixing and cooling of these fluids.

Distribution trays or plates of different types are described in the patent literature [16–20]. Well-designed distribution trays will uniformly redistribute the fluids and minimize the radial temperature gradients in the catalyst beds, thus assuring a stable operation and longer catalyst cycles.

Reactor diameter is determined by the desired mass velocity and acceptable reactor pressure drops. Commercial hydrocracking reactors may have interior diameters up to 4.5 m (15 ft). Depending on the design pressure and its diameter, the thickness of the reactor walls can be as much as 28 cm (11 in.).

The number of catalytic beds in a reactor and their respective lengths are determined from temperature rise profiles. The maximum acceptable temperature rise per bed defines the length of the catalyst bed. The acceptable temperature maximum, in turn, depends on the operating mode of the unit. For example, operations designed to maximize naphtha have a different maximum than those designed to maximize middle distillates. A typical reactor operated to maximize naphtha yields will have five or six beds with four or five quench zones. A typical

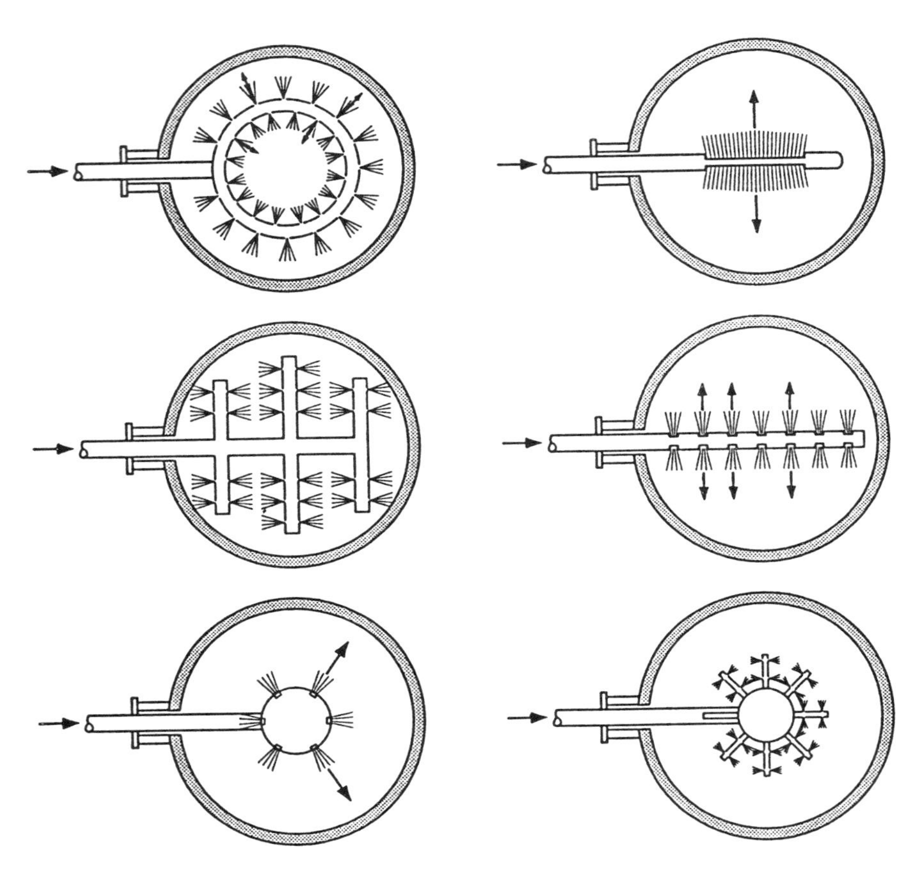

Figure 10.9 Different designs proposed for sparger in quench box.

reactor operated to maximize middle-distillate yields will have four beds with three quench zones. Commercial catalyst beds can reach lengths up to 6 m (20 ft).

A typical hydrotreating reactor used in hydrocracking processes has two beds with one quench zone. Feedstocks containing unsaturated material (e.g., coker gas oil, cycle oils) may require three or four beds with two or three quench zones to control the hydrogenative heat release. The maximum bed length may reach 12 m (40 ft). Trickle bed reactors have been described in detail in several publications [21–24]. The hydrocracking reactor is operated in plug flow, in a two-phase mode (reactants in vapor phase) or in a three-phase, trickle bed mode (reactants in vapor and liquid phase). The mode of operation depends on the boiling range of the feedstock. With lighter feedstocks, the reactor often operates in a two-phase mode. Heavier feedstocks involve a trickle bed operation because vapor/liquid mixtures pass through the reactor. Even distribution of fluids in the catalyst bed and even radial temperature distribution are key requirements for an efficient and safe operation.

The pressure drop in a fixed bed is very sensitive to bed void fraction. The void fraction can be controlled by particle size, shape, and method of catalyst loading. A range of particle sizes is not desirable because it results in smaller void fractions and can cause a significant pressure drop. A high void fraction results in low pressure drop and is desirable in some reactions. A high void fraction is also desirable to minimize bed plugging, e.g., in the hydroprocessing of residual feedstock.

There are two methods of catalyst loading: sock loading and dense loading. Sock loading is done by pouring the catalyst into a hopper mounted on top of the reactor and then allowing it to flow through a sock into the reactor. Dense loading or dense bed packing is done with the help of a mechanical device. Catalyst loaded by sock loading will have a higher void fraction than catalyst that was dense loaded. Dense bed packing and the resulting higher pressure drop provides a more even distribution of liquid in a trickle bed reactor. If diffusion limitations are negligible, dense loading is desirable in order to maximize the reaction rate per unit volume. This is often the case in hydrocracking reactors.

The other advantage of dense bed packing is that it orients the catalyst particles in a horizontal and uniform manner. This improves the vapor/liquid distribution through the catalyst beds. Catalyst particle orientation is important especially for cylindrically shaped extruded catalyst in vapor/liquid reactant systems. When the catalyst particles are oriented in a horizontal position in the catalyst bed, liquid maldistribution (channeling) is eliminated. This maldistribution tends to occur when the catalyst loading is done by the sock loading method, which generally causes the extrudates to be oriented in a downward slant toward the reactor walls increasing bed voids and creating liquid maldistribution.

During operation, the hydrocracking catalyst gradually loses some of its activity. In order to maintain the conversion constant, the average bed temperature is gradually increased. That temperature increase is in many cases less than 1°C/month (Figure 10.10). When the average bed temperature reaches a value close to the designed maximum, the catalyst has to be reactivated. Because the required temperature increase per unit time is relatively small, the reactor can be operated with the same catalyst for several years before replacement of the deactivated catalyst becomes necessary. Similar changes take place in the hydrotreating reactor.

Because the hydrocracking reactor is operated at high temperatures and high hydrogen partial pressures, it should be monitored and inspected on a regular basis. Only such monitoring and inspection can assure safe and continuous operation of the unit.

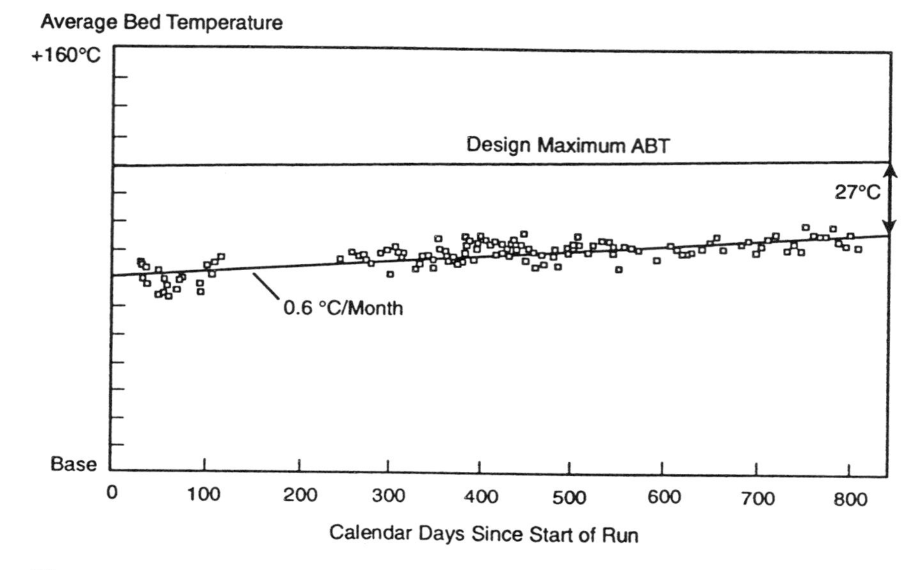

Figure 10.10 Variation of average bed temperature with time on stream. Process configuration: single-stage hydrocracking; feed: VGO; objective: maximize diesel product.

10.6. Kinetic Aspects

Kinetics is the study of the rates of reaction. The rates of reaction determine the key properties of a hydrocracking catalyst: initial activity, selectivity, stability, and product quality.

The initial activity is measured by the temperature required to obtain the desired product at the start of the run. In general, catalyst activity is a measure of the relative rate of feedstock conversion. In hydrocracking, activity is defined as the temperature required to obtain fixed conversion under certain process conditions. Hydrocracking conversion is usually defined in terms of change of endpoint:

$$\% \text{ conversion} = \left(\frac{\text{EP}^+{}_{\text{feed}} - \text{EP}^+{}_{\text{product}}}{\text{EP}^+{}_{\text{feed}}} \right) \times 100$$

where EP^+ indicates the fraction of material in the feed or product boiling above the desired endpoint (usually in wt % or vol %).

The rate of reaction is shown as follows:

$$\text{Rate} = k_0 \times \exp\left[\frac{-E_a}{RT} \right] \times (\text{EP}^+{}_{\text{feed}})^n$$

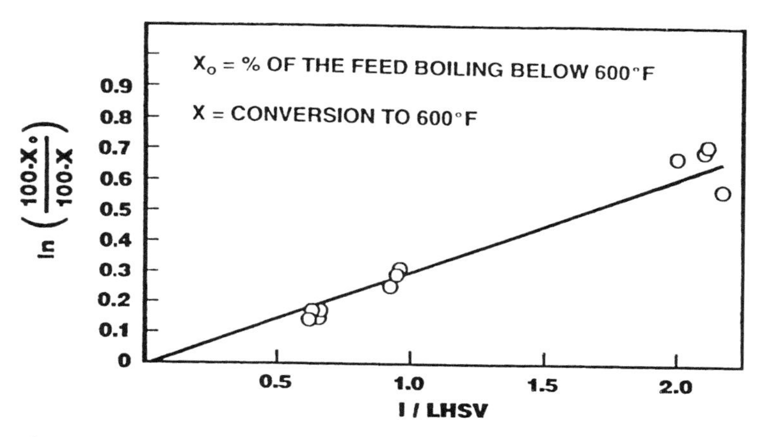

Figure 10.11 Vacuum gas oil hydrocracking first-order kinetics.

where k_0 is the rate constant, E_a is the activation energy, R is the gas constant, T is the absolute temperature, and n is the reaction order. Figures 10.11 and 10.12 show that the rate of hydrocracking of a vacuum gas oil (VGO) typically follows first-order kinetics ($n = 1$) and has an activation energy (E_a) of 47.7 kcal/gmol. For the usual case of a plug flow reactor, the relationship between conversion and

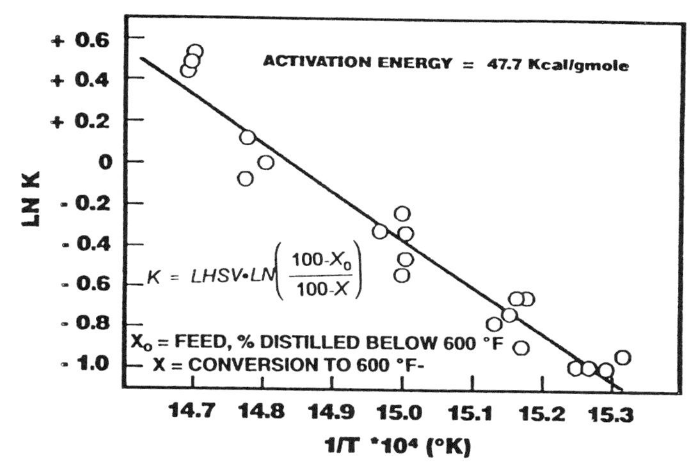

Figure 10.12 Arrhenius plot for first-order kinetics of vacuum gas oil hydrocracking.

temperature may be expressed by the following equations:

$$\text{Conversion} = 1 - \exp\left[\frac{-k}{\text{LHSV}}\right]$$

where LHSV is the space velocity, and

$$k = k_0 \times \exp\left[\frac{-E_a}{RT}\right]$$

Thus, as can be seen from Figure 10.12, a catalyst that is 15°C more active than a reference has a rate about twice as high as that of the reference at the same temperature (k_0 is twice as high). Therefore, a space velocity that is twice as high can be used to obtain the same conversion at the same temperature.

Although the reaction rates of individual feedstock components can be lumped together into an overall rate of boiling range conversion, these individual rates vary. In general, the relative reaction rates depend on the strength of adsorption of the reactants onto the catalyst surface. The adsorption strength decreases in the following order: heteroaromatic > multiring aromatic > monoaromatic > multiring naphthene > mononaphthene > paraffin. Figure 10.13 compares the reaction rates for individual feedstock components [25]. In this graph, the relative reaction rates are shown by the steepness of the slope of the lines. In addition, large molecules tend to react more rapidly for a given type of reactant, such as paraffins [26]. Other authors [27] have established a similar crackability ranking occurring during hydrocracking as shown in Figure 10.14. If removal of hetero

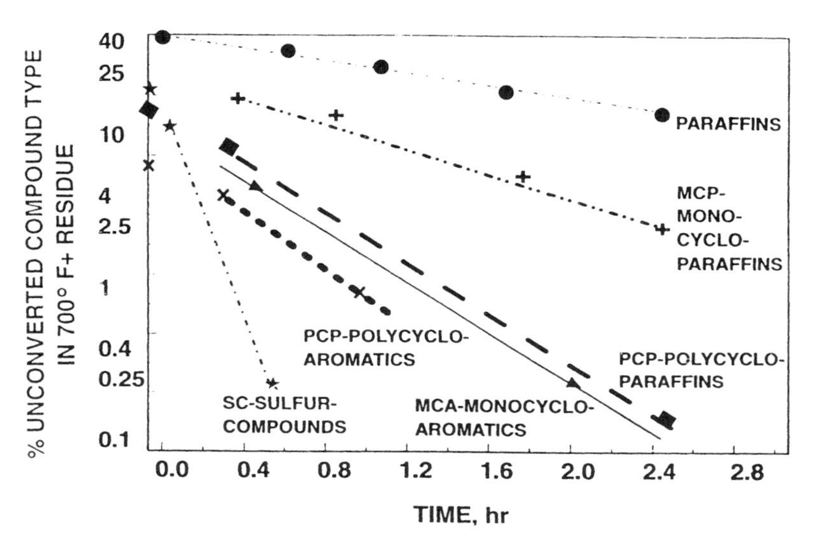

Figure 10.13 Hydrocracking of Kuwait vacuum gas oil [25].

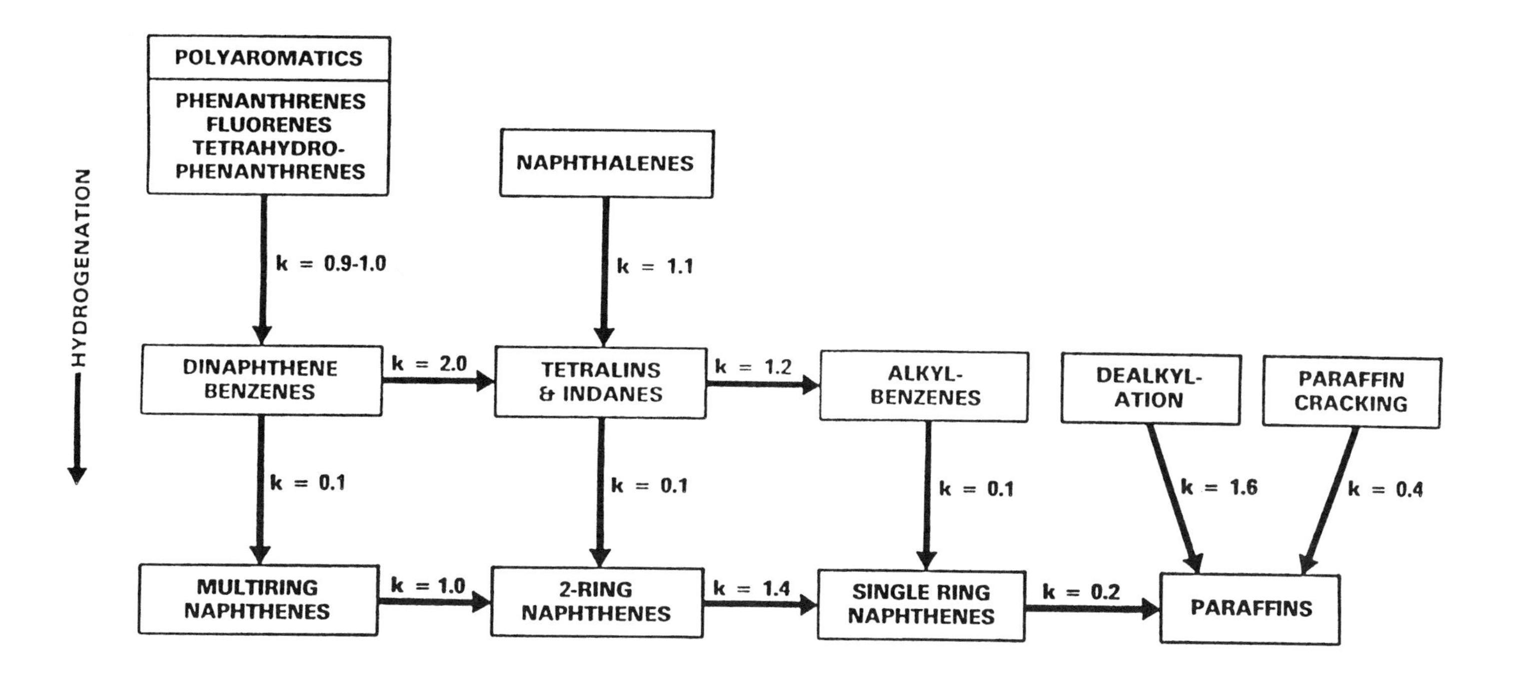

Figure 10.14 Relative rates of reactions under hydrocracking conditions [27].

Table 10.1 Effect of Feed Components on Activity

Feed component	Effect on activity
Nitrogen or NH_3	Poison for acid sites
Sulfur or H_2S	May poison noble metal some S needed for base metal
Oxygen or H_2O	May poison acid sites

atoms and metals is factored in, the ease of reaction in the hydrocracking process is as follows, in descending order: metal removal > olefin saturation > sulfur removal > nitrogen removal > saturation of rings > cracking of naphthenes > cracking of paraffins.

Table 10.1 shows the effect of heteroatomic species on hydrocracking activity. Nitrogen is the most severe poison of acidic sites. Sulfur poisons noble metal hydrogenation sites. Oxygen may poison acid sites. Figures 10.15 [28] and 10.16 [29] show the effect of feed nitrogen and reactor ammonia pressure on hydrocracking activity. Figure 10.15 shows that amorphous SiO_2-Al_2O_3 cracking activity is affected to a greater extent than crystalline zeolite cracking activity. This result may be due to the greater concentration and strength of acid sites on the zeolite (see Chapter 3).

Table 10.2 is a list of factors that influence hydrocracking activity. In general, increasing the effectiveness of the acid function or the metal function may increase hydrocracking activity. Operating conditions also influence hydro-

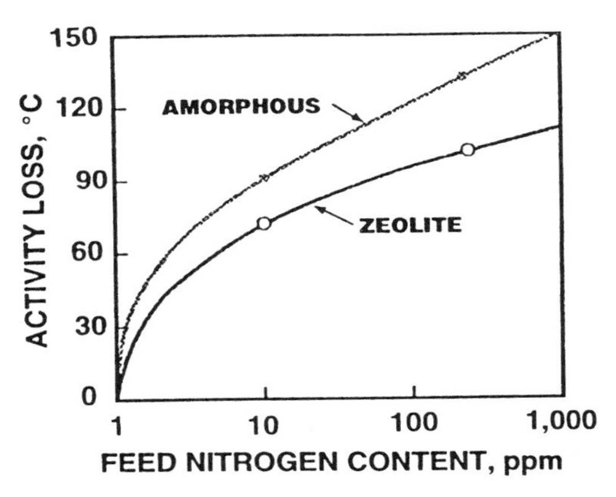

Figure 10.15 Effect of feed nitrogen on hydrocracker activity amorphous SiO_2-Al_2O_3 and zeolite catalysts [28].

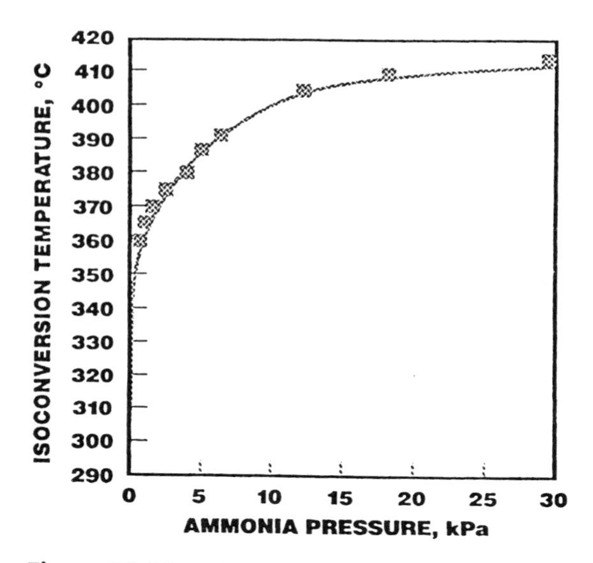

Figure 10.16 Effect of NH_3 hydrocracking activity [29].

cracking activity: more severe operating conditions require higher start-of-run temperatures.

Catalyst selectivity is a measure of the rate of formation of a desired product relative to the rate of conversion of the feed (or formation of other products). Hydrocracking selectivity is expressed as the yield of the desired product at a specific conversion. Yield is determined by the rate of formation of the desired

Table 10.2 Factors Increasing Hydrocracking Activity

Catalyst properties
- Increasing acid site strength
- Increasing acid site concentration
- Increasing metal site concentration (especially in the absence of NH_3)

Operating conditions
- Decreasing severity:
  Higher H_2 pressure,[a] higher recycle gas rate, lower conversion per pass, higher product EP, lower LHSV
- Feedstock components (e.g., aromatic vs. paraffins)

[a] H_2 pressure depends on total pressure, H_2 purity, recycle gas rate, and conversion per pass.

product relative to the feed rate. At 100% conversion, catalyst yield equals catalyst selectivity.

Figure 10.17 is a graph of the carbon numbers for various feed components as a function of boiling point [30]. This graph shows that a typical VGO feed may have components with an average of 30 carbon atoms. If the feed molecule cracks once, products are formed with an average carbon number of 15, which is the distillate range. This set of reactions may be expressed as:

$$\text{VGO} \xrightarrow{k_1} \text{distillate} \xrightarrow{k_2} \text{naphtha} + \text{light ends}$$

where k_1 and k_2 are the reaction rate constants for the two hydrocracking reactions. The selectivity to distillate products depends on the ratio of k_2/k_1. A typical ratio of k_2/k_1 is about 1/4 [25]. On the one hand, maximum distillate is produced by catalysts with a relatively low k_2/k_1 ratio. On the other hand, a catalyst designed for naphtha production will have a higher k_2/k_1 ratio.

Hydrocracking selectivity is also affected by the operating conditions. In general, more severe operating conditions cause higher selectivity for secondary products.

Stangeland [31] proposed an empirical equation for catalytic hydrocracking that allows the calculation of the rate of reaction, k, as a function of the true boiling point, T_b, expressed in °F:

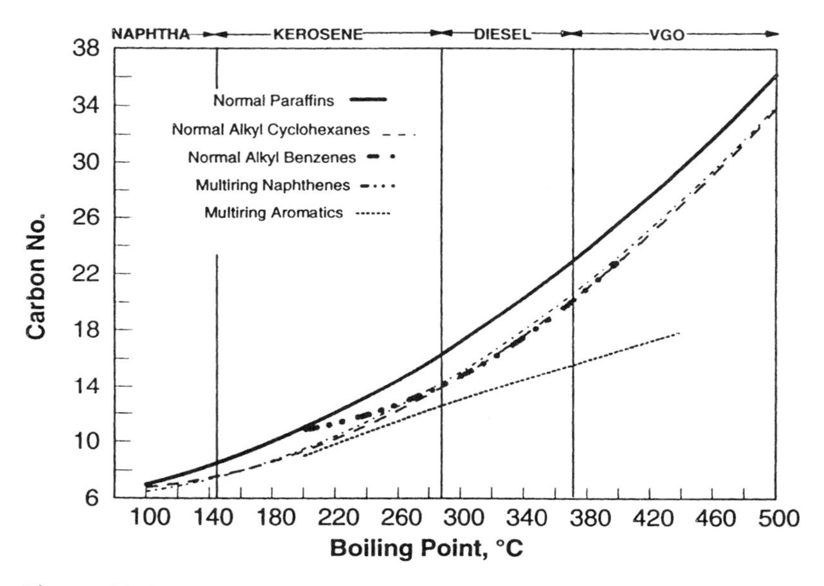

Figure 10.17 Carbon number distribution of petroleum fractions [30].

$$k(T) = k_0[T + A(T^3 - T)]$$

where $T = T_b/1000$, k_0 = base reactivity or limiting reactivity of the fraction under consideration, and A is an empirical constant, in the range $0 < A < 1$. $A = 0$ gives a linear relationship between rate and boiling point, whereas values of $A > 1$ give rapidly increasing reactivity with boiling point (Figure 10.18) [31]. Experimental data indicate an A value of about 0.5. The corresponding curve shows a strong increase in rate of reaction, k, with increasing true boiling point, T_b, i.e., with increasing molecular weight. Experimental data obtained for pure compounds show that the cracking rate decreases rapidly for hydrocarbons boiling below 120°C [32]. Other experiments have shown that the rate of hydrocracking for pure *n*-paraffins increases in the ratio 1:32:72:120 for $C_5/C_{10}/C_{15}/C_{20}$. Data obtained from the rate equation generally agree with these findings. However, due to the complex nature of crude oil–derived feeds, the correlation between the rate of reaction and true boiling point will vary from feed to feed.

The rate of reaction is also affected by the adsorption of the hydrocarbons on the catalyst. In general, heavier hydrocarbons adsorb more strongly and react faster than lighter ones.

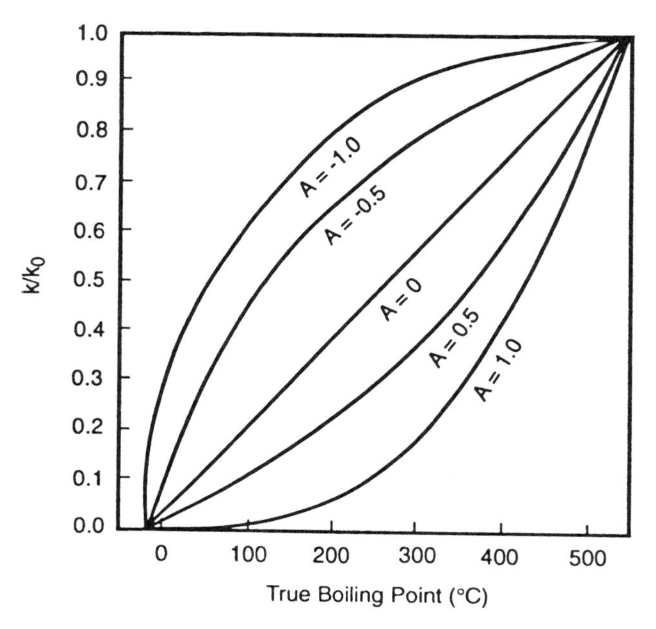

Figure 10.18 Relative rate constant for hydrocracking gas oil as a function of true boiling point, for different A values (see text) [31].

The rate equation described has been used for the following:

1. Estimating yields at a desired conversion by interpolation, using data obtained at other conversions
2. Optimizing yields in a two-stage operation by varying conversion levels
3. Estimating yields for feeds that have not yet been tested
4. Providing a consistent method for describing cracking rate constants for a variety of feeds
5. Indicating the effects of different catalysts and feeds on kinetics and yields

Examples of applications of the rate equation are given in Stangeland's publication [31]. One example shows the use of the rate equation for the optimization of jet fuel yields in a two-stage hydrocracking operation. A comparison of predicted and measured jet yields at different conversions in a two-stage hydrocracker is shown in Figure 10.19. Filled symbols indicate experimental data, with the corresponding jet yields in parentheses. Contour lines represent model predictions.

As the first-stage severity increases, the feed to the second stage becomes lighter. This in turn shifts the product distribution from each stage. The combined

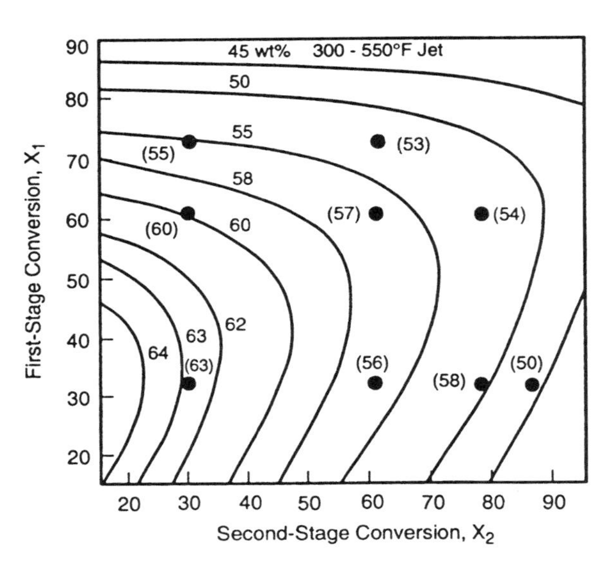

Figure 10.19 Predicted and measured jet yields at different conversions in a two-stage hydrocracker. Filled symbols indicate experimental data, with the corresponding jet yields in parentheses. Contour lines represent model predictions [31].

yields from such a two-stage operation were simulated. Many combinations of first- and second-stage conversions were used in order to build up the jet yield contours. The figure shows a fairly good agreement between predicted and measured yields.

More recently, Trytten and Gray [33] developed a method of estimating hydrocracking performance for an oil using a group additivity method. Data from ^{13}C-NMR, simulated distillation analysis, density and elemental analyses were used to calculate the average number of C-C bonds per molecule of oil. The percentage of C-C bonds actually cracked was calculated from the number of C-C bonds in the feed and in the products. Additional kinetic studies of hydrocracking of crude oil–derived feedstocks are listed in Section 6.2.6.

Catalyst stability is a measure of change of reaction rate over time. Hydrocrackers are typically operated in the constant conversion mode, with temperature adjustments made to maintain the desired conversion. Hydrocracking activity stability is defined as the temperature change required to maintain constant conversion. Changes in product yield over time on-stream have been reported for some zeolitic hydrocracking catalysts [34,35]. Hydrocracking yield stability is defined as the yield change with time at constant conversion and is usually expressed as a function of temperature change. In addition to the catalyst properties, the other factors that contribute to increased catalyst stability are feedstock properties and the operating conditions. Feedstocks with low polynuclear aromatics concentration and low nitrogen content contribute to increased stability. Decreasing operating severity such as running a higher H_2 partial pressure, higher recycle gas rate, lower conversion per pass, higher product endpoint, and lower LHSV contribute to increasing the catalyst stability.

The product quality is a measure of the ability of the process to yield products with the desired use specifications such as pour point, smoke point, or octane. Table 10.3 shows some of the important product quality measurements and the chemical basis for these measurements. In general, product quality for

Table 10.3 Chemical Basis for Product Quality

Quality measurement	Chemical basis
High smoke point	Low concentration of aromatics
Low aniline point	Low concentration of aromatics
Low pour point	Low concentration of *n*-paraffins
Low freeze point	Low concentration of *n*-paraffins
Low cloud point	Low concentration of *n*-paraffins
High octane	High ratio of iso/normal paraffins
	High concentration of aromatics
	High concentration of naphthenes

distillate fuels requires low aromatics and low *n*-paraffin concentrations. Because the production of either distillate or naphtha can be controlled by the choice of hydrocracking operating conditions and catalyst, hydrocracking can effectively balance seasonal demands for distillate production or naphtha for gasoline blending.

References

1. M. Skripek, D. A. Lindsay, and K. E. Whitehead, NPRA Annual Mtg., March 1988, San Antonio, TX, AM-88-55.
2. M. E. Reno, S. W. Shorey, and T. W. Tippett, NPRA Annual Mtg., March 1992, New Orleans, LA, AM-92-46.
3. A. G. Bridge, D. R. Cash, D. V. Law, and L. J. Scotti, 7th Annual Int. Refining Conf., May 9–12, 1994, Singapore.
4. M. G. Hunter and A. R. Gentry, Annual Int. Refining Conf., May 9–12, 1994, Singapore.
5. T. S. Chou, C. R. Kennedy, and S. S. Shih, U.S. Patent No. 4,997,544 (1991).
6. L. C. Gutberlet, S. G. Kukes, and F. T. Miller, U.S. Patent No. 4,875,991 (1989).
7. L. C. Gutberlet, J. C. Kelterborn, S. G. Kukes, and J. T. Miller, U.S. Patent No. 4,834,865 (1989).
8. L. C. Gutberlet, A. L. Hensley, Jr., and S. G. Kukes, U.S. Patent No. 4,797,196 (1989).
9. W. D. Gillespie, U.S. Patent No. 5,232,578 (1993).
10. J. L. Aderhold, Jr., A. L. Hensley, Jr., J. C. Kelterborn, and S. G. Kukes, U.S. Patent No. 4,954,241 (1990).
11. R. L. Howell, R. F. Sullivan, C.-W. Hung, and D. S. Laity, paper presented at the Petroleum Refining Conference, Tokyo, 1988.
12. W. R. Hennemuth and D. B. Carson, U.S. Patent No. 3,598,541 (1971).
13. P. M. Bhagat, R. M. Koros, R. D. Patel, J. M. Peruyero, and J. T. Wyatt, U.S. Patent No. 4,960,571 (1990).
14. R. M. Shirk, U.S. Patent No. 3,378,349 (1968).
15. L. Alcock, U.S. Patent No. 3,880,961 (1975).
16. Ch-Ch. J. Shih and B. A. Christolini, U.S. Patent No. 5,158,714 (1992).
17. F. A. Aly, R. G. Graven, and D. W. Lewis, U.S. Patent No. 4,836,989 (1989).
18. M. P. Grossboll, R. R. Edison, and T. Dresser, U.S. Patent No. 4,126,540 (1978).
19. W. R. Derr, Jr., et al., U.S. Patent No. 4,126,539 (1978).
20. F. V. Hanson and P. W. Snyder, Jr., U.S. Patent No. 3,541,000 (1970).
21. C. N. Satterfield, *AIChE J. 21*:209 (1975).
22. Y. T. Shah, *Gas–Liquid–Solid Reactor Design*, McGraw-Hill, New York, 1979.
23. P. A. Ramachandran and R. V. Chaudhari, *Three-Phase Catalytic Reactors*, Gordon and Breach, London, 1983.
24. A. Gianetto and P. L. Silveston, *Multiphase Chemical Reactors: Theory, Design, Scale-Up*, Hemisphere, New York, 1986.
25. R. Krishna, *Erd/uol und Kohle-Erdgas-Petrochemie 42*:194 (1989).
26. J. Weitkamp, in *Hydrocracking and Hydrotreating* (J. W. Ward and S. A. Qader, eds.), A.C.S. Symposium Series 20, 1975, p. 1.

27. V. A. Filimonov, A. A. Popov, V. A. Khavkin, I. Y. Perezghina, L. N. Osipov, S. P. Rogov, and A. V. Agafonov, *Int. Chem. Eng. 12*(1):21 (1972).
28. J. W. Scott and A. G. Bridge, *Adv. Chem. 103*:113 (1971).
29. P. Dufresne, et al., *Catalysts in Petroleum Refining 1989* (D. L. Trimm et al, eds.), Elsevier, Amsterdam, 1990, p. 301.
30. F. D. Rossini et al., *Hydrocarbons from Petroleum*, Van Nostrand Reinhold, New York, 1953.
31. B. E. Stangeland, *Ind. Eng. Chem. Proc. Des. Dev. 13*(1):71 (1974).
32. J. W. Scott and A. G. Bridge, A.C.S. Adv. Chem. Series 103, Washington, D.C., 1971.
33. L. C. Trytten and M. R. Gray, *Fuel 69*:397 (1990).
34. R. F. Sullivan and J. A. Meyer, in *Hydrocracking and Hydrotreating* (J. W. Ward and S. A. Oadar, eds.), A.C.S. Symposium Series 20, Washington, D.C., 1975, p. 28.
35. B. E. Stangeland and J. R. Kittrell, *Ud. Eng. Chem. Proc. Ser. Dev. 11*:15 (1972).

11

FACTORS AFFECTING PRODUCT YIELDS AND QUALITY

A typical characteristic of hydrocracking processes is feedstock and product yield flexibility. It allows the conversion of a wide range of feedstocks into a wide range of products, such as liquid petroleum gas (LPG), naphtha, middle distillates (jet and diesel fuels), fuel oil, lubricating oil base, petrochemical feedstocks, and fluid catalytic cracking (FCC) feedstock. The products obtained from hydrocracking processes have very low levels of sulfur and nitrogen, as well as olefins.

Changes in product yields and quality can often be brought about by changes in feedstock, in process configuration, in operating variables, and in catalyst design. These changes account for the considerable flexibility of the hydrocracking process.

11.1. Effect of Feedstock

Industrial feedstocks used in hydrocracking are complex mixtures of hydrocarbons and hetero compounds (see Chapter 1). Even with modern analytical techniques it is impossible to identify all of the individual compounds present in these feedstocks [1]. Therefore, feedstocks are usually characterized by gross properties such as gravity, aniline point, boiling range, hydrogen content, and hetero atom (mostly sulfur and nitrogen) content. Feedstocks can also be characterized by the proportion of different classes or types of hydrocarbons present, such as paraffins, naphthenes, and aromatics. At one extreme are highly paraffinic feedstocks, whereas at the other extreme are aromatic feedstocks. This is illustrated by the composition and properties of vacuum gas oil (VGO) feeds obtained from different crude oils, listed in Table 11.1 [2]. VGOs obtained from Bombay High or Minas crudes are highly paraffinic, whereas those obtained from Maya, Alaska North Slope, or California are highly aromatic.

Table 11.1 Variation of VGO Properties with Crude Source [2]

Crude source	Bombay High, India	Minas, Indonesia	Taching, China	Arabian Heavy, Saudi Arabia	Maya, Mexico	Alaska North Slope, USA	California, USA
LV % yield of 340–540°C VGO	26	37	37	24	23	31	24
API gravity	28	34	31	22	21	20	14
Mass spec.							
Group type, LV %							
Paraffins	46	44	37	17	11	8	1
Naphthenes	23	34	47	25	29	36	42
Aromatics	31	21	16	58	58	55	57
Sulfur, wt %	0.5	0.1	0.1	3.0	3.1	1.2	1.0
Nitrogen, ppm	410	610	690	150	1880	1510	6470

Table 1.2 lists the different feedstocks used and major products obtained in the hydrocracking process. The table illustrates the considerable flexibility of the hydrocracking process with regard to both feedstocks and products.

The ease of hydrocracking a specific feedstock depends on its composition and properties. Highly paraffinic feeds will require high temperatures for hydrocracking because normal paraffins are difficult to crack [2]. However, they do not contribute significantly to catalyst deactivation (low coke formation tendency). Therefore, catalyst selection for paraffinic feeds is generally dictated by the liquid yield distribution desired. Catalysts with a high hydrogenation-to-acidity ratio will produce higher C_5^+ liquid yields and a heavier liquid product distribution from paraffinic feeds. Catalysts with a low hydrogenation-to-acidity ratio will produce higher yields of lighter liquid product and light isoparaffins, particularly i-C_4.

Hydrocracking of aromatic feeds requires catalysts with good hydrogenation ability because aromatics can be cracked only after hydrogenation. Aromatic feeds also have a tendency to deposit coke on the catalyst surface and thus deactivate the catalyst. High hydrogenation activity catalysts with moderate-to-high acidity are needed to process aromatic feedstocks [2]. Diffusion effects also become important when processing feeds containing polyaromatic molecules.

In general, the heavier the feedstock (higher feed endpoint), the more severe the operating conditions required to obtain satisfactory product yields and quality. Heavier feeds are usually processed at higher temperatures, higher hydrogen partial pressures, and lower space velocities. Conversely, addition of a heavier feedstock to a lighter feedstock reduces conversion. The more severe process conditions required by heavier feedstocks result in higher light ends (C_1–C_4) and light naphtha yields, as well as lower middle-distillate yields. Catalysts with high hydrogenation activity and moderate-to-high acidity are needed to process heavy feedstocks [2].

Nitrogen-rich feedstocks (e.g., over 1000 ppm nitrogen) have to be processed at higher hydrogen partial pressure and/or lower space velocity during the hydrotreatment step in order to reduce the nitrogen content of the effluent to levels acceptable for hydrocracking [3]. Conversely, basic nitrogen compounds may be added to the feedstock during hydrocracking in order to control both catalyst activity and selectivity [4,5].

The content of metals, asphaltenes, and Conradson carbon in a feedstock affects catalyst stability and cycle length. For that reason, limits are often set for these contaminants to make a feedstock acceptable for hydrocracking. Most refiners use the following constraints: (Ni + V) <2 ppm; asphaltenes <500 ppm; Conradson carbon residue <1 wt %. In general, a more paraffinic feed produces a more paraffinic, higher quality product while consuming less hydrogen than a more aromatic feed.

Because feedstock compositions can show significant variations, product quality will depend strongly on feedstock used. This is illustrated in Table 11.2,

Table 11.2 Effect of Feedstock on Diesel Quality (Single-Stage Hydrocracking) [6]

Feed description	Middle East VGO	FCC HCO
Boiling range, °C	343–566	343–455
Gravity, °API	21	8.5
Sulfur, wt %	2.5	2.0
Nitrogen, ppm	900	900
Carbon residue, wt %	0.5	0.1
HC types, wt %		
Paraffins	22	11
Naphthenes	21	9
Monoaromatics	21	4
Polyaromatics	11	55
Sulfur compounds	18	15
O&N compounds	7	6
Diesel Properties		
Diesel boiling range, °C	160–350	177–343
Gravity, °API	44	34
Sulfur, ppm	16	10
Nitrogen, ppm	<1	<1
Aniline pt, °C	82	63
Cetane no.	60	50
Aromatics	8	17

which shows the composition of two hydrocracking feedstocks and the quality of diesel product obtained from these feedstocks [6]. The aromatics content of diesel fuel obtained from a Middle Eastern VGO is less than half that obtained from FCC heavy cycle oil (8 vs. 17 vol % aromatics). This is due to the significant difference in polyaromatics content of the two feedstocks: 11 vol % in VGO vs. 55 vol % in cycle oil. The higher aromatics content of diesel product obtained from FCC heavy cycle oil reduces the aniline point and cetane number: 63°C (145°F) vs. 82°C (180°F) aniline point, and 50 vs. 60 cetane number. The lower API gravity of the heavy cycle oil (8.5° API vs. 21° API) leads to a diesel product with lower API gravity: 34° API vs. 44° API.

11.2. Effect of Process Configuration

The single-stage configuration is the most common design and requires the lowest capital investment. The once-through configuration is used when the fractionator bottoms are also desired products. These unconverted materials can be FCC feedstocks, ethylene cracker feeds, or lube oil base stocks. However, the middle distillates obtained in this process are of lower quality due to a higher aromatics content. Because the once-through process is operated at high severity in order to obtain reasonable overall conversions, it produces higher yields of light ends compared to a recycle operation. The once-through process also has less flexibility than a recycle process with regard to product yield structure.

A single-stage recycle configuration is similar to the second stage and fractionation section of a two-stage configuration. It achieves complete conversion of feed to lighter products. However, it is less selective for liquid product than a two-stage configuration. It is generally used for feeds with low endpoints and for low-capacity plants.

A two-stage configuration offers more product yield flexibility than a single-stage operation because it allows the use of different hydrocracking catalysts in each stage. In a process in which middle distillates are the desired products, the use of a two-stage rather than a single-stage configuration results in a more complete hydrogenation of the products, with a significant reduction in aromatics and an increase in the smoke point of the middle-distillate fractions. However, the capital investment required for a two-stage plant is higher than for a single-stage plant.

11.3. Effect of Operating Variables

Operating variables such as temperature, hydrogen partial pressure, amount of ammonia present, conversion per pass, feed/recycle ratio, and fractionation conditions have a significant impact on the yield and quality of the resulting products. For example, a change in reaction temperature, distillation cut point, and conversion per pass can shift the maximum yield from naphtha to jet or diesel fuel. As the product objective changes from naphtha to jet fuel to diesel fuel, the reactor temperature is decreased and the naphtha endpoint reduced. The decrease in temperature decreases conversion per pass and increases the yield of heavier fractions while decreasing the yield of lighter fractions. Furthermore, a slight change in fractionation conditions, such as changes in product endpoint, will further enhance the yield of the desired product.

This is illustrated in Table 11.3 for a single-stage, recycle hydrocracking process. By decreasing the reactor temperature by 6°C (11°F) or 12°C (22°F) and by reducing the naphtha endpoint from 216°C (420°F) to 121°C (250°F) or 118°C (245°F), the maximum yield is shifted from naphtha to jet or diesel fuel, respectively. A zeolite-based catalyst was used. When operating in the naphtha mode, a yield of 90 vol % naphtha is obtained. When operating in the jet fuel mode, a yield of 69 vol % jet fuel is obtained, with a decline in naphtha yield. In the diesel mode, a yield of 62 vol % heavy diesel or 77 vol % total diesel is obtained, with a further reduction in naphtha yield. The properties of the products obtained are indicative of good-quality products.

Table 11.3 Illustration of Process Flexibility (Single-Stage Hydrocracking)

Feedstock		*Vacuum Gas Oil*	
Boiling range, °C		340–553	
Specific gravity, g/cm^3		0.9273	
Nitrogen, wppm		920	
Sulfur, wt %		2.6	
Aniline point, °C		82	
Product Objective	*Naphtha*	*Jet Fuel*	*Diesel*
Reactor temp., °C	Base	Base-6	Base-12
Yields, vol % FF			
C_4	11	8	7
C_5–82°C	25	18	16
82°C-plus naphtha	90	29	21
Jet	—	69	15
Diesel	—	—	62
Total C_4^+	126	124	121
Product Quality			
C_5–82°C			
RON, clear	79	79	80
Naphtha			
P/N/A	45/500/5	44/52/4	38/56/6
RON, clear	41	63	67
End Point, °C	216	121	118
Jet A-1			
Flash point, °C	—	38	—
Freeze point, °C	—	−48	—
Smoke point, mm	—	34	—
Aromatics, vol %	—	7	—
Endpoint, °C	—	282	—
Diesel			
Cloud point, °C	—	—	−15
Specific gravity, g/cm^3	—	—	0.8063
Cetane number	—	—	55
Flash point, °F	—	—	52
Endpoint, °C	—	—	349

In general, process severity, as reflected in reactor temperature and hydrogen partial pressure, increases with increasing feedstock boiling range. Higher temperatures and hydrogen partial pressures increase the reaction rates. Higher hydrogen partial pressures also reduce coke deposition and catalyst deactivation (see below).

When the objective is the production of a higher boiling range product, the severity of the operation is reduced. For example, a single-stage hydrocracking process can be used to convert VGOs either to distillates or to lube oil base stock. However, the processing conditions for lube oil base stock are much less severe because a heavier product is the major objective.

11.3.1. Reactor Temperature

It has already been shown that an increase in reactor temperature increases feedstock conversion and shifts yields to lighter products, whereas a decrease in temperature decreases conversion and enhances the yields of heavier products (Table 11.3). While hydrocracking catalysts can be kept on stream for several years, their gradual aging affects both activity and selectivity. However, the selectivity of amorphous catalyst is much less affected by catalyst aging than the zeolitic type of catalyst. In order to maintain conversion constant, the operating temperature is gradually increased to make up for the loss in activity. The increase in temperature results in a gradual shift in product distribution from heavier to lighter products (i.e., from diesel fuel to jet fuel to naphtha). Such a shift is due to the increased severity of the operation, which leads to products with a smaller average molecular weight (lower boiling range). A further increase in temperature will decrease the total C_5^+ liquid yield, as well as the yields of individual liquid products, while increasing the C_4^- gas yield. An increase in reactor temperature will also decrease the i-C_5/n-C_5 and i-C_6/n-C_6 ratios in the product [7]. From a thermodynamic point of view, an increase in operating temperature decreases the rate of aromatics hydrogenation [8]. This explains why the aromatic content of middle distillates increases with catalyst age.

11.3.2. Hydrogen Partial Pressure

Hydrogen partial pressure will affect both the degree of hydrogenation of unsaturated compounds in the feedstock and the rate of hydrocarbon cracking. Furthermore, it will affect the rate of hydrodenitrogenation and hydrodesulfurization of the feedstock.

An increase in hydrogen partial pressure increases the rate of aromatics hydrogenation. Hydrogen partial pressure has a dual effect on catalytic cracking and isomerization. On the one hand, an increase in pressure has a favorable effect due to enhanced hydrogenation of coke and cleaning of the catalyst surface. On the other hand, the rate of cracking and isomerization reactions decreases when hydrogen partial pressure increases. These two effects are in competition [8]. When processing aromatic feedstocks, the first effect will predominate. When processing paraffinic, low-coke-forming feedstock, the second effect will predominate. An increase in hydrogen partial pressure increases the rate of hydrodenitrogenation. The rate of hydrodesulfurization is less sensitive to hydrogen partial pressure. These effects are illustrated in the following examples.

The effect of hydrogen partial pressure on product quality obtained from the hydrotreatment of an Arabian Light VGO is shown in Table 11.4 [8]. Total pressure is 70 bars (1030 psig) or 140 bars (2045 psig). The data show that the hydrotreated product obtained at higher hydrogen partial pressure is lower in nitrogen and sulfur, as well as in aromatics. The data also show the stronger sensitivity to pressure of hydrodenitrogenation compared to hydrodesulfurization. The hydrotreated product obtained at higher hydrogen partial pressure leads to higher conversion during the subsequent hydrocracking step of the process.

The effect of hydrogen partial pressure on conversion and product quality obtained in the second stage of a hydrocracking process, using the same hydrotreated feedstock, is shown in Table 11.5 [8]. Total pressure is 120 bars (1755 psig) or 90 bars (1320 psig), and hydrogen flow varies in order to keep the partial pressures of hydrogen sulfide and ammonia constant. The hydrotreated feed is predominantly paraffinic and naphthenic. The data show that an increase in hydrogen partial pressure increases conversion and improves the quality of jet fuel, gas oil, and fuel oil products by reducing their aromatic content. The higher conversion at higher hydrogen partial pressure is due in this case to stronger suppression of catalyst coking (see also Chapter 12).

11.3.3. Ammonia Partial Pressure

The presence of ammonia in the reaction environment will affect both catalyst activity and selectivity. Catalyst activity decreases in the presence of ammonia. The required operating temperature of hydrocracking catalysts for a specified conversion is proportional to the logarithm of the partial pressure of ammonia in the reactor [9]. The higher operating temperature also produces more C_4^- gas. Furthermore, the middle-distillate selectivity increases significantly with increas-

Table 11.4 Effect of Hydrogen Pressure on Product Quality Obtained from Hydrotreatment of Arabian Light VGO [8]

	Feed VGO Arabian Light	Unconverted Fraction from First-Stage Hydrotreatment	
		Pressure 70 B	Pressure 140 B
Specific gravity	0.906	0.864	0.844
N (Wt ppm)	650	10	1
S (Wt ppm)	24300	200	40
Aromatics Total	47	31	9.6
UV (wt %) Mono	24	21	8.2
DI	11	6.1	1.0
TRI	12	4.4	0.4

Table 11.5 Effect of Second-Stage Hydrogen Pressure on Conversion and Product Quality for Same Hydrotreated Feedstock [8]

Total pressure (bars)		120	90
Hydrogen flow (L/L)		1000	735
Partial pressures (bars)			
Hydrogen		112.6	82.6
Hydrocarbon		6.3	6.3
H_2S		0.97	0.97
NH_3		0.13	0.13
Conversion (wt %)		78	73
	Jet fuel	3.3	4.7
Aromatic (wt %)	Gas oil	1.7	3.5
	Fuel oil	1.1	1.9

ing partial pressure of ammonia (Figure 11.1) [9]. The iso-to-normal ratio of light paraffins decreases with increasing ammonia partial pressure and decreasing catalyst activity. Therefore, an increase in ammonia partial pressure has a similar effect on catalyst activity, middle-distillate selectivity, and iso-to-normal paraffin ratio as the decrease in the level of cracking component.

The decrease in catalyst activity in the presence of ammonia is due to the poisoning of its acid sites, resulting in the weakening of the catalyst cracking function [3]. The increase in middle-distillate selectivity with increasing ammonia partial pressure is also related to the decrease in the total number of acid sites. Larger molecules of the feedstock can still be cracked, yielding products in the diesel and jet fuel boiling range; however, the cracking of medium-size and smaller molecules is reduced. This is in line with results obtained with model compounds, which showed that longer paraffinic chains can be cracked at lower severity than shorter paraffinic chains [10]. The decrease in iso-to-normal ratio of light paraffins with lower catalyst activity is due to the suppression of isomerization reactions (less carbenium ion formation) under these conditions.

11.4. Effect of Polynuclear Aromatics

Polynuclear aromatic compounds (PNAs) are polycyclic, condensed hydrocarbons that contain two or more fused aromatic rings. They are also called polyaromatic hydrocarbons (PAHs). The light PNAs, with two to six rings, are present in straight-run VGO. Their concentration increases during the hydrocracking process. This is illustrated in Table 11.6 [11]. The data were obtained by high pressure liquid chromatography (HPLC) analysis and show the composition of a residue-derived, desulfurized VGO at start-of-run (SOR) and at end-of-run (EOR)

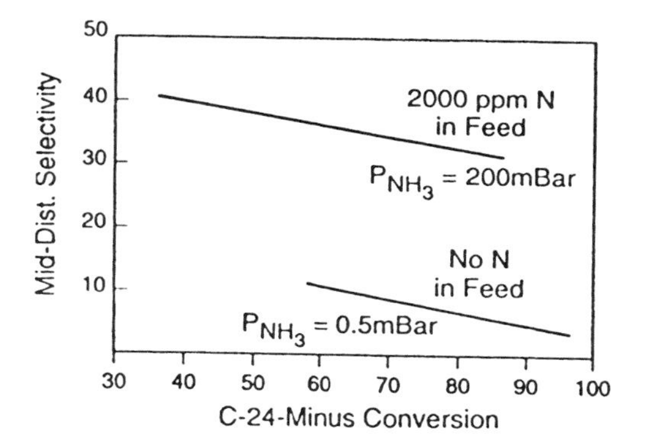

Figure 11.1 Effect of NH_3 partial pressure on middle-distillate selectivity [9].

in a two-stage hydrocracking operation. The results obtained show a significant increase (two- to four-fold) in PNAs (tri-, tetra-, and pentaromatics +) in the EOR stream as compared to the SOR stream.

Light PNAs also serve as precursors in the formation of larger PNAs. Most of the heavy PNAs (over six rings) are formed during the processing of heavy feedstocks at conversions close to 100% under severe operating conditions. The heavy PNAs (HPNAs) have a deleterious effect on the hydrocracking operation by forming deposits on the catalyst and in the unit.

Examples of PNAs encountered during the processing of heavy feedstocks

Table 11.6 Composition of Residue-Derived, Desulferized VGO at Start and End of Run [11]: Two-Stage Hydrocracking

	Conc. in VGO, wt %		Ratio
HPLC fraction	SOR	EOR	EOR/SOR
Saturates	57.8	53.0	0.92
Monoaromatics	27.50	28.46	1.03
Diaromatics	13.04	14.35	1.10
Triaromatics	1.198	2.862	2.39
Tetraaromatics	0.187	0.729	3.90
Pentaaromatics+	0.134	0.380	2.84
Tot. aromatics	42.05	46.78	1.11

SOR, start-of-run; EOR, end-of-run.

are shown in Figure 11.2. The HPNAs are formed in the hydrocracking unit by (a) condensation of two or more smaller PNAs present in the hydrocracker feed; (b) dehydrogenation of larger hydrogenated polycyclic compounds; or (c) cyclization of side chains on preexisting PNAs, followed by dehydrogenation. An increase in reactor temperature increases dehydrogenation.

The formation of HPNAs from light PNAs by cyclization of side chains can be described by the reaction pathway named "naphthalene zigzag" shown in Figure 11.3 [11]. The reaction scheme involves two types of ring closure reactions: the "ortho ring" closure is kinetically favored and involves a four-carbon atom addition; and the thermodynamically favored "peri ring" closure, which involves a two-carbon atom addition to a bay region of the molecule. It is assumed that in the process of forming larger PNAs the two types of ring closure alternate.

Compared to the PNAs in the feed, those in the fractionator bottoms not only are heavier but also are unsubstituted or contain shorter alkyl side chains. As side chains shorten or disappear, the solubility of HPNAs decreases.

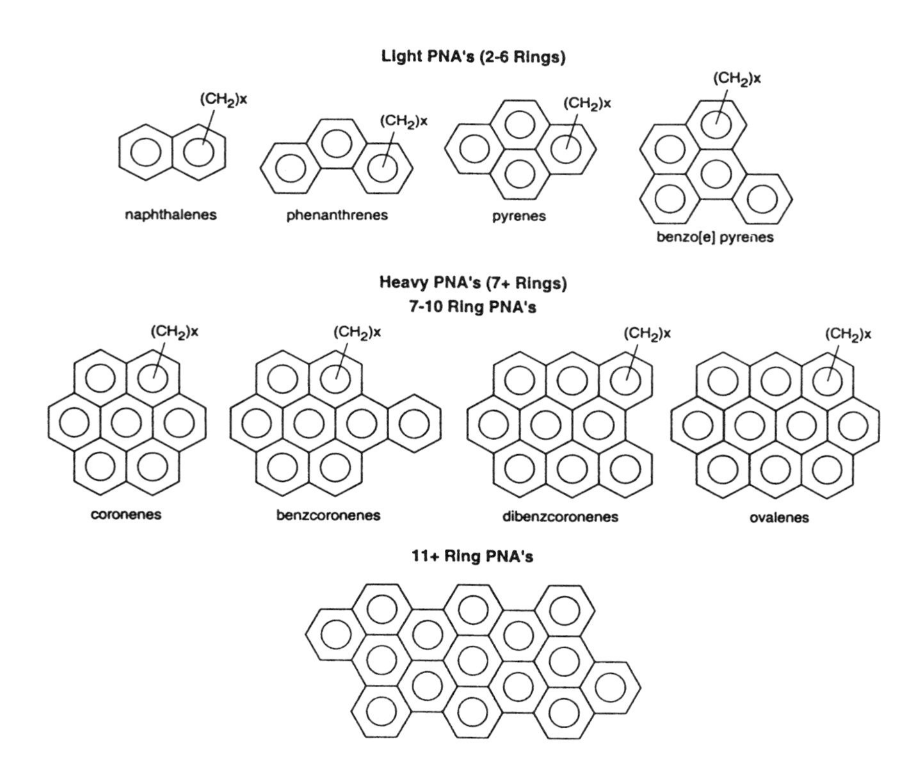

Figure 11.2 Polynuclear aromatics (PNAs) encountered during processing of heavy feedstocks.

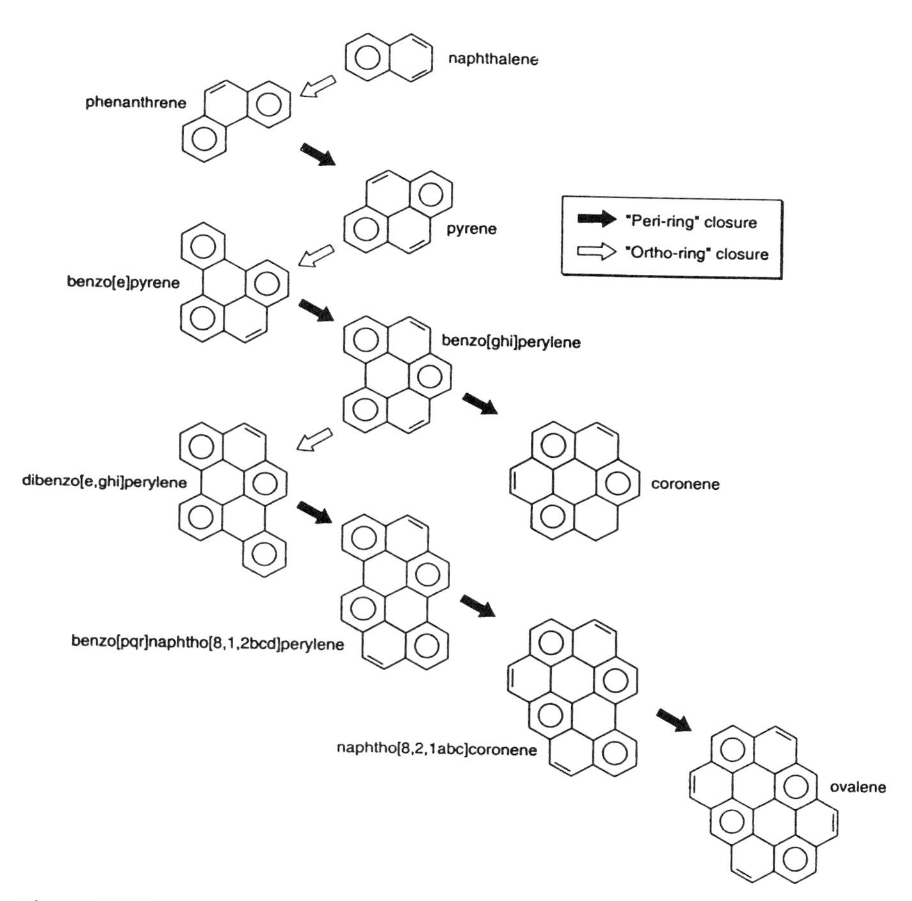

Figure 11.3 Reaction pathway for formation of heavy PNAs from light PNAs ("naphthalene zigzag" pathway) [11].

PNAs are preferentially adsorbed on the catalyst because they are more polar than most other feed components. The HPNAs deposited on the catalyst are coke precursors and accelerate catalyst deactivation.

HPNAs are high-boiling compounds and thus concentrate in the fractionator bottom stream, which in recycle hydrocracking is routed back to the reactor. HPNAs are difficult to crack and their concentration increases as new feed is added. As the buildup of HPNAs surpasses their solubility at temperatures used for the condensation and separation of the reactor products, some HPNAs precipitate and foul the reactor product condenser. Such fouling may impede heat transfer and force a reduction of conversion or feed rate. The analysis of these deposits has shown the presence of HPNAs with up to 30 aromatic rings in the molecule.

The formation of HPNAs depends on feedstock composition and boiling range. An increase in feedstock endpoint results in an increase in HPNA formation because heavier feedstocks contain more HPNA precursors. The higher reactor temperature required by catalyst deactivation also favors an increase in HPNAs because it enhances dehydrogenation and condensation. The temperature effect on the HPNA concentration becomes even more significant when conversion approaches 100%. According to some authors, under similar operating conditions, amorphous catalysts form fewer PNAs than zeolite-based catalyst [12]. This may be due to the higher cracking rate of bulky hydrocarbon molecules over amorphous solid acids than over zeolite catalysts (see Chapter 6).

In once-through processes, HPNAs do not accumulate and therefore do not cause operational problems. In a recycle operation, the buildup of HPNAs may cause operational problems, and "HPNA management" becomes necessary. Several approaches can be used to deal with this problem.

HPNAs can be partially hydrogenated and cracked over high-zeolite catalysts [13,14] or iron catalysts [15]. In addition to catalyst type, the HPNA conversion is affected by space velocity, reactor temperature, hydrogen partial pressure, and unit conversion level. However, only a fraction of the HPNAs can be converted by hydrocracking. The buildup of HPNAs in the recycle stream can also be reduced by lowering the feed endpoint or operating at less than 100% conversion. In the latter case, 5–10% of the fractionator bottom stream is removed prior to recycle. HPNAs can also be removed from the recycle oil via fractionation or adsorption (e.g., on activated carbon or by chromatography) [16–18]. Furthermore, "hot flash separation" can be used to separate lighter products from HPNAs at temperatures at which HPNAs will not precipitate. The lighter products are subsequently separated in the product condenser. Some PNAs can also be separated by precipitation with a paraffinic stream added to the heavy oil [19].

11.5. Effect of Catalyst Design

In addition to changing operating variables, more profound changes in the product slate can be obtained by using the appropriate catalyst. First-stage or single-stage hydrocracking catalysts require a high ratio of hydrogenation-to-cracking activity to saturate heavy aromatic feed molecules and remove hetero atoms. For heavy feedstocks (i.e., high endpoint) such as VGOs and deasphalted residual oils, the pore size of the catalyst must be large enough to minimize diffusion restrictions. Furthermore, the catalyst should be active in the presence of NH_3 and H_2S generated during first-stage conversion. Amorphous or low-zeolite catalysts are often used for that purpose.

Second-stage catalysts will process feeds that have been partially hydrogenated and from which contaminants and most hetero atoms have been removed. Such catalysts are required to crack all of the feed to materials boiling below the recycle cut point. To maximize naphtha yields, catalysts with low hydrogenation-to-acidity ratio are required. These catalysts will maximize aromatics and branched paraffins in the products, thus increasing their octane rating. Zeolite catalysts are commonly used for this purpose due to their strong acidity and stability.

To maximize middle-distillate yields, the second-stage catalyst requires a high hydrogenation-to-acidity ratio. The high ratio is also required to improve product properties, such as jet fuel smoke point and diesel cetane number. Amorphous catalysts and zeolite/amorphous catalysts with low-acidity zeolites are used for this purpose. Amorphous catalysts are used when diesel is the most sought after product, whereas low-acidity zeolite catalysts are employed when jet fuel needs to be maximized. The effect of catalyst design on catalyst performance is described in more detail in Chapter 7.

11.6. Process Flexibility

The hydrocracking process can be designed with the flexibility to change the main product slate, depending on market demands. A hydrocracker can be designed to operate either in a maximum naphtha mode or in a maximum middle-distillate mode. As already shown, such flexibility can be achieved by changes in operating variables and catalyst composition.

The quality of the products produced can be affected not only by the process variables but by the composition of the catalyst used. Table 11.7 illustrates quality data obtained for diesel fuel produced from a highly aromatic cracked feedstock over three different hydrocracking catalysts at identical hydrogen partial pressure [20]. Catalyst A, an amorphous catalyst, reduces the polyaromatic content but increases slightly the monoaromatic content. Catalyst B, a zeolite-based catalyst, reduces the monoaromatic content to 20 vol % and the polyaromatic content to a

Table 11.7 Catalyst Influence on Diesel Quality [20]

		Catalyst		
	Feed	A	B	C
Gravity, °API	26	32	36	41
Boiling range, °C	232–360	177–337	177–337	177–337
Aromatics, vol %				
Mono	35	39	20	3.5
Poly	17	3	4	2.5

level similar to that of catalyst A. Catalyst C, a high-zeolite catalyst, drastically reduces both the monoaromatic and polyaromatic content. The increased conversion of aromatics is related primarily to the increase in catalyst activity in the order $A < B < C$.

References

1. M. M. Boduszynski, *Energy and Fuels 2*:597 (1988).
2. R. L. Howell, R. F. Sullivan, C. W. Hung, and D. S. Laity, paper presented at the Petroleum Refining Conference, Tokyo, 1988.
3. P. Dufresne, A. Quesada, and S. Mignard, in *Catalysts in Petroleum Refining 1989* (D. L. Trimm et al., eds.), Elsevier, New York, 1990, p. 301.
4. N. Y. Chen, T. Sh. Chou, G. G. Karsner, C. R. Kennedy, R. B. LaPierre, M. G. Melconian, R. J. Quann, and S. S. Wong, U.S. Patent No. 5,100,535 (1992).
5. M. R. Apelian and C. R. Kennedy, U.S. Patent No. 5,062,943 (1991).
6. M. Skripek, D. A. Lindsay, and K. E. Whitehead, NPRA Annual Mtg., March 1988, San Antonio, TX, AM-88-55.
7. R. F. Sullivan and J. A. Meyer, in *Hydrocracking and Hydrotreating* (J. W. Ward and S. A. Qader, eds.), A.C.S. Symp. Series 20, 1975, p. 28.
8. P. Dufresne, P. H. Bigeard, and A. Billon, *Catal. Today 1*:367 (1987).
9. P. J. Nat, NPRA Annual Mtg., March 1988, San Antonio, TX, AM-88-75.
10. J. Weitkamp, in A.C.S. Symp. Series 20 (J. W. Ward and S. A. Qader, eds.), 1975, p. 1.
11. R. F. Sullivan, M. M. Boduszynski, and J. C. Fetzer, *Energy and Fuels 3*:603 (1989).
12. A. G. Bridge, D. C. Cash, and J. F. Mayer, NPRA Ann. Mtg., March 1993, San Antonio, TX, AM-93-60.
13. A. J. Gruia, U.S. Patent No. 5,139,644 (1992).
14. A. J. Gruia, U.S. Patent No. 5,007,998 (1991).
15. G. P. Hamner and C. W. Hudson, U.S. Patent No. 4,608,153 (1986).
16. J. C. Fetzer and D. G. Lammel, U.S. Patent No. 5,190,633 (1993).
17. A. J. Gruia, U.S. Patent No. 5,139,646 (1992).
18. P. J. Bosserman and V. T. Taniguchi, U.S. Patent No. 5,124,023 (1992).
19. R. W. Bachtel, D. R. Cash, J. C. Fetzer, D. G. Lammel, and J. M. Rosenbaum, U.S. Patent No. 5,232,577 (1993).
20. J. W. Ward, in *Catalysts in Petroleum Refining 1989* (D. L. Trimm et al., eds.), Elsevier, Amsterdam, 1990, p. 417.

12

MILD HYDROCRACKING

In addition to conventional hydrocracking, the refining industry has used a milder form of hydrocracking over the last several years [1–5]. Mild hydrocracking is a process used to produce relatively high yields of lighter products by using operating conditions similar to those used for vacuum gas oil (VGO) desulfurization. Generally, mild hydrocracking is a conversion of VGO desulfurization units being operated at more severe conditions, i.e., higher temperatures. The general features of conventional and mild hydrocracking are compared in Table 12.1 [5–7]. The major difference between the two processes is the hydrogen partial pressure at which they operate; mild hydrocrackers operate on a once-through basis whereas most conventional hydrocrackers recycle the unconverted material. While conventional hydrocracking is carried out at hydrogen partial pressures varying from 100–180 bars (1465–2625 psig), mild hydrocracking uses pressures of 30–70 bars (450–1030 psig). The lower hydrogen partial pressure in mild hydrocracking leads to lower conversion and to less hydrogenated products. It produces a large quantity of low-sulfur fuel oil and smaller quantities of middle distillates. Furthermore, the total conversion of feedstock that can be achieved with conventional hydrocracking cannot be achieved with mild hydrocracking (see also Section 11.3.2).

12.1. Conversion of VGO

Mild hydrocracking is a process very similar to hydrotreating. A VGO hydrotreater can often be operated in a mild hydrocracking mode by using a multiple catalyst bed arrangement for easy quenching and larger quantities of make-up hydrogen for deeper hydrogenation. The multiple-bed arrangement may involve single or multiple catalysts [4]. A VGO desulfurizer operating in the mild hydrocracking mode is shown schematically in Figure 12.1 [9]. VGO feed is mixed with make-up hydrogen, preheated, mixed with recycle gas, and passed through the reactor. The temperature rise caused by the exothermic reactions is controlled by using multiple catalyst beds and quenching with cold recycle gas (see Section

Table 12.1 Typical Process and Operating Conditions for Mild and Conventional Hydrocracking

	Mild hydrocracking	Conventional hydrocracking
Process	One-stage	One-stage or two-stage
Operating Conditions		
Conversion, wt %	20–70	70–100
Temp., °C	350–440	350–450
H_2 pressure, bar	30–70	100–200
LHSV, h^{-1}	0.3–1.5	0.5–2.0
H_2/oil, Nm^3/m^3	300–1000	1000–2000

10.5). The reactor effluent is cooled and is sent first to a high-pressure separator and then to a low-pressure separator where the liquid product is separated from the hydrogen. The hydrogen is recycled. The effluent from the low-pressure separator is sent to fractionation.

The effect of high- and low-pressure operation on product yields and quality is shown in Table 12.2 [6]. The data show that the low-pressure operation reduces hydrogen consumption (by about 20%) but results in poorer jet and diesel fuel quality. However, most often, the jet fuel smoke point and diesel cetane number are still considered acceptable.

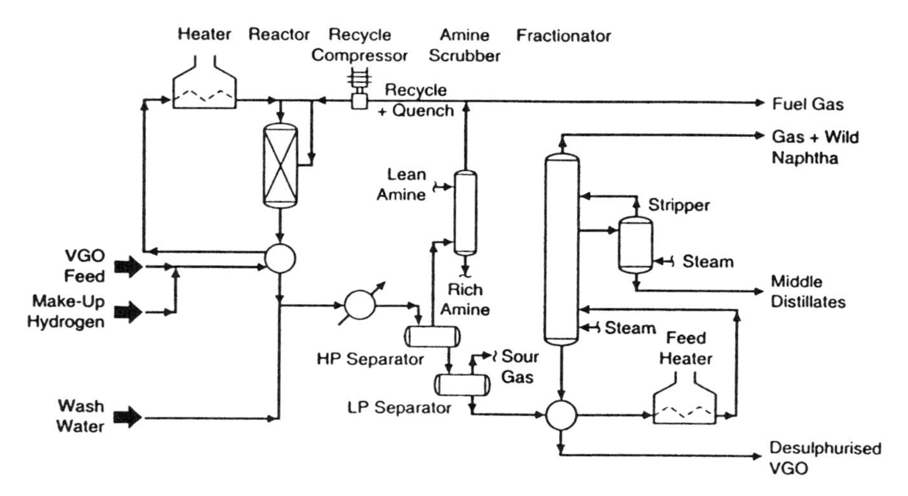

Figure 12.1 Flow diagram of VGO desulfurizer operating in a mild hydrocracking mode [9].

Table 12.2 Effect of High and Low Pressure on Product Yields and Quality [6]

	High pressure (140 bar)	Low pressure (70 bar)
Material Balance, wt %		
$H_2S + NH_3$	2.5	2.5
$C_1 + C_2$	0.5	0.5
$C_3 + C_4$	3.8	4.1
Naphtha	19.0	20.0
Jet fuel	34.8	34.0
Diesel fuel	41.7	40.6
	102.3	101.8
Product Properties		
Jet fuel smoke point	30	23
Diesel oil cetane index	68	58

The effect of pressure, temperature, and space velocity on conversion in a mild hydrocracking operation is shown in Fig. 12.2 [9]. For a given catalyst, conversion increases with the increase in pressure or temperature, and decreases with increasing space velocity.

The fuel oil obtained by mild hydrocracking is an excellent feedstock for catalytic cracking. However, the middle distillates obtained in this process are of lesser quality than those obtained by conventional hydrocracking due to the lower degree of hydrogenation. Thus, the diesel fuel obtained by mild hydrocracking has a lower cetane number and the jet fuel has a lower smoke point. Still, these middle distillates are of good enough quality, most of the time, to be used as components of the middle-distillate pool.

Mild hydrocracking encompasses a wide range of processing conditions as shown in Table 12.3 [10]. The process variables can affect both the cracking (acid) function and/or the hydrogenation (metal) function of catalyst. The cracking function is most strongly affected by reactor temperature and catalyst volume. A combination of low LHSV and high reactor temperature results in higher conversion levels. The hydrotreating function of the catalyst is most strongly affected by hydrogen partial pressure and hydrogen circulation rate. High hydrogen partial pressure and circulation rate provides for better hydrogenation of the products (e.g., desulfurization, denitrogenation, aromatic saturation) and helps to minimize catalyst deactivation. Typical catalyst cycle length for mild hydrocrackers is less than the cycle lengths achieved on conventional hydrocrackers; whereas cycle lengths of 2 or more years are achieved on hydrocrackers, the usual cycle life for a mild hydrocracker is 6–12 months.

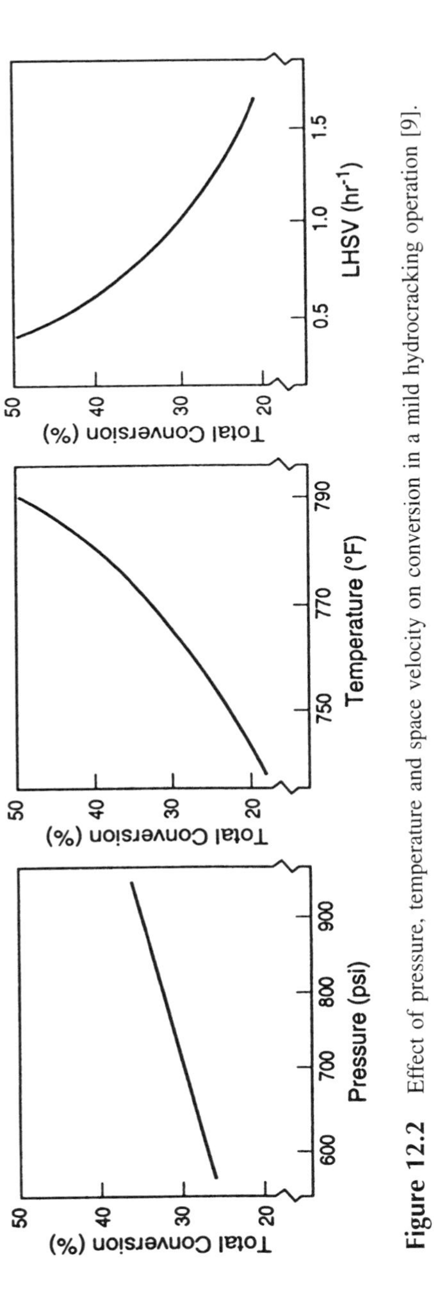

Figure 12.2 Effect of pressure, temperature and space velocity on conversion in a mild hydrocracking operation [9].

Table 12.3 Effect of Process Variables in Mild Hydrocracking [10]

Variable	Range	Conversion	Desulfurization
Rx. temp., °C	370–450	Higher More conversion	Higher Greater desulfurization More unstable
LHSV	0.3–2.0	Lower More conversion	Lower Greater desulfurization More stable
Hydrogen circulation, nm^3/m^3	250–850	Higher Quench to lower ΔT More conversion More stable	Higher Greater desulfurization More stable
Hydrogen partial pressure, bar	35–75+	Higher More stable	Higher Greater desulfurization More stable

More stable, low catalyst deactivation rate; more unstable, high catalyst deactivation rate.

The yield structure of the mildly hydrocracked products is relatively independent of the charge stock properties but more strongly a function of conversion severity level. As shown in Figure 12.3 [10], amorphous distillate hydrocracking catalyst can perform in a hydrocracking or mild hydrocracking mode of operation. Depending on the application, VGO can be processed over this catalyst to maximize either high-quality middle distillates or low-sulfur FCC feed with high yields for each extreme. Figure 12.4 [10] further illustrates the high selectivity to middle distillates achieved when processing a 650°F+ VGO over amorphous distillate hydrocracking catalyst under both hydrocracking and mild hydrocracking conditions. There is a fairly large impact of unit pressure on hydrogen consumption and product quality.

The catalysts used in mild hydrocracking are, in general, similar to those used in hydrotreating [11,12]. The difference is that catalysts used in mild hydrocracking are mildly acidic, thus allowing more conversion by carbenium ion cracking than the less acidic hydrotreating catalysts. These catalysts usually consist of cobalt or nickel oxide combined with molybdenum or tungsten oxide, supported on amorphous silica-alumina, alumina doped with halogen, or a mildly acidic zeolite. Amorphous distillate hydrocracking catalysts that contain both the hydrogenation (metal) and cracking (acid sites) functions can also be used in mild hydrocracking operations [10]. Catalysts containing medium-pore molecular sieves have also been described [8]. NiMoP on Al_2O_3 supports and NiMo catalysts on boron-promoted Al_2O_3 supports have also been reported to be effective in mild hydrocracking. The catalysts are often used in conjunction with conventional

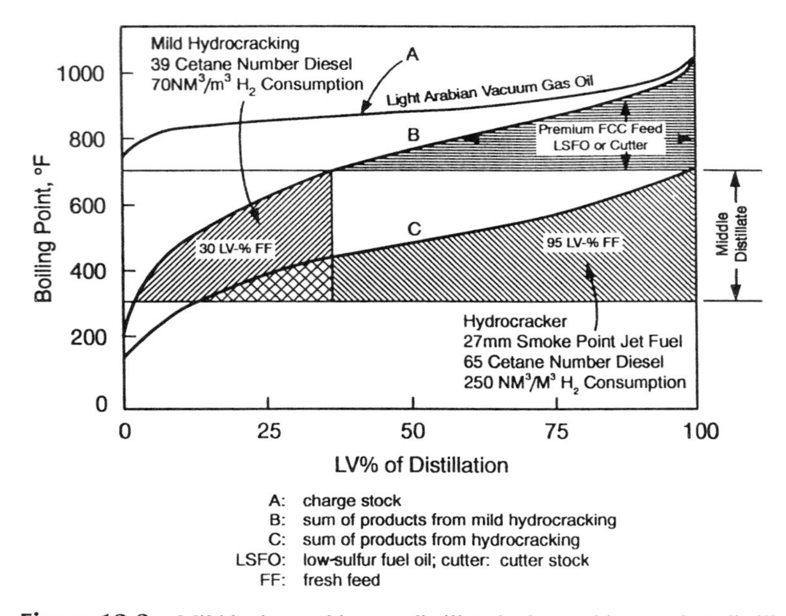

Figure 12.3 Mild hydrocracking vs. distillate hydrocracking product distillations [10].

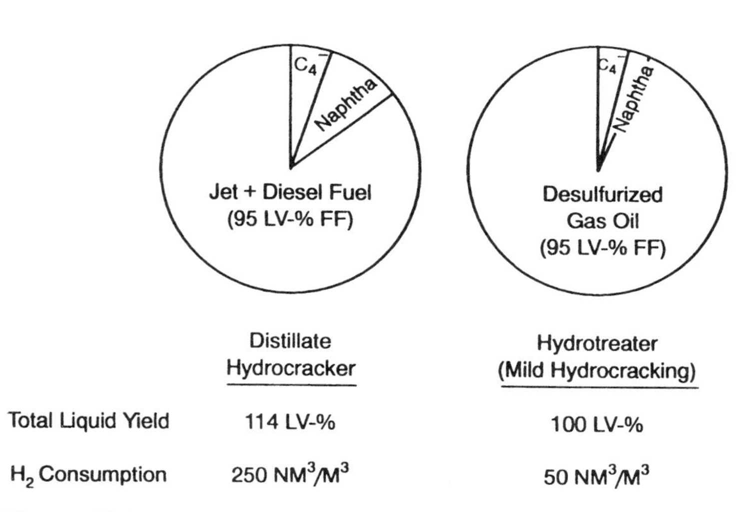

Figure 12.4 Gas oil processing flexibility [10].

hydrotreating catalysts placed upstream. In this case, the mild hydrocracking catalyst operates in the presence of relatively high hydrogen sulfide and ammonia partial pressures. Despite the relatively poor quality of the middle-distillate products in terms of hydrogen content, mild hydrocracking is economically very attractive due to lower hydrogen consumption and hydrogen pressure, as well as lower capital cost.

12.2. Conversion of Residue

Mild hydrocracking can be used for conversion of heavy feedstocks. The process has proven successful especially for conversion of the fraction boiling above 540°C (1005°F) [13].

The effect of catalyst composition and porosity on mild hydrocracking of heavy oils has been investigated by Dai and Campbell [14]. The authors used in their study a series of NiMo and NiMoP catalysts containing a variety of supports with different acidities and porosities (alumina, boria-alumina, magnesia-alumina, silica-alumina, titania-alumina, Y-zeolite + alumina). Mildly acidic alumina catalysts containing basic oxides, such as magnesia or lithia, as well as the boria-alumina catalyst showed the best performance. Catalysts with strongly acidic supports, such as silica-alumina or Y zeolite, were not suited for conversion of heavy oils in the mild hydrocracking mode because of high sediment formation. The sediment formation generally decreases with increasing macroporosity.

The kinetics of mild hydrocracking of residue has been analyzed by Yui [15] and Yui and Sanford [16]. The rate of reaction was calculated using the following power law expression:

$$\text{Rate} = -kC^n \text{pH}_2{}^{\beta}$$

where $n = 1.0$ and $\beta = 0.323$ (n and β are empirical constants); k = reaction rate constant (h^{-1}); C = concentration of reactant ($kmol/m^3$); and pH_2 = partial H_2 pressure (kPa). The activation energy for residue mild hydrocracking was found to be 82 kJ/mol. The relatively low value was attributed to a combination of thermal and catalytic cracking reactions. The same power law expression was used to analyze the kinetics of hydrotreating and of hydrogenation of aromatics [15,17]. For example, for hydrodesulfurization (HDS) $n = 1.5$ and $\beta = 0.86$, whereas for hydrodenitrogenation (HDN) $n = 1.0$ and $\beta = 1.39$. The power law expression has proven useful for interpolation, but it cannot give intrinsic kinetics or be used to extrapolate reactor performance [18].

References

1. J. W. Gosselink, W. H. J. Stork, A. F. de Vries, and C. H. Smit, in *Catalyst Deactivation 1987* (B. Delmon and G. F. Froment, eds.), Elsevier, Amsterdam, 1987, p. 279.

2. E. F. Galley, H. D. Dreyer, and W. F. Himmel, *Hydrocarbon Processing*, 1982, p. 97.
3. H. H. Free and H. J. Lovink, *Erdoel Kohle, Erdgas, Petrochem. Brennst. Chem, Hydrocarbon Technol. 40*:346 (1986).
4. J. W. Gosselink, A. Van de Paverd, and W. H. J. Stock, in *Catalysts in Petroleum Refining 1989* (D. L. Trimm et al., eds.), Elsevier, Amsterdam, 1990, p. 385.
5. P. J. Nat, J. W. F. M. Schoonhoven, and F. L. Plantenga, in *Catalysts in Petroleum Refining 1989* (D. L. Trimm et al., eds.), Elsevier, Amsterdam, 1990, p. 399.
6. P. Dufresne, P. H. Bigeard, and A. Billon, *Catal. Today 1*:367 (1987).
7. M. G. Hunter, D. A. Pappal, and C. L. Pesek, NPRA Annual Nat. Mtg., March 1994, San Antonio, TX, AM-94-21.
8. J. W. Ward, U.S. Patent No. 4,683,050 (1987).
9. G. E. Weismantel, in *Petroleum Processing Handbook*, Marcel Dekker, New York, 1992, p. 592.
10. T. N. Kalnes, P. R. Lamb, D. G. Tajbl, and D. R. Pegg, NPRA Ann. Mtg., March 1984, San Antonio, TX, AM-84-36.
11. J. W. Ward, U.S. Patent No. 4,686,030 (1987).
12. R. E. Galiasso, J. Lujano, N. P. Martinez, J. R. Velasquez, and C. Zerpa, U.S. Patent No. 4,689,314 (1987).
13. H. Sue, M. Yoshimoto, R. B. Armstrong, and B. Klein, Prepr. 4th UNITAR/UNDP Conf. Heavy Crude Tar Sands, Edmonton, Alberta, August 1988, Paper No. 86.
14. E. P. Dai and C. N. Campbell, in *Catalytic Hydroprocessing of Petroleum and Distillates* (M. C. Oballa and S. S. Shih, eds.), Marcel Dekker, New York, 1994, p. 127.
15. S. M. Yui, *AOSTRA J. Res. 5*:211 (1989).
16. S. M. Yui and E. C. Sanford, *Ind. Eng. Chem. Res. 28*:1278 (1989).
17. S. M. Yui and E. C. Sanford, *Can. J. Chem. Eng. 69*:1087 (1991).
18. M. R. Gray, *Upgrading Petroleum Residues and Heavy Oils*, Marcel Dekker, New York, 1994, p. 301.

13

CATALYTIC DEWAXING

Catalytic dewaxing is a particular hydrocracking process used to improve cold flow properties of middle distillates and lubricants by cracking normal and near-normal paraffins. Dewaxing is not always achieved by cracking; this can also be accomplished by isomerization as done by Chevron's Isodewaxing process. The properties targeted for improvement are pour point and viscosity of middle distillates and lubricants, cloud point of diesel fuel, and freeze point of jet fuel. The properties listed become important, especially at low temperatures.

The melting points of normal and branched paraffins are compared in Figure 13.1 [1]. Due to their high melting points, long-chain normal paraffins have the most detrimental effect on the low-temperature properties of middle distillates and lube oils. By reducing the amount or chain length of normal and minimally branched paraffins in these fuels and lubricants, their cold flow properties are improved. This can be accomplished by using a catalytic dewaxing process. Such a process can also be used to improve the flow properties of gas oils.

Typically, a single-stage, once-through hydrocracking process is used for catalytic dewaxing, with or without hydrotreating, depending on the sulfur and nitrogen content of the feedstock. The catalytic process is carried out as a trickle flow operation over a bifunctional, zeolite catalyst, under hydrogen flow. Though a noble metal catalyst can be used, the hydrocracking catalyst utilized in this process is frequently a nonnoble metal (e.g., nickel) supported on a medium-pore zeolite, such as ZSM-5 or erionite. The medium-pore zeolite ZSM-5 appears to be particularly suitable to obtain a high selectivity, as shown by the comparison between ZSM-5 and mordenite, a large-pore zeolite initially used for catalytic dewaxing (Table 13.1) [2].

In a catalytic dewaxing catalyst, the zeolite is combined with a binder, such as alumina, and the hydrogenation metal is supported on the composite. The mixture is calcined, exchanged with an ammonium salt solution to reduce the sodium content, and treated with a nickel nitrate solution to obtain the nickel-exchanged product. The catalyst is dried and calcined in air. Other zeolite catalysts, such as Pt or Pd on H-mordenite, H-ferrierite [3], silicalite [4], ZSM-22 [5],

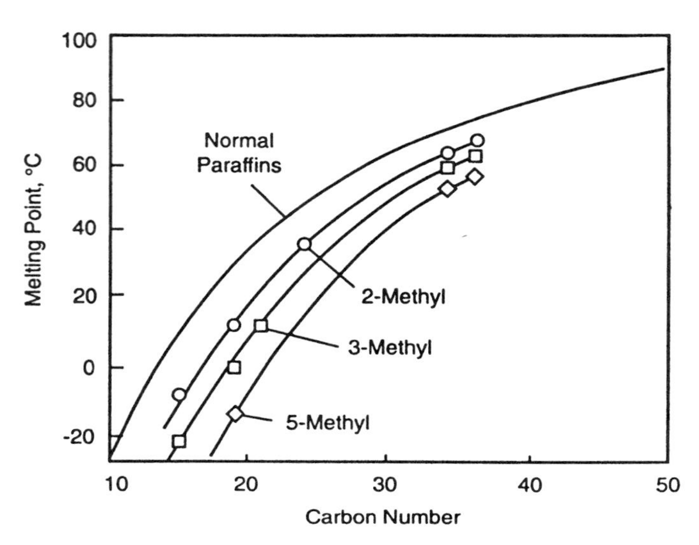

Figure 13.1 Melting points of normal and branched paraffins [1].

ZSM-23 [6], zeolite β [7,8], SAPO-11 [9], or zeolite L [10], have also been described in patents. The hydrotreating catalyst used in catalytic dewaxing is typically a nickel oxide/molybdenum oxide supported on alumina.

13.1. Reactions

The normal or near-normal paraffin chains enter the medium-pore zeolite and are cracked to smaller molecules. Oligomerization can also take place. The reaction within medium-pore zeolites is shape-selective because branched paraffins have lower diffusion rates in the zeolite pores and therefore their catalytic conversion is considerably reduced under these conditions. The decrease in concentration of long-chain *n*-paraffins in distillate after catalytic dewaxing is illustrated by the chromatograms in Figure 13.2 [11].

Table 13.1 Dewaxing With ZSM-5 vs. Mordenite [2]

Catalyst	Feed	ZSM-5		Pt/mordenite	
Conversion, wt %	0	20	23	27	36
Product pour point, °C	+35	−12	−26	−12	−26
Viscosity index	—	94.3	89.4	89.6	77.6

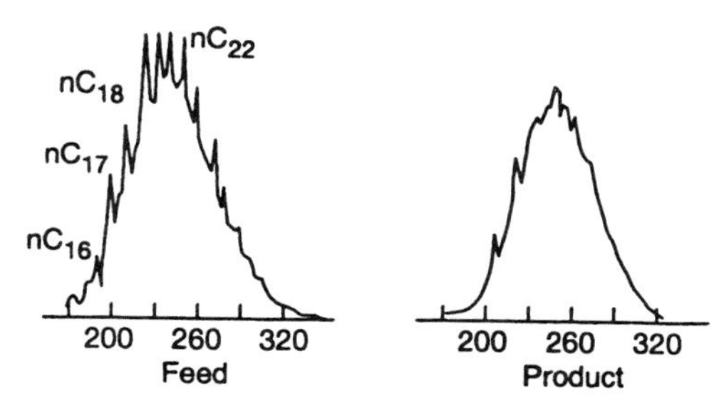

Figure 13.2 Shape-selective catalytic dewaxing over ZSM-5 zeolite [12].

The effect of shape selectivity on cracking rates over a ZSM-5 catalyst is shown in Figure 13.3 [12]. Among the different hexane isomers, *n*-hexane has the highest cracking rate due to its high diffusion rate. The suppression of isoparaffin cracking is beneficial because the isoparaffins are the preferred compounds in lube oils because of their relatively high viscosity index (VI) and low pour point. Normal paraffins have the highest VI but are undesirable due to their very high pour points. Aromatics generally have the lowest VI and poor oxidative stability, and are removed prior to dewaxing, usually by solvent extraction. Naphthenes have intermediate VI and very low pour points (Table 13.2).

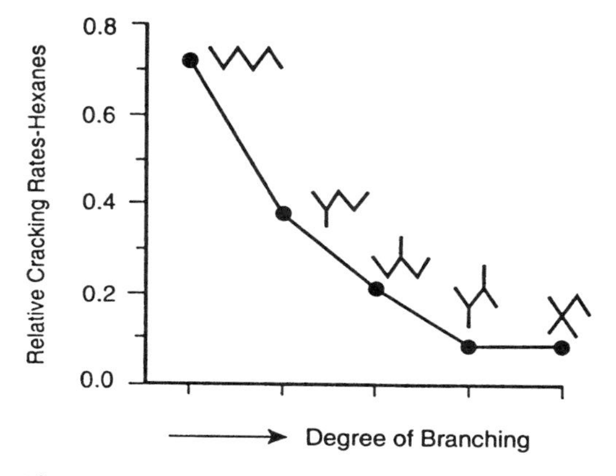

Figure 13.3 Effect of shape selectivity on cracking rates of different hexane isomers over ZSM-5 zeolite [12].

Table 13.2 Viscosity Index of Hydrocarbons

Type of hydrocarbons	Viscosity index
n-Paraffins	175
i-Paraffins	155
Mononaphthenes	142
Dinaphthenes+	70
Aromatics	50

13.2. Distillate Dewaxing

Catalytic dewaxing is commonly used to upgrade distillates and lube oils. Figure 13.4 shows the flow diagram for the Mobil distillate dewaxing process (MDDW) that uses a fixed bed of ZSM-5 catalyst [1]. The heated waxy distillate is passed with hydrogen over the zeolite catalyst in the dewaxing reactor. The products are separated into gases and liquid, and the desired dewaxed distillate is obtained by fractionation. Typical operating conditions are 260–430°C (500–805°F) reactor temperature, 2–5 MPa (290–725 psi) reactor pressure, 250–450 m^3 (1500–2650 scf/B) of H_2 per m^3 of oil, 2–6 months operating cycle length between catalyst regenerations.

A typical product yield and change in chemical composition resulting from dewaxing a heavy gas oil is shown in Figure 13.5 [11]. In this example, the heavy gas oil, containing 30% paraffins (8% normal) with 21°C (70°F) pour point, is converted by catalytic dewaxing to 86% yield gas oil with −26°C (−15°F) pour

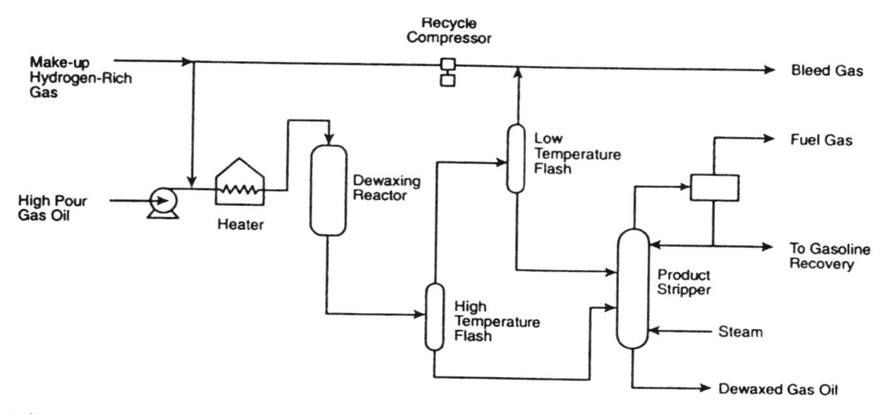

Figure 13.4 Flow diagram of Mobil distillate dewaxing process (MDDW) [1].

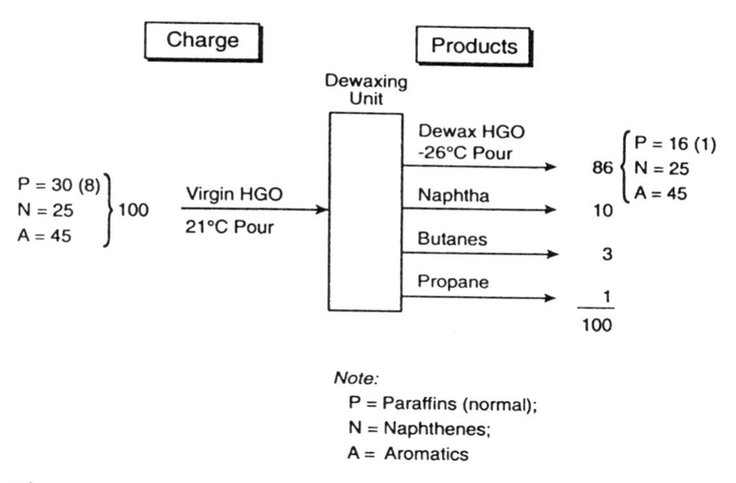

Figure 13.5 Typical product yields and changes in chemical composition resulting from dewaxing of heavy gas oil [11].

point and 16% paraffins (1% normal). Other products obtained in the process are naphtha, butanes, and propane. The naphtha is usually of good quality (high octane number) and can be used as a blending component in the gasoline pool. The butanes and propane can be used as heating fuel.

In addition to improving distillate quality, the dewaxing process makes possible an increase in distillate yield by extending the distillate endpoint, which otherwise would not be possible due to unacceptably high pour and cloud points of the resulting fuel. The heavy gas oil fraction from residue, with a high pour point, can be partially converted by dewaxing to more valuable, low-pour-point gas oil, to be used as blending stock or for further processing. This is illustrated in Table 13.3, which shows the effect of dewaxing of heavy gas oils derived from different crudes [11]. The pour point of the heavy gas oil fractions varies from 20°C to 35°C (70–95°F). The product 165°C+ (330°F+) gas oil yields vary from 54 to 91 vol % feed, and the pour point varies from −12°C to −7°C (10–20°F). The cloud point of the dewaxed gas oils is also reduced. Sulfur- and nitrogen-containing feedstocks can be used in the catalytic dewaxing process because the zeolite can selectively exclude most sulfur and nitrogen compounds, which otherwise may act as catalyst poisons.

The feedstock can be hydrotreated before or after dewaxing. Figure 13.6 [13] shows the process flow of a typical Unicracking/dewaxing unit, in which hydrotreating precedes dewaxing. It is a once-through, fixed-bed catalytic process that can be used to lower the pour point of both distillates and lube oil stocks. The feedstock can be deasphalted oil, vacuum gas oil, atmospheric gas oil, or a waxy

Table 13.3 Dewaxing of Heavy Gas Oils Derived from Three Crudes

	Crude source					
	A		B		C	
Feed Properties						
Gravity, °API		26.1		26.0		34.2
Pour point, °C		20		30		35
Cloud point, °C		22		TD		36
Sulfur, wt %		2.5		0.17		0.14
Nitrogen, ppm		450		610		110
ASTM Dist., °C						
10 vol %		375		377		334
50 vol %		390		390		370
90 vol %		403		411		390
95 vol %		415		431		402
Product yields	Vol %	Wt %	Vol %	Wt %	Vol %	Wt %
$C_1 + C_2$	—	0.1	—	0.6	—	0.6
C_3	—	1.8	—	4.2	—	4.3
C_4	2.8	1.9	8.0	5.2	10.8	7.4
C_5	2.0	1.4	6.4	4.6	10.6	8.0
C_6–165°C Naphtha	3.4	2.6	11.6	9.4	26.8	22.5
165°C + gas oil	91.4	92.2	71.3	76.0	54.3	57.2
	99.6	100.0	97.3	100.0	102.5	100.0
Product Properties						
165°C + gas oil						
Gravity, °API	24.2		24.0		23.7	
Pour point, °C	−12		−10		−7	
Cloud point, °C	−2		−7		−7	
C_6^+ naphtha						
Gravity, °API	64		63		65	
Composition, vol %						
Paraffins	44		35		32	
Olefins	48		52		61	
Naphthenes	6		8		5	
Aromatics	2		5		2	
C_5^+ Naphtha						
Octane no., clear	86		87		89	

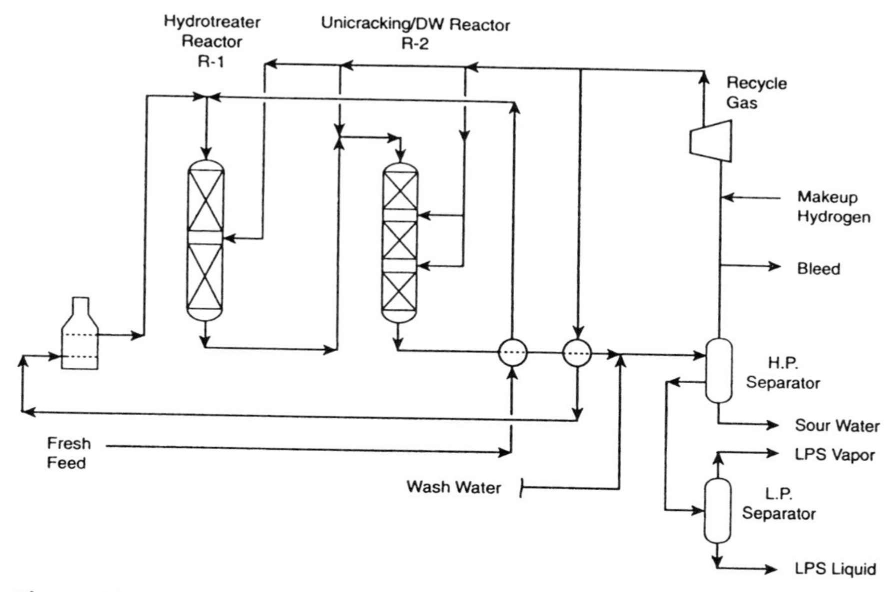

Figure 13.6 Simplified flow diagram of Unicracking/dewaxing process [13].

middle-distillate fraction. The feed enters the first reactor, R-1, for hydrotreating over a desulfurization-denitrogenation catalyst, where most sulfur and nitrogen is removed as H_2S and NH_3, respectively. The effluent from reactor R-1 passes directly to reactor R-2, which contains a nonnoble metal, zeolite-based dewaxing catalyst, which can function well in the presence of H_2S and NH_3. The dewaxed effluent from reactor R-2 passes through heat exchange trains to high- and low-pressure separators, where gases and liquid are separated. Vapors from the high-pressure separator are combined with make-up hydrogen and recycled to the reactors. Liquid product is sent to a fractionator.

A typical cycle in the dewaxing unit can last 2–4 years between regenerations. The unit can be operated to yield a low-pour-point product with minimum viscosity reduction.

13.3. Integration with Other Processes

Catalytic dewaxing can be integrated with other refining processes in order to obtain upgraded transportation fuels. The more important integrated processes are dewaxing/hydrogenation, dewaxing/FCC, and hydrocracking/dewaxing.

13.3.1. Dewaxing/Hydrogenation

Catalytic dewaxing can be combined with product hydrogenation to produce high-quality jet fuels from a wide range of feedstocks. Hydrogenation reduces the aromatics content of the jet fuel and increases its heating value. Catalytic dewaxing also makes it possible to convert a portion of heavier distillate to jet fuel. Table 13.4 illustrates the conversion of kerosene derived from a waxy Libyan crude to a high-quality, military jet fuel JP-7 by dewaxing and hydrogenation [1].

13.3.2. Dewaxing/FCC

Catalytic dewaxing can be integrated with a catalytic cracking process, especially for processing gas oils from very waxy crudes. Two methods can be used: (a) catalytic dewaxing of FCC cycle oils or (b) diversion of the light portion of catalytic cracker feed to the catalytic dewaxing unit, with concurrent processing of the heavier feed in the FCC unit. The second process is shown schematically in Figure 13.7 [14]. Instead of processing all of the 350–550°C (660–1020°F) feedstock in the FCC unit, the lighter fraction (350–415°C) (660–780°F) is submitted to catalytic dewaxing, whereas the heavier fraction (415–550°C) (780–1020°F) is processed in the FCC unit. Such a process is especially desirable when the demand for middle-distillate fuels exceeds that for gasoline.

13.3.3. Hydrocracking/Dewaxing

Catalytic dewaxing can be combined with conventional hydrocracking in order to improve the quality of the products by reducing their *n*-paraffin content. An example of such a combined hydrocracking/dewaxing process is the Paragon process. A flow diagram for the Paragon process is shown in Figure 13.8 [15]. In this process, the effluent from the first-stage reactor (where most sulfur and

Table 13.4 Catalytic Dewaxing Followed by Hydrogenation, Libyan 177–260°C Kerosene [1]

Property	Charge stock E	Dewaxed[a] product	Then[b] hydrogenated	JP-7 specification
Gravity, °API	48.1	47.2	47.5	44–50
Freezing point, °C	−33	−53	−54	−45 max
Aromatics, vol %	91	11.9	4.0	<5
Heating value, Btu/lb	18,710	18,655	18,835	18,750 min

[a]Zn/H-ZSM-5, 343°C, 34 atm, 15 H_2/HC mol ratio, 24 LHSV. 78.5 wt % yield of dewaxed product.
[b]Ni/Kieselguhr, 34 atm, 315°C.

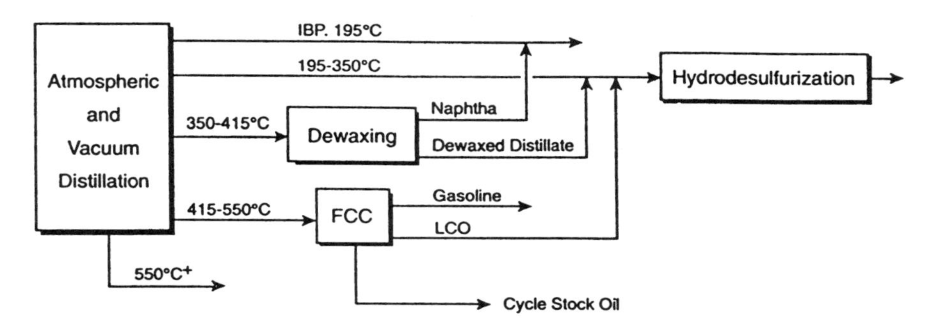

Figure 13.7 Integration of catalytic dewaxing unit with catalytic cracking unit: light fraction of FCC feed is diverted to catalytic dewaxing unit [14].

nitrogen is removed) enters a low-pressure Paragon reactor, filled with a shape-selective ZSM-5/Al_2O_3 catalyst. In this reactor, heavy *n*-paraffins are converted to lighter paraffins and olefins; some isomerization also takes place. The effluent from the reactor is passed through a separator and submitted to fractionation, where olefinic product (gases and C_5–150°C (C_5–300°F) liquid) is separated from

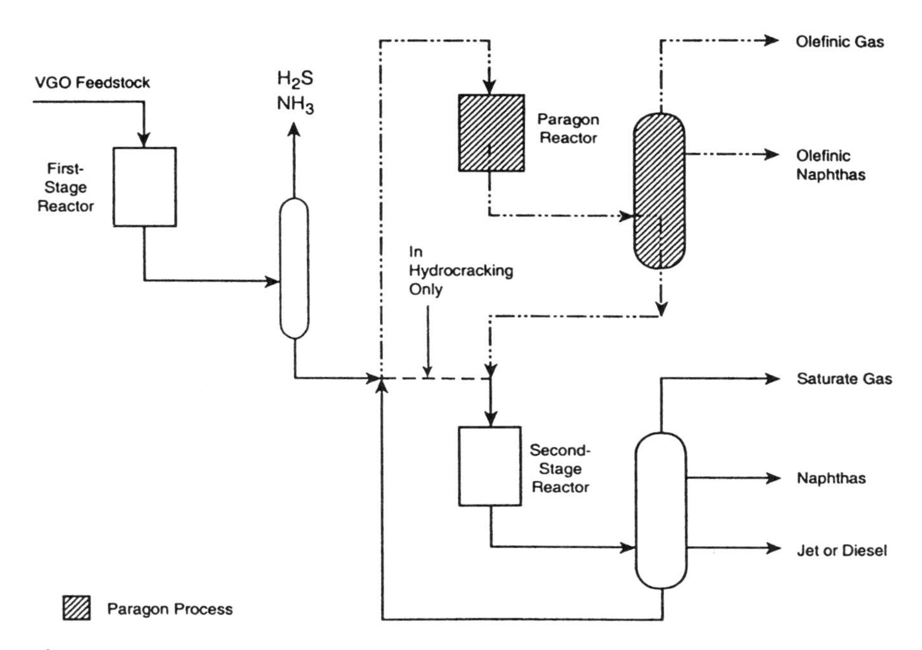

Figure 13.8 Flow diagram of conventional hydrocracker and the Paragon process [15].

unreacted (150°C+) (300°F+) liquid. This liquid is passed to the second-stage reactor of the hydrocracker for further processing. Although the products obtained by this process are of superior quality and considerable amounts of valuable olefins are generated, the process has not been commercialized due to the high cost of combining high-pressure and low-pressure equipment.

Simultaneous hydrocracking and hydrodewaxing of heavy oils, by using a catalyst containing zeolite β and a hydrogenation component, has also been described in a patent [7]. In this process, fused aromatic rings are hydrogenated and hydrocracked while dewaxing occurs simultaneously by hydrocracking long-chain, waxy paraffins.

13.4. Lube Oil Dewaxing

A wide variety of lubricating products are produced by the refining industry: automotive and aviation lubricants, gear and transformer oils, cutting and metal-working oils, hydraulic oils, and so on. Among these products, automotive oils play a dominant role. Viscosity, which is an indicator of an oil's resistance to flow, is considered the most important property of a lubricating oil. It is typically measured as kinematic or absolute viscosity. Kinematic viscosity is determined by the time required for an oil to flow through a capillary tube and is expressed in centistokes (cSt) or Saybolt universal seconds (SUS) at a specified temperature (typically 40 or 100°C (104 or 212°F)). Absolute viscosity measures the force required to shear the oil, is determined in a rotary viscometer, and is expressed in centipoise (cP).

During use, a lubricating oil may experience significant variations in operating temperatures. Higher temperatures tend to reduce the oil's viscosity, whereas lower temperatures increase it. Different lubricating oils are affected to different degrees by temperature. The relationship between viscosity and temperature for a certain oil is shown by the VI. The higher an oil's VI, the less effect temperature has on its viscosity. Since it is important for an oil to maintain viscosity at varying operating temperatures, higher VI values are generally desired.

The VI is dependent on the molecular composition of the oil. High concentrations of paraffins and single-ring naphthenes result in high VI (Table 13.2). Although *n*-paraffins have higher VI values than isoparaffins, their high pour points restrict their use in lube oils. It is therefore necessary to remove *n*-paraffins by dewaxing. Lube oil dewaxing is based on the same principle as distillate dewaxing, i.e., the removal of linear and slightly branched paraffins in order to reach an acceptable pour point with a minimum loss in yield.

To further improve the quality of lube oils, a refining step is added in conjunction with dewaxing. Refining processes commonly used are solvent extraction, acid/clay treatment, hydrogenation, or hydrocracking. Hydrocracking is done before dewaxing, whereas hydrofinishing is usually done after dewaxing. In

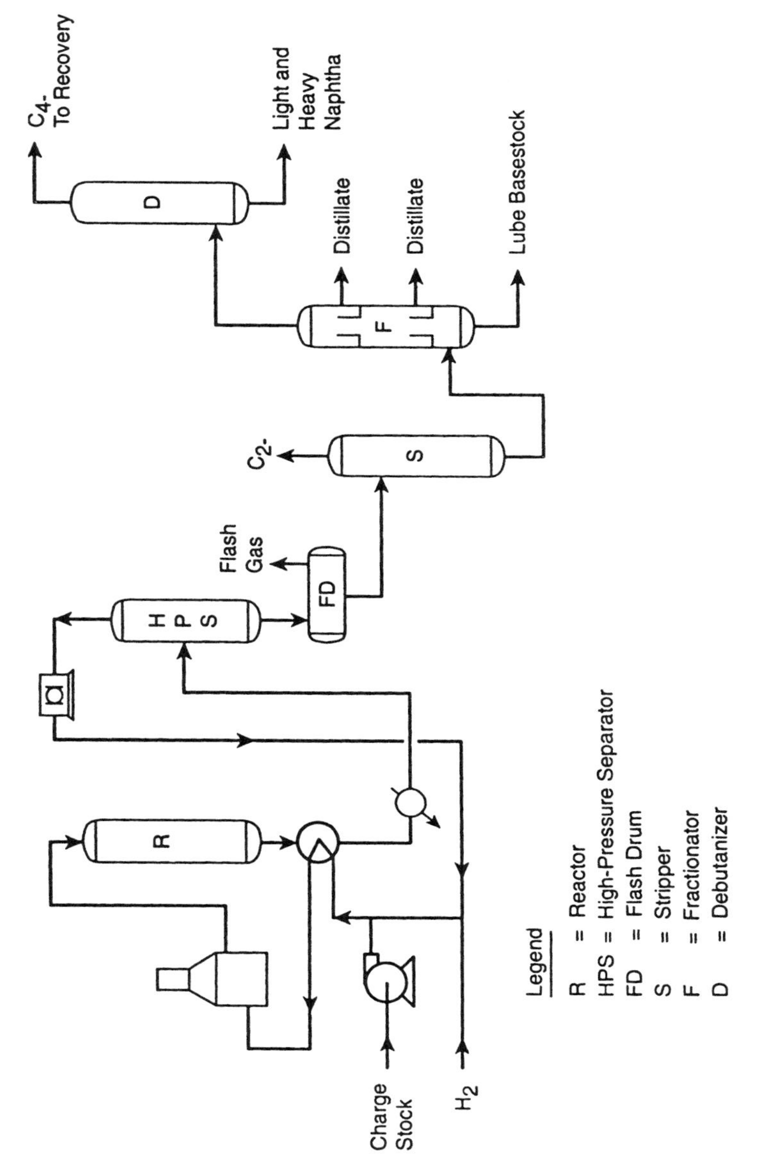

Figure 13.9 Flow diagram of single-stage Unicracking process used for lube basestock production.

the past, lube oil dewaxing was carried out primarily by solvent dewaxing. Such a process produces good-quality lube oils and wax, both products of significant value. However, the conventional solvent dewaxing process is expensive in comparison to catalytic lube oil dewaxing.

The catalysts used for lube oil dewaxing are dual-function, zeolite-based catalysts, similar to those used for distillate dewaxing. Different lube oil dewaxing processes have been designed and commercialized. The Unicracking process used for lube oil dewaxing is shown schematically in Figure 13.9. Reaction temperature is adjusted continually as the catalyst deactivates to maintain a particular target pour point. The lube oil dewaxing temperature is lower than the distillate dewaxing temperature in order to avoid viscosity losses that would occur at higher temperatures. The catalyst is regenerated when the maximum operating temperature is reached. A hydrotreating step is commonly used before or after catalytic dewaxing. Yields and properties of lube oils obtained in a commercial catalytic dewaxing process are shown in Table 13.5.

A lube oil hydrocracking process, which includes a catalytic dewaxing step, has also been commercialized [16]. In this process, waxy hydrocracked distillates

Table 13.5 Unicracker Yields and Lube Basestock Properties Obtained with VGO Feedstock

		Unicracker 315°C+ lube basestock	
	VGO feed	Low conversion	High conversion
°API	22.8	34.5	34.6
Distillation, °C:			
IBP	440	315	315
50%	480	465	455
EP	520	515	510
Viscosity at 38°C, SUS	264	161.7	127.4
Waxy VI	84	100	120
Dewaxed VI	—	95	115
Yields, wt %:			
NH_3	—	0.10	0.10
H_2S	—	2.59	2.59
C_1–C_4	—	0.99	1.49
C_5	—	0.40	0.70
C_6	—	0.59	0.70
C_7–177°C	—	2.97	5.06
177–315°C	—	14.85	23.83
315°C+	—	79.21	67.52
		101.70	101.99

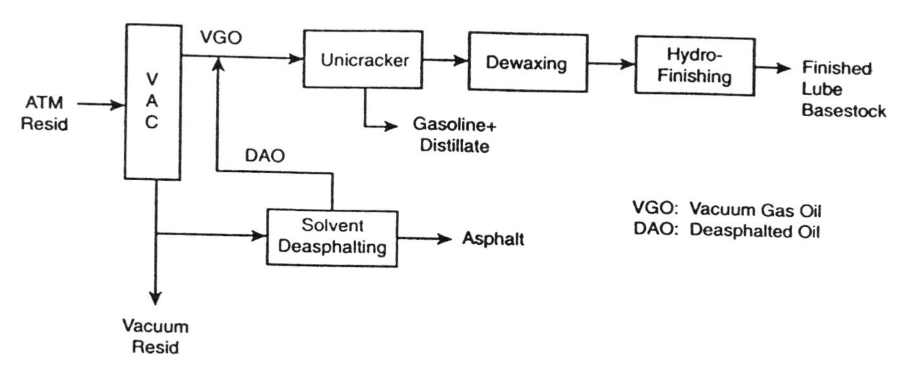

Figure 13.10 Flow diagram of Unicracking lube oil process, including solvent extraction and hydrofinishing.

are catalytically dewaxed, producing high-quality lube oils from crudes with few high-VI components (i.e., with few branched paraffins and single-ring naphthenes). The hydrocracking process creates high-VI components and allows their separation from low-VI components such as multiring aromatics. Subsequent dewaxing further improves cold flow quality of the lube oil by removing the waxy component. The combined hydrocracking/dewaxing process allows the production of lube oils from low-quality crudes.

In some processes, catalytic dewaxing is preceded or followed by solvent extraction and/or hydrotreating. The solvents (e.g., furfural, *n*-methyl-2-pyrrolidone) increase the VI of lube oils by extracting low-VI components. Hydrotreating reduces the sulfur content of the product and hydrogenates aromatics. A hydrocracking/dewaxing process preceded by solvent deasphalting and followed by hydrotreating (hydrofinishing) is shown schematically in Figure 13.10.

13.5. Other Applications

Catalytic dewaxing can be used to improve the flow properties of a synthetic crude oil in order to facilitate its flow in a pipeline. The high pour point of high-nitrogen synthetic crude can be reduced by using a once-through, hydrotreater/dewaxer combination. The hydrotreating catalyst is a typical nickel molybdate-alumina catalyst. The dewaxing catalyst is a nonnoble metal supported on a medium pore molecular sieve. By using such a process, the nitrogen content, viscosity, and pour point are reduced (Table 13.6) [17]. The product obtained can be further processed.

Catalytic dewaxing can also be used to improve the flow properties of gas oils by using a once-through, single-stage hydrotreating/dewaxing process. A high

Table 13.6 Properties of Crude Feedstock and Product [15]

	Feed	Product
Gravity, °API	25.2	39.2
Distillation		
IBP, °C	C_5	C_4
Max, °C	590	540
Pour point, °C	24	−4
Viscosity cSt at 100°F	25	4
Nitrogen, wt %	1.7	90 ppm
Con Carbon, wt %	2.1	<0.01

yield of product with low pour point, low viscosity, and low sulfur/nitrogen content can be obtained by this process.

References

1. N. Y. Chen, W. E. Garwood, and F. G. Dwyer, *Shape Selective Catalysis in Industrial Applications*, Marcel Dekker, New York, 1989.
2. J. G. Bendoraitis, A. W. Chester, F. G. Dwyer, and W. E. Garwood, in *Proc. 7th Int. Zeol. Conf.*, Tokyo, 1986, p. 669.
3. E. Bowes and B. P. Pelrine, U.S. Patent No. 4,388,177 (1983).
4. P. J. Bredael and J. F. Grootjans, U.S. Patent No. 4,810,356 (1989).
5. A. W. Chester, W. E. Garwood, and J. C. Vartuli, U.S. Patent No. 4,574,043 (1986).
6. W. E. Garwood, N. LeQuang, and S. S. Wong, U.S. Patent No. 5,037,528 (1991).
7. R. B. LaPierre and R. D. Partridge, U.S. Patent No. 5,284,573 (1994).
8. P. J. Angevine, K. M. Mitchell, S. M. Oleck, and S. S. Shih, U.S. Patent No. 4,612,108 (1986).
9. S. J. Miller, U.S. Patent No. 4,921,594 (1990).
10. C. Olavesen, U.S. Patent No. 4,421,634 (1983).
11. N. Y. Chen, R. L. Gorring, H. R. Ireland, and T. R. Stein, *Oil Gas J.* 75:165 (1977).
12. I. E. Maxwell, *Catal. Today 1*:385–413 (1987).
13. R. Bertram and F. Danzinger, NPRA Annual Mtg., San Antonio, TX, March 1994, AM-94-50.
14. R. H. Perry, Jr., F. E. Davis, Jr., and R. B. Smith, *Oil Gas J. 76*(21):78 (1978).
15. D. J. O'Rear, *Ind. Eng. Chem. Res. 26*:2337 (1987).
16. T. R. Farrell and J. A. Zakarian, *Oil Gas J. 84*(20):47 (1986).
17. J. W. Ward, in *Catalysts in Petroleum Refining 1989*, Elsevier, Amsterdam, 1990, p. 417.

14

OTHER APPLICATIONS OF HYDROCRACKING

14.1. Hydrocracking/Catalytic Cracking Integration

The integration of a hydrocracker and a fluid catalytic cracker results in a highly effective and versatile processing combination [1]. While catalytic cracking is highly suitable for cracking paraffinic feedstock into high-octane gasoline, it is far less suitable for cracking highly aromatic feedstocks, particularly into middle-distillate products. However, hydrocracking excels in converting highly aromatic feedstocks (e.g., cat cracker cycle oils and coker gas oils) into high-quality middle distillates such as jet and diesel fuels. The lighter products obtained are high-quality reformer feedstock and gasoline pool blending stock.

Two major processing schemes are in use:

1. The hydrocracking of gas oil to improve the quality of FCC feedstock and to produce other high-value products.
2. The hydrocracking of FCC cycle oils to gasoline and diesel fuels.

14.1.1. Improvement of FCC Feedstocks

Highly aromatic feedstocks, such as heavy vacuum gas oils (VGOs) or coker gas oils, are ill suited for catalytic cracking. The quality of such feedstocks can be significantly improved by hydrocracking. The process is usually carried out with nonnoble metal, zeolite-based catalysts. Typically, the heavy feedstock is fed to a once-through, partial conversion hydrocracker (Figure 14.1). In addition to superior catalytic cracker feedstock, high yields of middle distillate are obtained. The quality of these middle distillates is significantly better than that produced in the FCC in terms of cetane number, stability, and sulfur content.

In addition to hydrogenation of aromatics and cracking to lower boiling products, hydrodesulfurization and hydrodenitrogenation also take place. The resulting catalytic cracker feedstock has a lower boiling range and is low in sulfur, nitrogen, and aromatics.

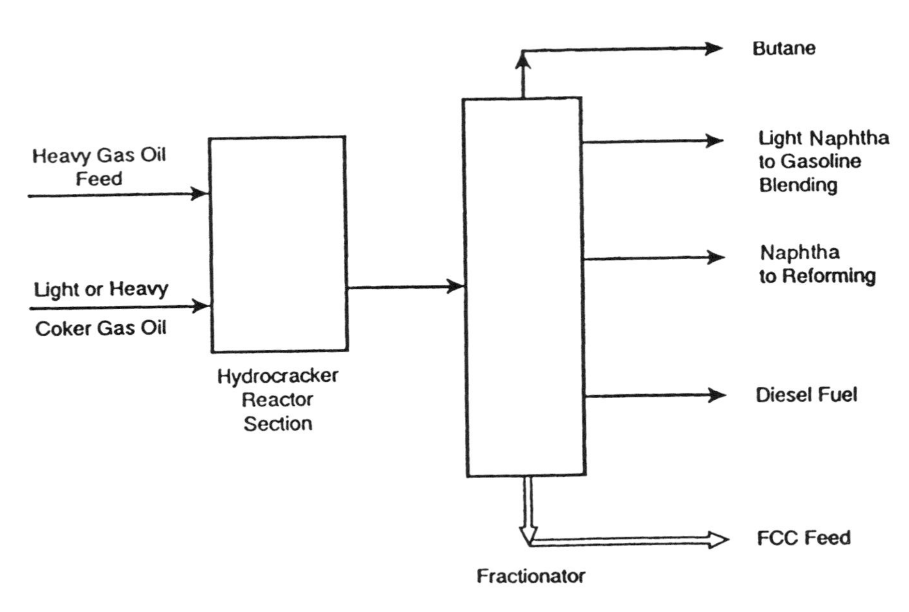

Figure 14.1 Simplified flow diagram of once-through hydrocracking process used for FCC feedstock upgrading.

The effect of hydrocracking on FCC feedstock is illustrated in Table 14.1 [1]. The data show that the hydrocracked feedstock is low in aromatics, sulfur, and nitrogen. The FCC product yields and conversions obtained with the corresponding feedstocks are shown in Table 14.2. By using the hydrocracked feedstock, both conversion and gasoline yield are significantly improved.

14.1.2. Upgrading FCC Cycle Oils

Hydrocracking can be used to convert cat cracker light and heavy cycle oils to naphtha and diesel fuel [2]. The naphtha produced is a high-quality reformer feed, whereas the diesel product has a significantly higher cetane number than the cycle oil and contains essentially no sulfur. The properties of diesel fuel obtained by hydrocracking a blend of FCC cycle oils are shown in Table 14.3.

14.2. Production of Ethylene Cracker Feedstocks

This process converts unconverted oil from a hydrocracker to a high-quality ethylene cracker feedstock by reducing the concentration of polynuclear aromatics, naphthenes, and heterocycles, and by increasing the paraffinic component of the feedstock. Hydrocracker unconverted oil has an analytical composition

Table 14.1 Hydrocracking for FCC Feedstock Upgrading [1]

Feedstock properties	Untreated feed	Once-through hydrocracking
		(379°C + product)
Specific gravity, g/cm^3	0.919	0.857
Distillation, ASTM D-1160		
10% point/endpoint, °C	402/555	418/555
Nitrogen, wt %		
Total	0.144	0.0006
Basic	0.052	0.0001
Sulfur, wt %	1.8	0.002
Refractive index @67°C	1.4957	1.4619
Conradson carbon residue, wt %	0.3	<0.1
Hydrocarbon types, wt %		
Paraffins	25	48
Mononaphthenes	11	23
Polynaphthenes	12	16
Monoaromatics	18	4
Polyaromatics	12	9
Hetero atom compounds	22	0
Product Yields from Once-through Hydrocracking at 60% Conversion		
Gas, Nm^3/m^3	9.8	
Liquid yields	*Vol % feed*	
Butane	4.3	
C_5–85°C (light naphtha)	10.4	
85–177°C (heavy naphtha)	22.3	
177–379°C (diesel)	34.6	
379°C+ (FCC feed)	39.4	
Hydrogen consumption, Nm^3/m^3	180	

Table 14.2 Effect of FCC Feedstock Upgrading on Conversion and Product Yields (vol %)

FCC feed component:	Untreated VGO feed to FCC	Once-through hydrocracked 379°C+ product to FCC
Conversion	72.8	86.9
C_5^+ gasoline	56.3	66.4
Light cycle oil	19.5	10.4
C_3/C_4 olefins	13.5	19.3

Table 14.3 Once-Through Hydrocracking to Produce Diesel Product from FCC Cycle Oil Blend

	Feedstock[a] properties	Diesel product properties
Gravity, °API	28	41
Sulfur, wppm	5400	10
Nitrogen, wppm	240	<1
Boiling range, °C	249–357	177–338
Cetane index (D-976)	22	59
Aromatics, vol %	48	6

[a]Blend of light and heavy cycle oils

significantly different from straight-run fractions of the same boiling range. The ratio of paraffins, naphthenes, and aromatics is quite different from that of conventional feedstock (Table 14.4) [3]. It is also known that polycondensed hydrocarbons, often with long-chain paraffinic groups in the side chain, are characteristic of such fractions. The latter ones cause the very high olefinic yields obtained from such fractions upon pyrolysis in comparison with low-boiling hydrocracking distillate and straight-run gas oil fractions (Table 14.5) [3]. In the upgrading process polycyclic aromatics from the feedstock are hydrogenated and multiring naphthenes undergo hydrodecyclization, cracking, and hydrogenation, resulting in the formation of paraffins. The process is usually single stage. The feedstock is passed first over hydrotreating catalyst in order to remove the hetero compounds. Subsequently, the feedstock is passed over a nonnoble metal, zeolite-based cracking catalyst and the products are separated by fractionation. The heavy fraction (e.g., 350–565°C cut) (660–1050°F) is used as ethylene cracker feedstock. Naphtha and middle distillates are also obtained in this process. The resulting naphtha is a good reformer feedstock.

The Bureau of Mines Cracking Index (BMCI) is considered a major factor

Table 14.4 Analytical Characterization of Pyrolysis Feedstocks[a] [3]

Hydrocarbon groups (wt %)	Naphtha	AGO	HAGO	VGO	Hydrocracker unconverted oil
Paraffins	67	47	54	22	57
Naphthenes	23	26	41	27	32
Aromatics	9	27	4	46	9
Not identified	—	—	—	5	2

[a]Feedstocks used in the production of ethylene by pyrolysis.
AGO, atmospheric gas oil; HAGO, heavy atmospheric gas oil.

Table 14.5 Pyrolysis Products from Different Feedstocks [3]

Products (wt %)	Naphtha	AGO	HAGO	Hydrocracker unconverted oil
Gas	72	55	73	71
Gasoline	24	21	20	22
Pyrolysis oil	4	24	7	13
C_2H_4	27	22	28	29
C_3H_6	17	11	16	16
C_4H_8	5	3	6	6.5
Coke	0.3	1.2	0.4	0.6

AGO, atmospheric gas oil; HAGO, heavy atmospheric gas oil.

in evaluating ethylene cracker feedstocks. The lower the BMCI, the higher the ethylene yield. A high content in paraffins and long-chain alkyl cyclics in the hydrocracked bottoms leads to good performance in ethylene cracker furnaces. The correlation between BMCI and the ethylene yield for a great variety of feedstocks shows a linear dependence (Figure 14.2) [3]. This is not the case, however, if the ethylene yields of all hydrocracking fractions, starting with naphtha up to hydrocracking residue, are correlated with their corresponding BMCI. The potential of ethylene formation is significantly higher for the lowest and highest boiling fractions, whereas the middle distillate gives only very low ethylene yields (Figure 14.3) [3]. This is due to the different structure of hydrocarbons in the fractions in question.

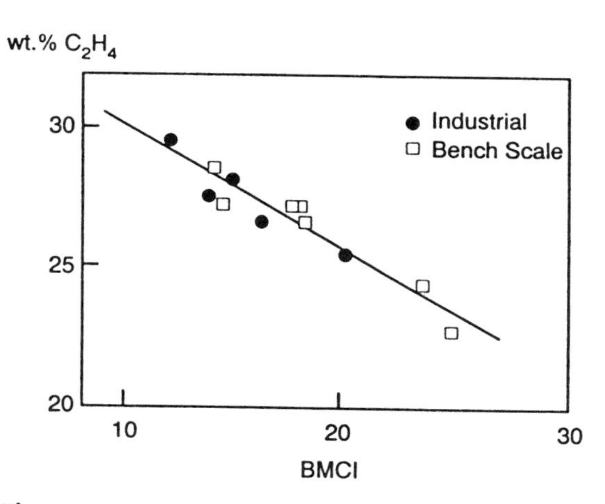

Figure 14.2 Yields of ethylene from pyrolysis of different feedstocks as a function of BMCI [3].

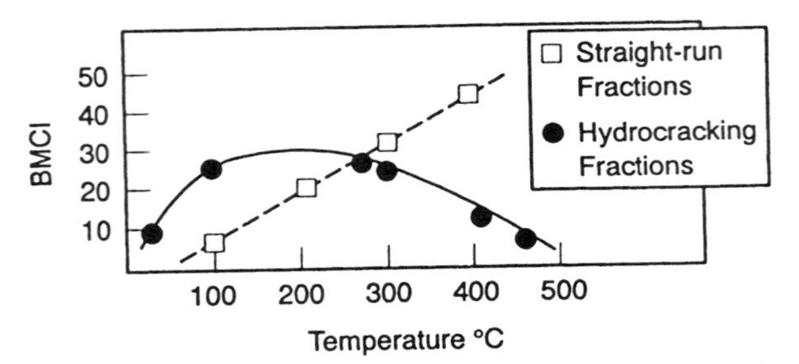

Figure 14.3 BMCI of straight-run and hydrocracking fractions of different boiling range [3].

Feedstock properties, reactor operating conditions, and product yields and properties obtained at low and high conversion (32 vs. 69 vol % conversion) are shown in Table 14.6. The data show that at low conversion the yield in ethylene cracker feed is significantly higher than at high conversion (67.8 vs. 31.1 vol %), but its quality is inferior (17 vs. 9 BMCI).

By choosing the proper catalyst and process parameters, the yield and quality of the naphtha and middle distillates obtained in the process can be optimized. For example, a high-activity, high-zeolite catalyst will enhance the naphtha yield but will give low yields of poor-quality middle distillates. By contrast, a lower activity catalyst, containing a low-unit cell size zeolite, will enhance the yield and quality of middle-distillate fractions. The unconverted oil is used as feedstock for the ethylene cracker.

14.3. Production of Petrochemical Feedstocks

Hydrocracking of heavy oil fractions (e.g., VGO) is an effective method to prepare petrochemical feedstocks, such as liquefied petroleum gas and naphtha. Whereas straight-run naphthas are a common source of BTX products, hydrocracking of a VGO can produce naphthas with a considerably higher content in BTX precursors (C_6–C_9 molecules). In the process, polynuclear aromatics are hydrogenated to the corresponding naphthenes and cracked. The single ring in naphthenes and aromatics is maintained intact, whereas its side chains and associated rings are hydrocracked. Thus, heavy polycyclic compounds are cracked to molecules in the C_6–C_9 range. The higher the content in cyclic compounds in the feedstock, the higher the yield of BTX precursors.

Depending on process conditions, the petrochemical naphtha yield can be maximized, or a mixture of petrochemical naphtha and middle distillates can be obtained. Changes in product distribution can result from changes in reactor temperature and fractionation conditions, as illustrated in Table 14.7 [4]. At a high

Table 14.6 Once-Through Hydrocracking of VGO for Producing Ethylene Cracker Feed

Feed Properties		
Gravity, °API (g/cm³)	22 (0.920)	
Nitrogen, wppm	0.08	
Sulfur, wt %	2.4	
Distillations (D-1160), °C (°F)		
IBP	367 (693)	
10%	414 (777)	
50%	454 (849)	
90%	514 (957)	
FBP	567 (1053)	
Reactor Operating Conditions		
	Low conversion	*High conversion*
Conversion, vol % of feed	32	69
H_2, psi (Bar)	1840 (123)	1840 (123)
Average reactor temp., °C (°F) (in hyrocracker)	Base	+22 (+40)
Product Yields and Product Quality		
	Low conversion	*High conversion*
Product yields, vol % of feed		
Lt. naphtha, C_5–85°C (C_5–185°F)	2.6	8.2
Hvy. naphtha, 85–160°C (185–320°F)	8.1	24.0
Jet Fuel, 160–260°C (320–500°F)	13.6	26.8
Diesel Fuel, 260–350°C (500–662°F)	14.5	20.0
Ethylene cracker feed, 350°C+ (662°F+)	67.8	31.1
Hydrogen consumption, Nm³/m³ (scf/bbl)	1070 (180)	1360 (230)
Product quality		
Ethylene cracker feed		
Gravity, °API	34	37
g/cm³	0.857	0.837
Hydrogen, wt %	14.1	14.4
BMCI[a]	17	9

[a]Bureau of Mines Cracking Index

Table 14.7 Hydrocracking Yield Response to Reactor Temperature Nonnoble Metal Catalyst [4]

Feedstock:	343–540°C Gas oil		
Gravity, °API	23.6		
Nitrogen, wt %	1250		
Sulfur, wt %	2.0		
Aniline point, °C	85		
Unicracker reactor avg. temp., °C	*376*	*367*	*360*
Product objective:	*PC Naphtha*	*Turbine fuel*	*Diesel*
Yield, vol % feedstock			
C_4	19.6	8.9	4.2
C_5–60°C	21.7	11.3	6.3
60°C+ naphtha	87.0	45.7	28.5
149°C+ distillate	—	54.1	75.2
Total C_4^+	128.3	120.0	114.2
N + A[a]			
C_6–C_8	29.7	18.7	13.2
C_6–C_9	37.9	23.1	16.5
C_6	41.0	23.04	16.6
Naphtha ASTM EP, °F	335	320	325
H_2 consumption, scf/bbl	2215	1635	1415
149°C+ distillate			
Flash point, TCC, °C		53	—
Flash point, PMCC, °C		—	62
Freeze point, °C		<−60	—
Smoke point, mm		26.7	—
Cetane number		—	53.8
D-86 endpoint, °C		272	373

[a]In 50°C+ naphtha.

reactor temperature (376°C), the yield in petrochemical naphtha is maximized (87 vol %) and its C_6–C_9 naphthenes plus aromatics content is high (37.9 vol %). A lower reactor temperature and a reduction in naphtha endpoint enhances the yields of turbine fuel and diesel while reducing the naphtha yield and C_6–C_9 content.

14.4. Production of Lube Oil Base Stock

The combination of hydrocracking and dewaxing has been described in Section 13.3.3. However, it should be noted that hydrocracking in and by itself can produce high yields of premium quality lubricating base stocks. The primary reactions that occur during hydrocracking result in a significant increase in the viscosity index of unconverted product. Viscosity index is a measure of the change in viscosity of an oil as a function of temperature and a high viscosity index is desirable for lube oil base stock. The higher an oil's viscosity index, the less effect temperature has on its viscosity. Because one of the most important characteristics of an oil is its ability to maintain viscosity at varying operating temperatures, a higher viscosity index is generally desired. However, as shown in Figure 14.4 [5], an increase in the conversion level in the hydrocracker results in a decrease in the viscosity of the unconverted portion of the feedstock, which is the lube base stock. Since the value of a lube stock is determined by both viscosity index and viscosity, a suitable conversion level must be chosen to satisfy both the desired viscosity index and viscosity characteristics.

Either amorphous- or zeolitic-based catalysts can be used in hydrocrackers for the production of lube oil base stocks. The simplified flow diagram of a hydrocracker used to produce lube oil base stocks is similar to that shown in Figure 13.9. This flow scheme is very similar to that of a typical partial conversion single-stage hydrocracker used for the production of distillate range material (see Section 10.2).

When compared to solvent extraction, hydrocracking produces higher yields of better quality lube stocks that exhibit excellent response to commercial additive packages. The process is also less dependent on the quality of the feedstock and can reduce, or in some cases eliminate, the need for additional downstream finishing.

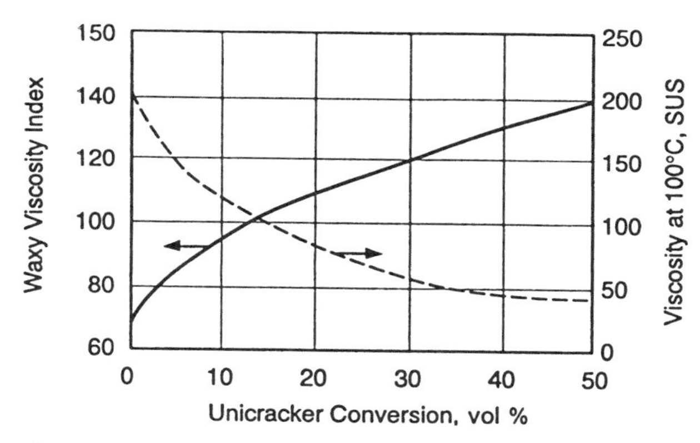

Figure 14.4 Effect of Unicracker conversion on lube base stock viscosity and viscosity index [5].

14.5. Petroleum Residue Upgrading

14.5.1. Feedstock Characteristics

The term petroleum resid (or residue) encompasses a broad range of materials typically boiling above ~345°C (650°F). These include atmospheric tower bottoms (ATB), heavy gas oil (HGO), vacuum gas oil (VGO), and vacuum tower bottoms (VTBs). Most crude oils contain 10–30% residue, although heavy crude oils can have residue contents exceeding 40% of the crude. Upgrading of these residues to higher value, distillable products is becoming an important objective of many refiners.

The chemical composition and boiling range of residues are significantly different from the lighter gas oil components of a crude. Residues contain oils, resins, and asphaltenes. In heavier crudes, the asphaltenes may represent a considerable portion of the residue fraction. In addition, because asphaltenes consist of condensed heterocyclic and aromatic rings that contain hetero atoms (S, N, O, V, Ni, etc.), their presence can have a deleterious effect on hydroprocessing catalysts, with the rate of catalyst deactivation being proportional to the concentration of asphaltenes in the feedstock [6].

When compared to the crude's lighter gas oil components, residues contain all of the metals (Ni, V, Fe, Cu, Na) present in the crude as well as significant amounts of sulfur and nitrogen. In addition, residues have a lower hydrogen content ($H/C < 1.5$), contain most of the crude's Conradson carbon, and have lower API gravities and higher viscosities. The properties of residues derived from light, medium, and heavy crudes are shown in Table 14.8 [7].

14.5.2. Upgrading Technologies

Residue processing is of considerable importance to the refining industry. The low value of fuel oil and increasing government-mandated restrictions on sulfur levels in residual fuel have created strong incentives to process residues to produce low-sulfur fuel oils or to upgrade the residue to higher value products. A detailed

Table 14.8 Properties of Typical Residues [7]

Property	Light (Nigerian)	Medium (Arabian)	Heavy (Mayan)
API gravity	22.0	17.0	11.0
Sulfur, wt %	0.26	3.0	4.1
Metals, ppm (Ni + V)	15	36	500
Conradson carbon, wt %	2.8	7.5	16.3

description of residue-upgrading technologies has been provided in a monograph by Gray [8], at an AIChE symposium [9], and at an ACS symposium [10]. Some of the technologies applied to the upgrading of petroleum residues were developed initially for processing other heavy feedstocks, such as coal liquids, shale oils, and tar sand bitumen.

In general, residue upgrading involves one or more of the following objectives [8]:

1. Removal of hetero atoms (mostly S and N) in both the residue and resulting distillate range products.
2. Conversion of high molecular weight residue components into lower molecular weight distillable products. This process involves the breaking of C-C, C-S, and C-N bonds present in the residue.
3. Increasing the H/C ratio of the residue and distillate products. For the distillate products, a high H/C ratio is desirable for ignition characteristics of the resulting jet and diesel fuels. If the residue portion of the product is to be further processed in a downstream conversion unit, the higher H/C ratio results in improved yields and lower coke production.

Processes used for residue upgrading can be classified in several ways. One classification divides these processes into carbon rejection and hydrogen addition processes [11].

Another classification used for categorizing residue-upgrading technologies divides the processes into solvent extraction, thermal and catalytic processes (Table 14.9). Some of these upgrading technologies require a pretreatment of the feedstock prior to processing in conventional units.

The major reactions that occur during residue upgrading include:

Hydrotreating (HDS, HDN, HDM)
Thermal and catalytic cracking
Hydrocracking
Hydrogenation
Coke formation

Table 14.9 Heavy Feed Upgrading

Solvent Extraction	Thermal Processes	Catalytic Processes
Deasphalted oil	Visbreaking	Catalytic cracking
	Coking	Hydrotreating-hydrocracking
	Steam cracking	Mild hydrocracking

Selection of the appropriate residue upgrading configuration depends on a variety of factors, including cost, desired product slate, and byproduct (solid or gas) uses.

14.5.3. Hydrotreating/Hydrocracking Processes

A variety of hydroprocessing technologies are being used for upgrading residual materials [8]. Depending on the process used, these technologies can generally be classified into fixed-bed, ebullating bed, moving bed, and slurry phase processes [12].

Figure 14.5 [11] compares three different reactor technologies used for hydroprocessing residue. Noteworthy are the significant differences in catalyst volume, catalyst particle size, and interparticle distance for the three reactor types. While significant amounts of catalyst are used in fixed-bed and ebullating bed reactors, only a minor amount of catalyst (1% of reactor volume) is used in the slurry phase reactor. The small size of the catalyst used in slurry phase reactors

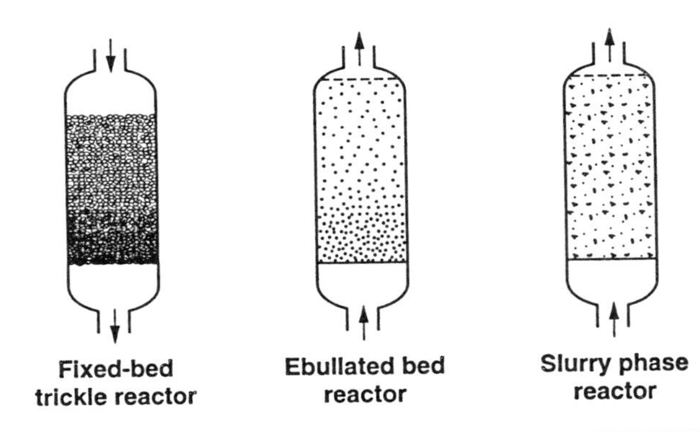

	HDM/HYCON (Shell) ARDS and VRDS (Chevron) RCD Unicracking (UOP) Exxon Idemitsu HYC etc.	H-Oil (Texaco) LC-Fining (Amoco)	CANMET Aurabon (UOP) VebaCombicracking
Vol% catalyst in reactor	~60	~40	~1
Catalyst size, mm	1.5 x 3	0.8 x 3	0.002
Particles/cm^3	~120	~250	2.4 x 10^9
Interparticle distance, mm	–	~1.6	~0.008

Figure 14.5 Comparison of reactor technologies [11].

permits a very high degree of catalyst dispersion to be achieved. This results in intimate contact between catalyst and feedstock and high conversions to be achieved while maintaining product stability.

Fixed-bed (or trickle bed) processes have been used extensively to produce low-sulfur fuel oil or as pretreats for downstream coking and catalytic cracking units. These processes utilize fixed beds of catalysts, typically operating at moderate to high hydrogen partial pressures to process the residue which contains metals, sulfur, and nitrogen. Examples of fixed-bed residue-upgrading technology are UOP's RCD Unionfining, Chevron's ARDS and VRDS, and Idemitsu's HYC process.

Hydroprocessing by employing an ebullating (expanded) bed such as Texaco's H-Oil process or Amoco's LC-Fining process is well established and has been proven commercially. The H-Oil process was originally developed in the 1960s and has been improved since then [13–17]. It is used to process residue and heavy oils to produce upgraded petroleum products such as LPG, gasoline, middle distillates, gas oil, and desulfurized fuel oil. A simplified flow diagram of the H-Oil residue-upgrading process is shown in Figure 14.6 [13].

The basic principles of the ebullated bed reactors are illustrated in Figure 14.7 [8]. Fresh feed, recycle oil, and high-pressure hydrogen are contacted with the ebullating bed of catalyst in an upward flow motion. Hydrocracking takes place at temperatures of 420–450°C (790–840°F), at liquid hourly space velocities of 0.1–1.5 h^{-1}, and at pressures varying from 100 to 200 bars (1500 to 3000 psig), depending on feedstock. Liquid and vapors are separated, acid gas is removed, and the hydrogen-rich gas is recycled to the process along with make-up hydrogen. Liquid product is passed through a light-component steam stripper and submitted to fractionation at atmospheric pressure and under vacuum. The vacuum bottoms obtained in this process can be recycled.

Ebullating bed reactors have several advantages over fixed-bed reactors [8]:

1. The ebullating bed is not plugged by solids present in some heavy feeds.
2. The liquid recycle gives good mixing within the reactor, thus ensuring that temperature gradients are minimized.
3. Catalysts can be added and withdrawn continuously, allowing long operating runs without shutting down the reactor.

The catalyst pellets should be small enough (usually $\leq$1 mm) to facilitate suspension by the liquid phase in the reactor. The rate of catalyst addition and withdrawal is such as to maintain constant conversion. Usually this amounts to about 1% of the catalyst contained in the reactor per day.

The products obtained are sent to hydrotreating in order to improve their quality. The virgin stock may also require hydrotreating prior to processing in the H-Oil unit. An upgraded version of the H-Oil process, the T-Star process, designed to process very difficult feedstocks, has been described more recently [18]. The

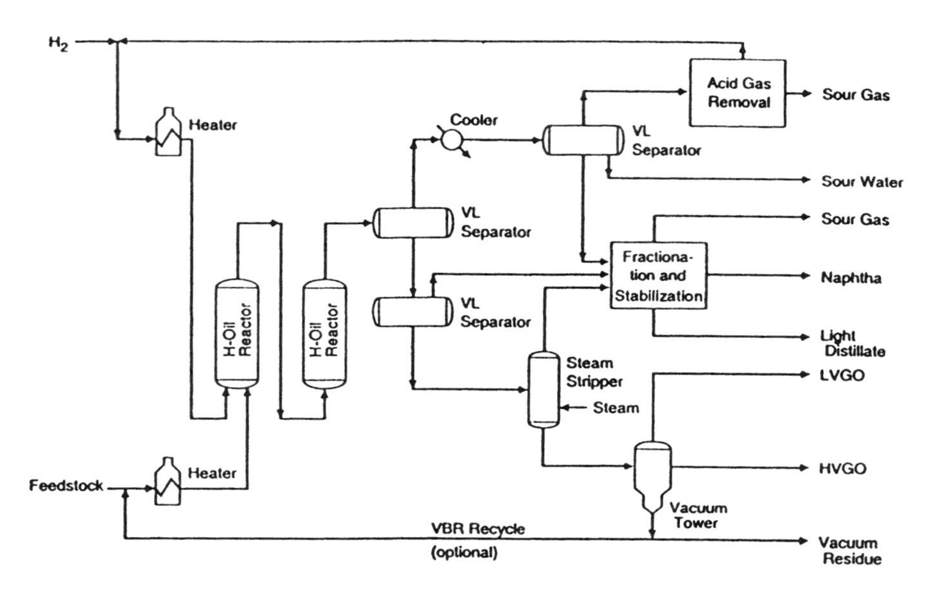

Figure 14.6 Simplified flow diagram of H-Oil residue upgrading process [13].

new process combines the conventional H-Oil process with a fixed-bed hydrotreating/hydrocracking process.

Typical quality of products obtained from a Heavy Arabian crude at high and low conversion, as well as from a coking operation, is shown in Table 14.10. The data show that the quality of products obtained by the H-Oil process is superior to that of products obtained by coking.

The LC-Fining process [19,20] also uses ebullating bed reactors. A flow diagram of the LC-Fining process is shown in Figure 14.8 [19]. The process is similar to the H-Oil process, from which it differs primarily in details of equipment design.

Shell's Hycon process is a commercial, moving bed process. Fresh or regenerated catalyst is added at the top of the reactor and spent catalyst is withdrawn from the bottom. Regeneration of the catalyst outside the reactor avoids the need for frequent shutdowns for regeneration that a fixed-bed process would require and enables the use of catalysts under severe (i.e., fast deactivation) process conditions. Specially designed hydrodemetallization and hydrodesulfurization catalysts are used in this process. Recently introduced cocurrent flow of catalyst and residue results in a more uniform buildup of metals on the catalyst and increases catalyst useful life [21]. The effluent from the reactor is submitted to atmospheric and vacuum fractionation.

The reaction severity of this process is high [over 200 bar pressure and over

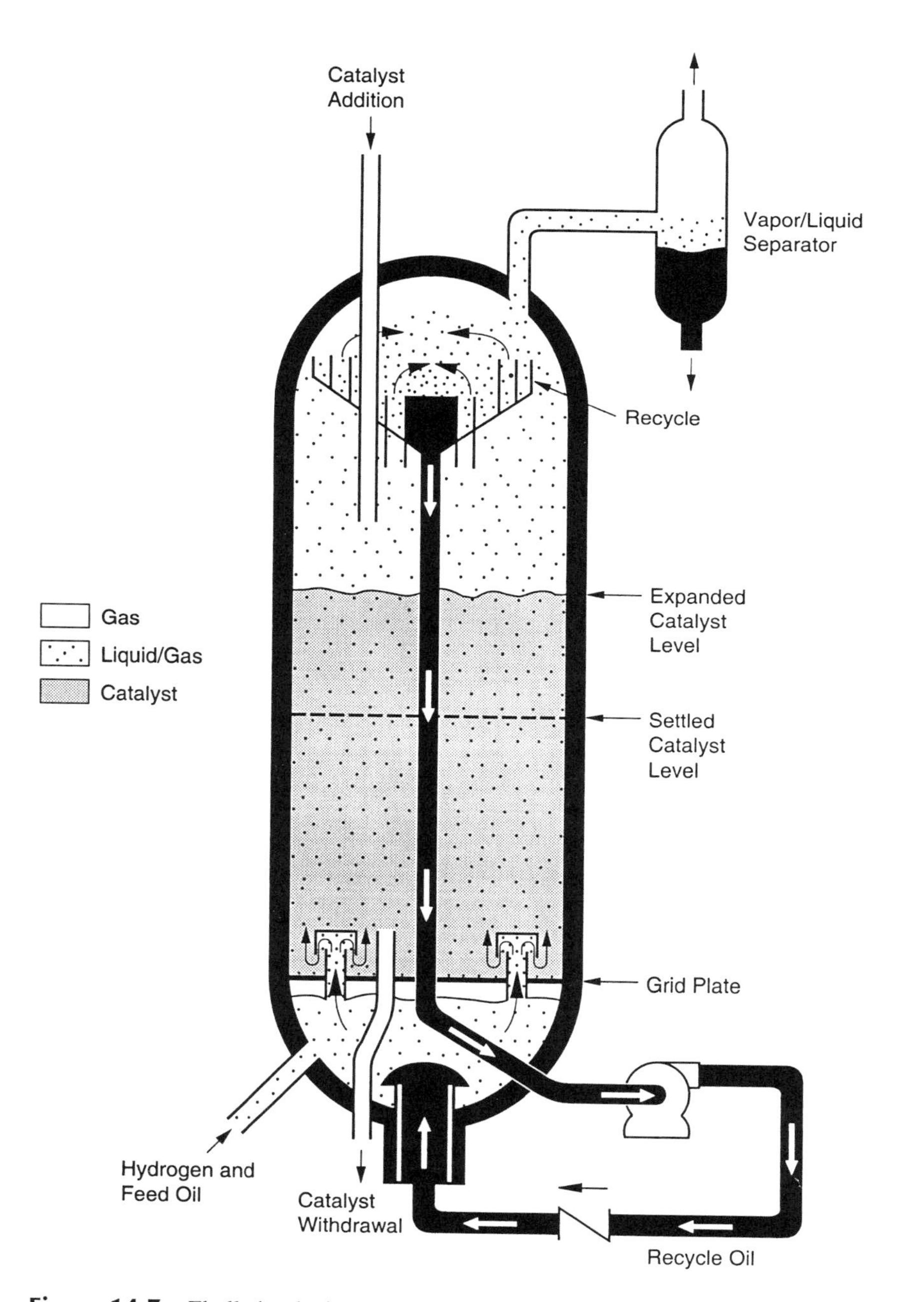

Figure 14.7 Ebullating bed reactor [8].

Table 14.10 H-Oil Processing of Arabian Crude: Typical Product Properties

		H-Oil Products		
Fraction/property	Virgin crude	High conv. 90 vol %	Low conv. 70 vol %	Coking
Naphtha				
Gravity, °API	63.5	62.0	62.2	61.0
Sulfur, wt %	0.02	0.15	0.06	1.4
Nitrogen, ppm	—	240	170	130
PONA	73/–/20/7	66/3/21/10	63/2/26/9	NA
Middle distillate				
Gravity, °API	37.0	34.4	34.5	28.0
Sulfur, wt %	1.2	0.59	0.26	2.36
Cetane index	53	43	43	37
Vacuum gas oil				
Gravity, °API	22.5	16.4	19.8	17.0
Sulfur, wt %	2.7	1.55	0.71	5.09
Nitrogen, ppm	1500	3480	1370	2700

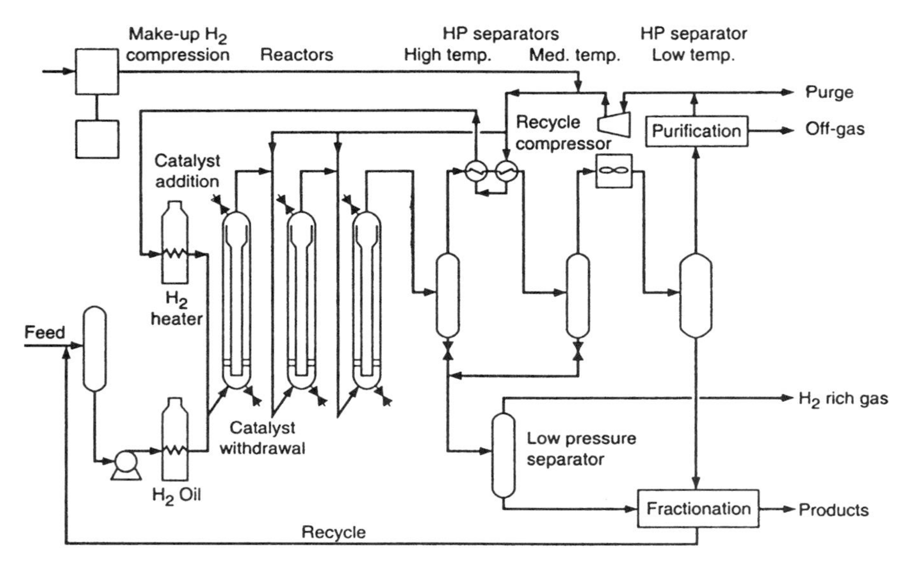

Figure 14.8 Simplified flow diagram of LC-Fining process [19].

400°C (750°F) reaction temperature] [22] and is designed to achieve up to 90% conversion of the 540°C+ (1005°F+) fraction. The properties of some heavy-residue feedstocks, the yields of products obtained from these feedstocks by the Hycon process, and the properties of some of these products are shown in Table 14.11 [22]. The data were obtained at the Shell residue hydrocracker at Pernis, Holland, which went on-stream in 1989. The major product is a low-sulfur VGO that can be used as FCC feedstock. Significant yields of kerosene/gas oil are also produced. The vacuum residue 575°C+ (1065°F+) obtained in the process has only 1.5% sulfur.

Slurry phase hydroprocessing is based on the modernization of the old German coal liquefaction technology. High temperatures (450–480°C) (840–895°F) and high pressures (135–205 bars) (2000–3000 psig) are used to crack the

Table 14.11 Upgrading of Vacuum Residues by the Hycon Process [22]

	Arab Heavy straight run	Arab Heavy flashed and cracked	Mayan straight run
Feedstocks			
Yield on crude	25	18	38
Hydrogen, wt %	10.1	8.8	9.6
Sulfur, wt %	5.6	6.0	5.2
N + V, ppm	240	340	765
CCR, wt %	24	32	31
Viscosity, cSt @ 150	440	2800	2900
Product Yields			
C_1–C_4	3.5	3.6	3.2
Naphtha (C_5–155°C)	4.6	4.3	4.5
Kerosine/gasoline (155–355°C)	21	19	20.5
VGO (355–575°C)	41	39	41
Short residue (575°C+)	27	31	27.5
H_2 consumption, wt % on feed	2.3	2.8	2.2

Product Properties	C_5–155°C C_5–310°F	155–285°C 310–545°F	285–355°C 545–670°F	355–575°C 670–1065°F	575°C+ 1065°F+
Sulfur	15 ppm	30 ppm	0.06%	0.3%	15%
Nitrogen	50 ppm	250 ppm	0.05%	0.1%	
Vanadium				0.5 ppm	20 ppm
Nickel					15 ppm
CCR				0.3%	17.4%
Cetane			42		
Cloud point, °C			−5		
Freezing point, °C		15 ppm			
Smoke point		50 ppm			

residue in the presence of finely dispersed powder. The powder is called, variously, additive or catalyst. Because of the high processing severity, it is possible to achieve over 90% conversion in a once-through mode at relatively high space velocities. The solid additive or catalyst inhibits coke formation at high conversion. These processes have been used so far in small, commercial scale plants or large pilot plants [11,21].

The hydroconversion mechanism in slurry phase processes is conceived as a thermally induced free-radical cracking reaction of heavy feedstock components, such as asphaltenes and resins, followed by catalytic hydrogenation of the unstable radicals to oil range molecules [23]. The highly dispersed catalyst favors the rapid stabilization of free-radical intermediates through hydrogenation, thus preventing coke forming condensation and polymerization reactions.

$$\begin{array}{l} \text{Heavy molecules} \\ \text{(asphaltenes,} \\ \text{resins)} \end{array} \xrightarrow{\Delta T} \begin{array}{l} \text{unstable} \\ \text{intermediates} \\ \text{(radicals, olefins)} \end{array} \xrightarrow[\text{cat.}]{+H_2} \begin{array}{c} \text{stabilized} \\ \text{``oil''} \\ \text{molecules} \end{array}$$

The catalyst may also facilitate the partial hydrogenation of heavy feedstock molecules prior to cracking.

Several slurry processes are available. These include CANMET, Veba Combicracking, Super Oil Cracking, Intevep's HDH process, Idemitsu Kosan (Kellogg's) MRH process, and UOP's Aurabon process [24].

References

1. T. Tippett, paper presented at the Unocal Technology Symposium, Kyoto, Japan, Nov. 1991.
2. W. R. Derr Jr., P. J. Owens, and M. S. Sarli, U.S. Patent No. 4,985,134 (1991).
3. G. Bach, B. M. Gräfe, D. Herschelmann, J. Kõlter, R. Krant, and K. H. Nestler, Industrial and Scientific Experience in Pyrolizing High-Boiling Hydrocarbon Fractions, Deep Crude Oil Processing '90 Symposium, Litvinov, Czechoslovakia, 1990.
4. J. W. Ward, in *Catalysts in Petroleum Refining 1989*, (D. L. Trimm, S. Akashah, M. Absi-Halabi, A. Bishara, eds.), Elsevier, Amsterdam, 1990, p. 417.
5. O. Genis, K. P. McCormick, S. W. Shorey, and E. J. Houde, Unicracking for Lubestock Production, Indian Institute of Petroleum Conference, New Delhi, 1994.
6. J. G. Speight, in *Catalysis on the Energy Scene* (S. Kaliaguine and A. Mahay, eds.), Elsevier, Amsterdam, 1984.
7. H. Qabazard, R. Adarme, and B. L. Crynes, in *Catalysts in Petroleum Refining 1989*, Elsevier, Amsterdam, 1990, p. 61.
8. M. R. Gray, *Upgrading Petroleum Residues and Heavy Oils*, Marcel Dekker, New York, 1994.
9. M. C. Oballa and S. S. Smith, (eds.), *Catalytic Hydroprocessing of Petroleum and Distillates*, Marcel Dekker, New York, 1994.
10. *Prepr. Div. Petrol. Chem. ACS 38*(2):1993.

11. F. M. Dautzenberg and J. C. De Deken, *Catal. Rev. Sci. Eng. 23*(3&4):421 (1984).
12. B. L. Schulman, NPRA Ann. Mtg., San Francisco, March 1989, AM-89-18.
13. R. P. Van Driesen and L. M. Rapp, NPRA Ann. Mtg., San Antonio, TX, March 1969, AM-69-34.
14. R. P. Van Driesen and L. M. Rapp, World Petroleum Congress, April 1967, I.P. No. 32.
15. K. G. Tasker et al., Japan Petroleum Institute, Petroleum Refining Conference, Tokyo, Japan, Oct. 1988.
16. R. M. Eccles, A. M. Gray, and W. B. Livington, *Oil Gas J. 80*(15).
17. G. Nongbri et al., in *Catalytic Hydroprocessing of Petroleum and Distillates* (M. C. Oballa and S. S. Shih, eds.), Marcel Dekker, New York, 1994, p. 55.
18. W. F. Johns, G. Clausen, G. Nongbri, and H. Kaufman, NPRA Ann. Mtg., San Antonio, TX, March 1993, AM-93-21.
19. B. Robson, K. Chinnery, C. Herster, C. Vieus, and J. Caspers, *Proc. of the C. E. Lummus Montreal Area Refiner's Seminar*, May 21, 1980.
20. J. D. Potts and H. Unger, *Proc. Intersoc. Energy Conv. Eng. Conf. 15*(3):1832 (1980).
21. B. Schnetze and H. Hofmann, *Hydrocarbon Proc. 63*(5):60 (1984).
22. A. Carreau, *Pétrole et Techniques 360*:33 (1990).
23. R. Bearden and C. L. Alderidge, *Energy Prog. 1*(1–4):44 (1981).
24. F. A. Adams, J. G. Gatsis, and J. G. Sikonia, *Proc. First Int. Conf. on Future of Heavy Crude Oils and Tar Sands* (R. F. Meyer and C. T. Steele, eds.), McGraw-Hill, New York, 1981, p. 632.

15

RECLAMATION OF SPENT CATALYSTS

The catalysts used in the hydrocracking process gradually deactivate, due primarily to coke deposition and, in the case of hydrotreating catalysts, contamination with nickel, vanadium, iron, copper, silica, and arsenic. Cycle length may vary from 1 to 5 years. Catalyst regeneration and, in the case of noble metal catalysts, rejuvenation restores most of the catalyst activity (see Chapter 8). However, after two or three cycles the catalyst can no longer be used due to irreversible loss of catalytic and/or mechanical properties. The spent catalyst is removed from the unit and replaced with a load of fresh or regenerated catalyst.

The disposal of spent petroleum-refining catalysts is a growing concern for refiners. In the past, most of the spent catalysts were sent to landfills. Metal extraction was limited mostly to noble metals. However, increasingly stringent environmental regulations have restricted the use of landfills due to long-term environmental risks. This has forced the industry to develop disposal routes that emphasize reclamation or recycling that convert the spent catalyst into useful products.

15.1. Composition of Spent Catalysts

Based on the type of metals present, the spent catalysts generated in the hydrocracking process can be divided into two categories: (a) noble metal catalysts and (b) base metal catalysts.

The noble metal catalysts used in hydrocracking typically contain between 0.4 and 1.0 wt % palladium, a metal of significant economic value. Well-established technologies are applied to the recovery of this metal.

As was shown in previous chapters, the base metal catalysts used in hydrocracking processes typically contain nickel or cobalt oxides associated with molybdenum or tungsten oxides on a support. In hydrotreating catalysts the support is usually γ-alumina, whereas in hydrocracking catalysts it consists of amorphous silica-alumina, of zeolite plus alumina, or some combination of these components (see Chapter 3). Some catalysts also contain phosphorus, titania, or other elements. Spent catalysts contain, additionally, carbon from coke (up to about 20%) and sulfur (up to about 10%) from the active phase of the sulfided catalysts. In spent hydrotreating catalysts, sulfur is also associated with contaminant metals, such as nickel and vanadium, deposited on the catalyst from the feedstock. Contaminating iron originates from corrosion of tanks, pipes, and other equipment. Silica results from the decomposition of antifoaming agents. Arsenic exists in small concentrations in some heavy crudes. The typical composition of spent Ni-Mo and Co-Mo hydrotreating catalysts and of a Ni-W hydrocracking catalyst is shown in Table 15.1 [1].

15.2. Reclamation Technologies

The following options are available for the disposal of spent catalysts:

1. Partial or complete metal recovery, with the subsequent reuse of the recovered metals.
2. Encapsulation and stabilization (fixation) of spent catalyst before land disposal [2,3]. Encapsulation involves surrounding the catalyst particles with an impervious layer of sealant (e.g., paraffin wax, polyethylene) [4], whereas stabilization converts the catalyst into an inert, nonleachable product by reaction with certain chemicals (e.g., silicic acid) or materials (e.g., cement, various glasses) [5,6].

Table 15.1 Typical Analyses of Spent Hydroprocessing Catalysts (wt %) (Calcined Basis) [1]

	Catalyst type		
Element	NiMo	CoMo	NiW
Al	33.1	28	17.7
Mo	12.3	6.2	0.47
V	0.05	10.5	<0.04
Ni	2.60	3.1	2.5
Co	<0.04	2.1	<0.05
W	<0.05	<0.05	15.2
Pb	<0.05	<0.05	<0.05
Si	0.30	1.2	13.2
Fe	0.12	0.7	0.18
As	<0.05	0.4	<0.05
P	3.2	0.05	0.2
Na	0.03	0.5	0.04

3. Disposal in landfills, without chemical pretreatment. As already mentioned, this approach is becoming increasingly difficult to implement due to stringent environmental regulations.

Noble metal catalysts are commonly processed in order to recover and reuse the metals. For base metal–containing catalysts, metal recovery or encapsulation/stabilization prior to land disposal is used. The chosen method depends on economics. When the catalyst contains more than 10 wt % metals, they are usually recovered. Spent hydroprocessing catalysts fall into that category.

15.2.1. Noble Metal Reclamation

Noble metals can be recovered from spent catalysts by hydrometallurgical methods or by a combination of pyro- and hydrometallurgical methods. The hydrometallurgical methods used for reclamation of noble metals from supported catalysts can be divided into two categories: (a) methods in which the noble metals and the support are dissolved, followed by separation of the noble metals from solution; and (b) methods in which the noble metals are separated from the support and subsequently recovered.

Methods of the first category are applied, for example, to alumina-supported catalysts. When recovering noble metals from a catalyst with a partially or totally insoluble support, as in a typical hydrocracking catalyst, the noble metals are separated from the support by dissolution in chlorinated hydrochloric acid or aqua regia [7]. The metals are then precipitated with sulfur dioxide in the presence of tellurium dioxide. If the support is partially soluble and contains an acid-soluble component, such as alumina, that component is extracted with dilute sulfuric acid prior to the solubilization of the noble metals. Traces of noble metals that may be dissolved during the sulfuric acid treatment are recovered by cementation with aluminum scrap in the presence of tellurium dioxide. The noble metals are separated from tellurium dioxide by extracting the latter with tri-*n*-butyl phosphate. Tellurium is stripped from the extract with concentrated hydrochloric acid and reused as collector.

The noble metals may also be leached from the spent catalyst with a cyanide solution at high temperature [8]. Soluble metal-cyanide complexes are formed. After separation from the support, the solution containing the cyanide complexes is heated, resulting in the precipitation of the metals.

When a mixture of platinum and palladium is present, the two metals are separated from the solution obtained with chlorinated hydrochloric acid or aqua regia. After dilution, platinum can be precipitated as ammonium chloroplatinate by addition of ammonium chloride [9]. The precipitate is dissolved for further purification from base-metal contaminants. Alternatively, palladium can be precipitated with dimethylglyoxime in the presence of platinum. Economics makes high yields and high purity critical factors in noble metal recovery processes.

A widely used pyro/hydrometallurgical process for noble metal recovery from spent catalysts involves smelting of the catalyst and accumulation of the noble metals on collector metals, such as iron, lead, or copper [7,9]. The collector is then acid-leached, leaving a noble metal concentrate behind. Smelting is often used to recover palladium from spent hydrocracking catalysts.

When using copper as a collector, the crushed and ground catalyst is mixed with required fluxes (silica, lime, iron oxide), as well as with a copper compound, and submitted to smelting. The charge is allowed to settle and the slag is skimmed. The noble metals are concentrated in the molten copper. The latter is air- or water-atomized to provide a high surface for leaching. Copper is leached with aqueous sulfuric acid using air as an oxidant. The noble metals remain as a residue. Copper is subsequently precipitated as a basic carbonate with soda ash and reused. The noble metals are submitted to separation and/or purification. Copper smelters may be used for the process described.

Plasma fusion is another pyro/hydrometallurgical method used for noble metal recovery [7]. Very high temperatures (over 2000°C) are used to melt the catalyst. In this case, no fluxes are needed to lower the melting point of catalyst components. Iron is added as a collector of noble metals. The catalyst is charged to the reactor and fused. The fused charge is settled to allow the slag and metal phases to separate. The iron alloy is atomized to provide a high surface for leaching. Sulfuric acid and air are used to dissolve the iron alloy, leaving behind the noble metal residue. The solution is separated from the residue and neutralized.

A dry chlorination process may also be used to recover noble metals [7]. In this process, the noble metals are converted to soluble chloro complexes, which can be readily leached with hot water. A mixture of crushed catalyst and sodium chloride (necessary to ensure the formation of chloro complexes) is introduced in the chlorinator and calcined in air in order to burn off any coke and other volatile components. Calcination is then continued in a CO stream in order to reduce the noble metals to elemental form. This calcination is followed by chlorination with chlorine gas at 600–700°C. At this stage, the noble metals are converted to chloro complexes (e.g., Na_2PdCl_4). The compounds are leached by flushing the reactor repeatedly with a mixture of hot water and steam. The noble metals can subsequently be precipitated from solution with sulfur dioxide in the presence of a tellurium dioxide collector. A simplified flow diagram of this process is shown in Figure 15.1. The major advantages of this method are low reagent cost and complete conversion of noble metals to their respective chlorides.

The method chosen for noble metal recovery depends on economics and on the required purity of the recovered metals.

15.2.2. Base Metal Reclamation

A variety of metal recovery methods have been described. In many of these methods, the initial calcination (roasting) of the catalyst to remove coke and sulfur

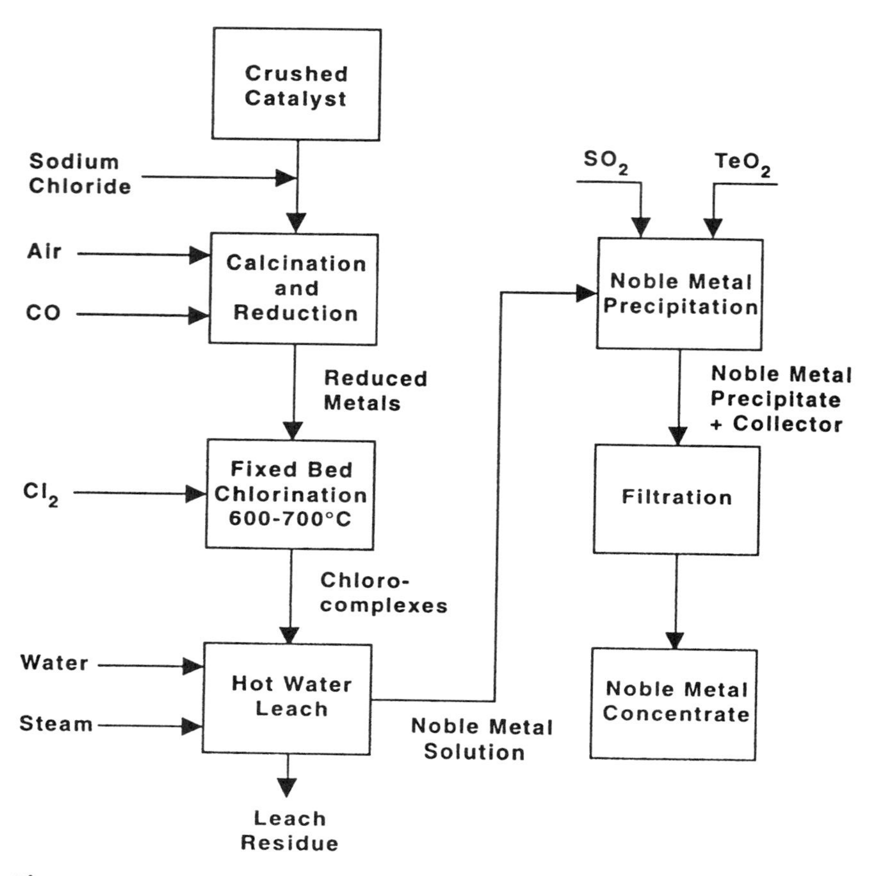

Figure 15.1 Noble metal recovery by dry chlorination process [7].

is followed by another calcination in the presence of a chemical additive that converts some of the metal oxides to water-soluble salts. These are subsequently leached with water and separated. For example, calcination in the presence of caustic or sodium carbonate converts molybdenum to water-soluble sodium molybdate. After dissolving in water, addition of nitric acid precipitates molybdic acid [10].

Other methods consist of leaching some of the metals from the calcined catalyst with solutions of oxalic acid, alone or in combination with other reagents [11–14]. The formation of metal oxalates during leaching allows the selective removal of nickel and vanadium from the catalyst. Dilute mineral acids have also been used for metal extraction.

Calcination in chlorine gas, leading to the formation of soluble metal oxychlorides, has been recommended for the removal of nickel, cobalt, and vanadium [15]. Chlorination was also used to produce volatile metal chlorides that can then be separated [15]. Selective solvent extraction has also been recommended for metal separation [17].

The technologies used commercially for total or partial reclamation of components from spent hydroprocessing catalysts can be divided into two categories: hydrometallurgical and pyrometallurgical technologies. In some instances a combination of the two technologies is also used. The metals are reclaimed for sale or use, whereas the remaining materials are sold, reused, or disposed of as nonhazardous waste.

a. *Hydrometallurgical Process.* This process involves the selective extraction of metals from the spent catalyst by using caustic or acid-leaching agents. In some processes, the spent catalyst is roasted prior to leaching in order to remove carbon and sulfur. The flow diagram of such a hydrometallurgical reclamation process (CRI-MET process) applied to a spent hydrotreating catalyst is shown in Figure 15.2 [18]. It may also apply to mild hydrocracking catalysts. When processing Ni-Mo or Co-Mo spent catalysts, the solution obtained from the first leaching with weak caustic contains molybdenum, vanadium, and some impurities. The extract is submitted to purification and separation, resulting in concentrates of the corresponding metals. Molybdenum and vanadium can be separated

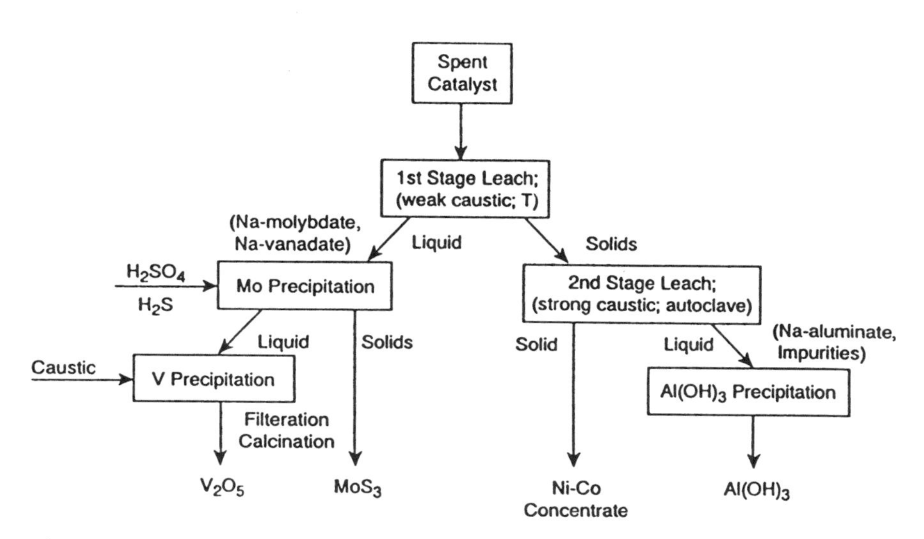

Figure 15.2 Flow diagram of CRI-MET hydrometallurgical reclamation process [18].

by precipitation or over an ion exchange resin. For example, in the CRI-MET process, molybdenum is precipitated as MoS_3 and vanadium as hydrated V_2O_5. In the Eurecat process, molybdenum is separated from vanadium by ionic exchange. Overall recovery of these two metals is over 90%.

The solids from the first-stage leaching, sometimes called tailing, are submitted to a second leaching, this time with strong caustic, in order to separate nickel and cobalt from the aluminum compound. Alumina is dissolved as aluminate and Ni-Co concentrate is obtained, with nearly 100% recovery of these metals. Aluminum is subsequently precipitated from solution as hydrated Al_2O_3. Alumina recovery varies from 75% to 90%, depending on the silica content of the catalyst. A more detailed description of the hydrometallurgical process used by CRI-MET is given in [19]. Other hydrometallurgical reclamation processes are also being used [9,20].

b. Hydro/Pyrometallurgical Process. Some processes combine the hydrometallurgical method with a pyrometallurgical method. For example, in the Eurecat process, the solid obtained after the first leaching step is submitted to a pyrometallurgical treatment in which the solid is fused in an arc furnace at over 1500°C [1]. Such a treatment produces a metal concentrate of nickel and cobalt, whereas aluminum and residual impurities form a slag. The metal concentrate can either be sold to specialty steel producers or further processed for separation of the two metals. Nickel and cobalt are separated from each other by solvent extraction and the respective solutions are subsequently submitted to electrolysis. The final products are high-purity, electrolytic nickel and cobalt. The alumina slag is nonhazardous and can be used as an aggregate for roadbase; it also can be purified for metallurgical or chemical applications. A flow diagram of the Eurecat hydro/pyrometallurgical reclamation process is shown in Figure 15.3 [1].

In another hydro/pyrometallurgical process, developed by Gulf Chemical and Metallurgical Corporation in conjunction with Sadacem [9,21], molybdenum and vanadium are converted to their sodium salts by roasting the catalyst with sodium carbonate. The calcined material is quenched with water, which dissolves the salts. The resulting solution is separated from the tailings by decantation. Ammonium chloride is added to the solution to precipitate ammonium vanadate, which is converted by calcination to vanadium pentoxide. The remaining solution is heated and acidified to precipitate molybdic acid. Upon calcination of the precipitate, molybdic oxide is obtained. The tailings are treated by a proprietary pyrometallurgical method to recover nickel and cobalt.

The endproducts obtained by some hydrometallurgical and hydro/pyrometallurgical processes are listed in Table 15.2.

c. Pyrometallurgical Process. This process involves smelting in an electrical furnace, designed to reclaim metals from steel wastes, pickling solutions, plating sludges, and nickel-copper ores. The spent catalyst may be added to

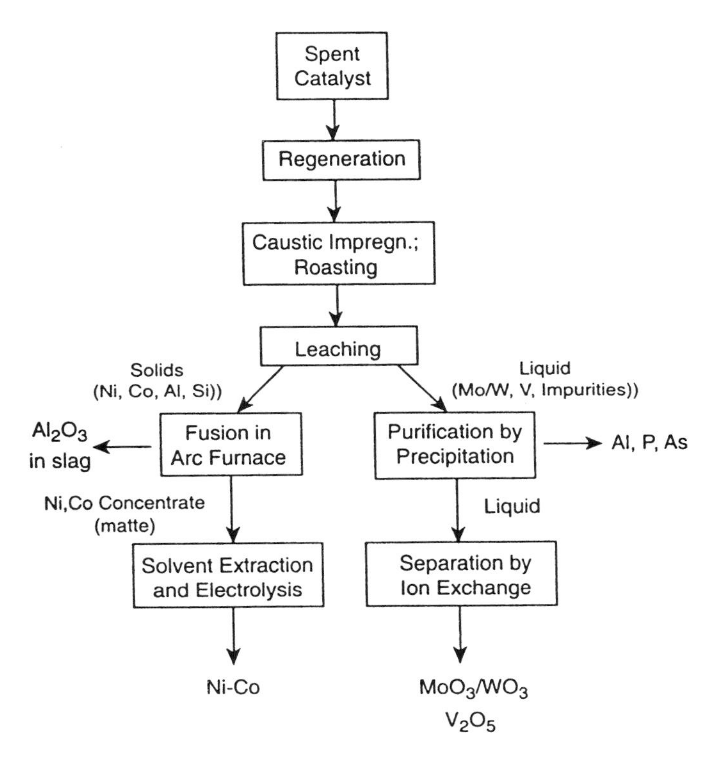

Figure 15.3 Flow diagram of Eurecat hydro/pyrometallurgical reclamation process [1].

the other materials processed in the smelter. The pyrometallurgical process typically consists of the following steps:

1. Feed preparation, blending, agglomeration and pelletizing
2. Roasting for carbon and sulfur elimination
3. Smelting in an electric furnace, resulting in a metal concentrate (matte) and slag
4. Metal extraction and casting, or further metal refining

The endproducts from such a process typically are stainless steel alloys, purified metal concentrates, electrolytic nickel and cobalt, MoO_3, V_2O_5, alumina slag, and/or some precious metals.

The technologies described have been commercialized and are available for reclamation of spent hydroprocessing catalysts. A more detailed discussion of hydroprocessing catalyst reclamation is given in [20].

Table 15.2 Differentiation of Reclamation Methods by Endproducts

Company	Products
CRI-MET	• Molybdenum trisulfide (MoS_3) • Vanadium pentoxide (V_2O_5) • Alumina trihydrate ($Al_2O_3 \cdot 3\ H_2O$) • Nickel-cobalt concentrate
Gulf Chemical and Metallurgical Corp.	• Molybdenum oxide (MoO_3), technical grade • Vanadium pentoxide flakes • Tailings containing nickel, cobalt, and alumina
Eurecat	• Molybdenum oxide (MoO_3) or ammonium molybdate • Vanadium pentoxide • Electrolytic nickel metal • Electrolytic cobalt metal • Alumina slag
TNO/Metrex	• Sodium molybdate, molybdenum sulfide, or Fe-Mo • Sodium vanadate • Nickel chloride or metal • Cobalt chloride or metal

References

1. G. Berrebi, P. Dufresne, and Y. Jacquier, *Environ. Prog. 12*(2):97 (1993).
2. R. B. Posajek (ed.), *Toxic and Hazardous Waste Disposal Vol. 1*, Ann Arbor Science Publishers, Mich., 1979.
3. R. B. Posajek, *Environ. Sci. Technol. 12*:382 (1978).
4. H. R. Lubowitz and C. C. Wiles, in *Catalyst Deactivation* (B. Delmon and G. Forment, eds.), Elsevier, Amsterdam, 1987, p. 189.
5. P. Colombo and R. M. Neilson, Jr., *Nucl. Technol. 32*:30 (1977).
6. Z. Dlouhy, *Disposal of Radioactive Waste*, Elsevier, Amsterdam, 1982.
7. J. E. Hoffman, *J. Metals*, June 1988, p. 40.
8. G. B. Atkinson, D. P. Desmond, and R. J. Kuczynski, U.S. Patent No. 5,160,711 (1992).
9. *Chem. Eng.*, Feb. 16, 1987, p. 25.
10. H. Castagna, G. Gravey, and A. Roth, U.S. Patent No. 4,075,278 (1978).
11. J. O. Hernandez, Symposium on recovery of spent catalysts, Div. Petr. Chem., A.C.S., Kansas City, 1982.
12. B. G. Silbernagel, R. R. Mohan, and G. H. Singhal, U.S. Patent No. 4,272,400 (1981).
13. H. Beuther and R. A. Flinn, *Ind. Eng. Chem. Prod. Res. Dev. 2*:53 (1963).
14. D. F. Farrell and J. W. Ward, U.S. Patent No. 4,089,806 (1978).
15. M. Ioka and T. Shimizu, *Jpn. Kakai Tokyo Koho. 79*:101, 794 (1978).
16. G. J. Gravey, C. LeGoff, and Goniv, U.S. Patent No. 4,182,747 (1980).
17. G. L. Hubrod and D. A. Van Leirsburg, U.S. Patent No. 4,434,141 (1984).

18. R. W. Goerlich, C. A. Vuitel, and J. G. Welch, NPRA Annual Mtg., San Antonio, TX, March 1987, AM-87-74.
19. T. La Rue, Ray Tinnin, E. Wiewiorowski, and R. Crnojevich, paper presented at the AIChE Mtg., New Orleans, LA, March 1986.
20. D. L. Trimm, *Catalysts in Petroleum Refining 1989*, Elsevier, Amsterdam, 1990, p. 41.
21. G. W. Walker, *Management of Spent Catalyst by GCMC*, paper presented at the AIChE Mtg., August 1990.

16

CONCLUSION

In nearly half a century since hydrocracking science and technology has evolved, significant progress has been made in the area of hydrocracking catalysts and processes. The use of zeolites as major components of hydrocracking catalysts has led to dramatic changes in both activity and selectivity. The more recently introduced dealuminated Y zeolites have allowed a further fine tuning of catalyst selectivity.

The use of a variety of mixed oxides as matrix components has been investigated. New physical characterization methods developed during the last decades have been used to obtain a better understanding of the nature of different hydrocracking catalysts. The reaction mechanism of hydrocarbon conversion over hydrocracking catalysts is now better understood. Some basic correlations between catalyst composition, catalyst activity, and catalyst selectivity have been established. In recent years an increasing number of hydrocracking catalysts have been tailor-made by catalyst manufacturers to meet the requirements of different refiners.

Hydrocracking processes have also seen dramatic changes over the last decades. Catalytic dewaxing processes using zeolite-based catalysts are now widely used by the refining industry. Hydrocracking is used to produce petrochemical feedstocks. New processes for upgrading residues have been developed and have undergone continual improvement. Hydrocracking processes have been successfully integrated with other refining processes, such as hydrotreating and catalytic cracking.

The need to convert heavy feedstocks to lighter, more valuable products assures that hydrocracking will continue to play an important role in the petroleum industry. Hydrocracking allows the refiner to meet the high fuel quality standards demanded by federal and state regulations, such as lower aromatic and sulfur levels in jet and diesel fuels, and lower sulfur levels in fuel oils. Increasing demand for distillates in the near future and heavier crude slates may lead to further increases in hydrocracking capacity. Residue-upgrading technology is likely to continue the trend of changing from thermal processes to more flexible hydroconversion processes.

The great flexibility of hydrocracking processes and the diversity of commercial hydrocracking catalysts allows the processing of a wide variety of feedstocks to a series of high-quality products. Such flexibility regarding both feedstocks and products, as well as government mandate fuel specifications, assures the continuing application and expansion of hydrocracking technology in the modern refinery.

GLOSSARY

ABD Average or apparent bulk density of a catalyst. Generally, the weight of catalyst per unit volume. Usually expressed in gm/cm^3. See also Compacted Density (CD).

Activity See Catalyst Activity.

Additive Substance used in small amounts to impart new or to improve existing characteristics of oils and catalysts.

Alkylate Product obtained by the acid-catalyzed coupling of light olefins, usually propylene and butylene (sometimes also amylene), with isobutane. Catalysts: sulfuric acid or hydrofluoric acid. Contains branched chain hydrocarbons and is a high-octane blending component for gasoline.

Alkylation Unit Feed (Alky Feed) Isobutane, butylene, and propylene produced in different oil refining operations and used as feed to the alkylation unit.

Aniline Point The minimum temperature for complete miscibility of equal volumes of aniline and test sample. Usually correlated to aromatics content: the lower the temperature the more aromatic the sample.

API American Petroleum Institute.

°API Gravity An arbitrary gravity scale for oils, widely used in the oil industry. Defined as:

$$°API = \frac{141.5}{\text{specific gravity } 60/60°F} - 131.5$$

Aromatics Class of hydrocarbons distinguished by the presence of one or several benzene rings. Present in crude oil and in a variety of oil products.

The definitions given are those used mostly in the petroleum industry, in petroleum refining literature, and by catalyst manufacturers. However, the definition of some terms can vary depending on the source. This applies especially to the boiling ranges of different petroleum fractions.

ART Process Asphalt Residual Treating process, developed by Engelhard Co. for processing residue.

Asphaltenes Broadly defined as the nonvolatile fraction of a feedstock that is insoluble in *n*-pentane and *n*-hexane. Complex, colloidal substances found in the residue obtained from crude oil distillation. Consist essentially of three-dimensional clusters of polycyclic aromatics, aliphatic chains, and naphthenic rings. Contain also hetero atoms (S, N, O, and metals). Generally soluble in benzene.

ASTM American Society for Testing and Materials. Founded in 1898 to develop standards on characteristics and performance of materials, products, and systems. Specifications and test methods described in ASTM standards are subject to periodic reviews and revisions.

ASTM Distillation Standardized laboratory test carried out at atmospheric pressure (e.g., D-86) or under vacuum (e.g., D-1160), without fractionation. Temperatures are recorded at regular intervals of collected distillate, expressed as vol % of the original sample.

ATB Atmospheric tower bottoms. The residue left at the bottom of the atmospheric crude oil fractionator. Boiling range: over 430°C (~800°F). Typically fed to a vacuum tower for further operation.

Atmospheric Gas Oil See Gas Oils.

Barrel U.S. petroleum industry standard of 42 U.S. gallons at 60°F and 14.696 psia.

Basic Nitrogen See Nitrogen and Contaminants.

Bifunctional Catalyst Catalyst having two catalytic functions. For example, a hydrocracking catalyst has a cracking function (due to the acid component) and a hydrogenation function (due to hydrogenation metals or metal sulfides).

Bitumen That portion of petroleum, asphalt, and tar products that will dissolve completely in carbon disulfide (CS_2).

Blending In refining, it is the operation in which two or more components are mixed together to obtain a desired range of properties in the finished product.

Blending Octane Numbers Calculated or estimated octane numbers used in practice by refiners so that linear blending can be achieved. They are different from "pure" octane numbers and may differ from refinery to refinery.

Blending Stock An intermediary hydrocarbon stream from a processing unit used in product blending to a final marketable stream.

BMCI Bureau of Mines Correlation Index; see Correlation Index.

Bottoms General term used for higher boiling residue that is removed from the bottom of a fractionating column.

Bromine Number A number, determined from a test, that is indicative of the degree of unsaturation in a sample. Often used to characterize the olefinicity of gasolines.

Btu British thermal unit. The heat required to raise the temperature of 1 pound of water by 1°F. Equivalent to 252 calories. The caloric value of crude oils is usually expressed in Btu.

BTX Abbreviation for mixtures of benzene, toluene, and xylenes. Important in the manufacture of petrochemicals, plastics (e.g., nylon, polyurethane, polyesters), insecticides, drugs, and dyes. Also used as solvents.

Bunker Fuel Oil The heaviest fuel oil sold, often used as fuel by ships and power plants. Also referred to as No. 6 fuel.

Canmet A slurry hydrocracking process operating in a three-phase mode consisting of a solid additive, liquid feed, and hydrogen.

Carbenium Ion Organic ion with a positive charge on a tricoordinated carbon atom (e.g., CH_3-CH_2-C^+H-CH_3).

Carbocation Organic ion with a positive charge on a carbon atom. Includes carbenium and carbonium ions.

Carbon Disulfide (CS_2). In hydrocracking, used as feedstock additive for catalyst presulfiding.

Carbon on Catalyst The actual carbon content of the carbonatious deposit on the catalyst following either reaction (carbon on spent catalyst) or regeneration (carbon on regenerated catalyst). Measured by a standard laboratory oxidation test.

Carbon Rejection Process (e.g., FCC, coking) usually applied to heavy oils or residues to increase the hydrogen/carbon ratio in the product without addition of hydrogen, by forming coke or pitch and a liquid product with a higher hydrogen/carbon ratio than the initial feedstock.

Carbon Residue The carbonaceous residue formed after evaporation and pyrolysis of a petroleum product. Tests for the determination of carbon residue (Conradson carbon, Ramsbottom carbon) provide an indication of the relative coke-forming tendency of the product.

Carbonium Ion Organic ion with a positive charge on a pentacoordinated carbon atom (e.g., CH_3-CH_2-C^+H_3-CH_3).

Catalyst A substance that changes the rate of a chemical reaction but is itself not consumed in the process.

Catalyst Activity In hydrocracking, the ability of a catalyst to convert a specified feedstock to lighter products under specific operating conditions. In commercial hydrocracking, catalyst activity is expressed as temperature required to obtain a specified conversion under certain process conditions. Lower temperature is indicative of higher activity.

Catalyst Bed A uniform section of catalyst volume in the reactor.

Catalyst Deactivation Decrease in catalytic activity and/or selectivity with time on stream. Can be caused by coke, by structural changes in the catalyst, or by catalyst poisons (contaminants). Can be reversible, e.g., when due to coke; or irreversible, e.g., when due to certain structural changes.

Catalyst Matrix Component of hydrocracking catalyst, in which the zeolite is embedded. Usually contains a synthetic component (e.g., alumina and/or amorphous silica-alumina). Affects primarily the physical properties of the catalyst, but may also have a catalytic role.

Catalyst Poisons Substances that have a deleterious effect on catalyst activity and/or selectivity when adsorbed on the catalyst (e.g., organic sulfur and nitrogen compounds). Catalyst poisoning can be reversible or irreversible. In hydrocracking, potential catalyst poisons are commonly removed by hydrotreating the feedstock prior to hydrocracking.

Catalyst Presulfiding Conversion of catalyst metal oxides to metal sulfides prior to catalytic process. Can be done in situ (in the hydroprocessing reactor) or ex situ (outside the hydroprocessing reactor). Is done with a H_2S/H_2 gas mixture, a regular feedstock containing organic sulfur compounds (nonspiked feedstock), or a feedstock to which an organic sulfur compound has been added (spiked feedstock).

Catalyst Reclamation Conversion of spent catalyst into useful products. The commercial processes used for catalyst reclamation can be hydrometallurgical, hydro/pyrometallurgical, or pyrometallurgical.

Catalyst Selectivity Describes the relative rates of two or more competing reactions on a catalyst. In hydrocracking, it is a measure of the vol % (or wt %) of feedstock converted to desired product or products. Product selectivities obtained with different catalysts are often compared at the same conversion.

Catalyst Stability See Stability.

Catalyst Support Component of hydrocracking catalyst that supports hydrogenation component and contains the cracking component(s) of the catalyst. Has acidic properties and consists of amorphous oxides (e.g., amorphous silica-alumina) or of zeolite embedded in metal oxide (e.g., alumina) or in mixture of oxides.

Catalyst Support Impregnation In hydrocracking, addition of solution of hydrogenation metal precursors to calcined catalyst support. Usually done by pore saturation (volume of solution added equals that of catalyst support pores) or by dipping (use of excess solution) and draining the excess. Followed by drying and calcining.

Catalytic Cracking The refining process of breaking down the larger, heavier, and more complex hydrocarbon molecules into simpler and lighter molecules accomplished through the use of a catalyst.

Catalytic Cycle Stock See Cycle Oil.

Cetane Number Percentage of pure cetane (*n*-hexadecane, $C_{16}H_{34}$) in a blend of cetane and α-methyl naphthalene ($C_{11}H_{10}$) which matches the ignition quality of a diesel fuel sample. Higher numbers indicate better quality over the range of about 30–65.

Characterization Factor The factor developed by Watson is used to characterize hydrocarbon streams. The Watson characterization factor *K* (or UOP *K* factor) is defined by the equation:

$$\text{Watson } K = \text{UOP } K = \frac{\sqrt[3]{\text{mean average boiling point, °R}}}{\text{specific gravity } 60/60\text{°F}}$$

where °R = degree Rankine
(°R = °F + 460)

The *K* factor ranges from 12.5 for very paraffinic stocks to 10.0 or less for highly aromatic stocks. The crackability of a feedstock increases with its *K* factor.

Chlorides See Contaminants and Organic Chlorides.

Clear Unleaded, in reference to gasoline. For example, the clear octane number of gasoline is the octane number of unleaded gasoline.

Cloud Point Temperature at which turbidity is first noticed when cooling an oil sample. It is indicative of the presence of paraffin wax in the oil.

Coke (1) In catalytic hydrocarbon conversion, a carbonaceous material deposited on the catalyst during reaction. A complex mixture of molecules, with a high content in polycyclic aromatics, containing 5–8% hydrogen and minor amounts of sulfur and nitrogen. (2) Nonvolatile product obtained from different coking processes of residues or heavy oils (petroleum coke). Produced in the form of sponge or needles. Used as fuel or in the manufacture of electrodes, graphite, and various carbides.

Coker Gas Oils Thermally cracked product from a coker unit.

Coker Products Any of a full range of products from light gases to gas oil, largely from the coking of crude and vacuum residue. Coker gas oil is a potential hydrocracking feedstock.

Coking Thermal cracking process for the conversion of heavy, low-grade oils (fractionator bottoms, reduced crude) to coke and lighter products, such as light gases, naphtha, and gas oil.

Cold Flow Properties Flow characteristics of the hydrocarbon stream when it is exposed to low temperature.

Combicracking A slurry three-phase mode residue conversion process developed by Veba and based on the coal liquefaction process.

Combined Feed Rate The rate of total feed (including all recycle streams) to the unit.

Compacted Density (CD) Generally the weight of a unit volume of compacted catalyst, expressed in g/cm^3. See also ABD.

Conradson Carbon Residue (CCR) The wt % carbonaceous residue remaining after the evaporation and pyrolysis of an oil under specific conditions. A value of about 0.25 or lower usually indicates a feed with low coke-forming tendency.

Constraint Index (CI) The ratio of the cracking rate constants for *n*-hexane and 3-methylpentane. Used to characterize the shape selectivity of zeolites. For large-pore zeolites (e.g., X, Y, L) CI < 1; for medium-pore zeolites (e.g., ZSM-5) CI $= 1$–12; for small-pore zeolites (e.g., A) CI > 12. Higher CI values are indicative of higher shape selectivity.

Conversion The vol % (or wt %) of fresh feed changed to middle-distillate, gasoline, and lighter products in the catalytic hydrocracking process. Also defined as product fraction (in wt % feed) boiling below a certain temperature. Conversion can be calculated by the following equation:

$$\text{Conversion (volume \%)} = \frac{V_{\text{feed}}^{\text{EP+}} - V_{\text{prod}}^{\text{EP+}}}{V_{\text{feed}}^{\text{EP+}}} \times 100$$

where $V_{\text{feed}}^{\text{EP+}}$ = vol % feed boiling above desired endpoint;
$V_{\text{prod}}^{\text{EP+}}$ = vol % product boiling above desired endpoint;

May also be expressed as wt %.

Conversion Per Pass Amount of feedstock converted to product materials in one pass through the reactor.

Coronenes Polynuclear aromatic compounds containing seven aromatic rings. Unsubstituted coronene: $C_{24}H_{12}$. Present in the residue obtained from crude oil distillation. Formed during hydrocracking of heavy feedstocks. Strong coke precursors with a deleterious effect on the hydrocracking process.

Correlation Index (BMCI) The index, developed by the U.S. Bureau of Mines, relates the average boiling point of a distillation fraction to its specific gravity. It is defined by the following empirical equation:

$$\text{BMCI} = 48{,}640/K + 473.7/d - 456.8$$

where K = mid-boiling point of the fraction, in °K and
d = specific gravity of the fraction, at 60/60°F.

The index can be used for classification of crude oils: lower values (0 to 15) are indicative of paraffinic oils, while high values (over 50) are indicative of aromatic oils. Can also be correlated to ethylene yield from steam cracker.

Cracking The breaking down of higher molecular weight hydrocarbons to lighter components by heat (thermal cracking) or by heat in the presence of a suitable catalyst (catalytic cracking). In petroleum technology the term cracking also includes secondary reactions of the products formed (isomerization, hydrogen transfer, condensation, etc.) leading to other hydrocarbons and coke.

Crude Oil (Petroleum) Naturally occurring liquid mixture, consisting primarily of hydrocarbons, but also containing organic compounds with hetero atoms such as sulfur, nitrogen, oxygen, and metals. Contains varying amounts of water, inorganic matter, and gas. Can be broadly classified on the basis of relative amounts of waxes and asphaltic material (light paraffinic, paraffinic, naphthenic, or asphaltic, aromatic and mixed-based crudes), or on the basis of specific gravity (light and heavy crudes).

Crude Oil Distillation The separation, through distillation, of the hydrocarbons from crude oil into individual streams with boiling ranges suitable for further processing. Typical crude oil fractions are fuel gas, propane, butane, light straight-run gasoline (LSR), heavy straight-run gasoline (naphtha) (HSR), kerosene, light gas oil (LGO), heavy gas oil (atmospheric gas oil) (HGO), atmospheric residue, vacuum gas oil (VGO), and vacuum residue. The initial boiling points and endpoints of the fractions vary, depending on refinery configuration and specification. The volume of each fraction varies, depending on type of crude oil.

Cut Portion of a crude oil, gasoline, middle distillate, or other liquid product boiling within certain temperature limits, usually on a true boiling point basis.

Cutter Stock An intermediate refinery hydrocarbon stream used to reduce viscosity of heavier hydrocarbon streams.

Cycle Oil (Cycle stock) Portion of a catalytic cracker effluent that is not converted to gasoline or lighter products. Material usually boils above ~200°C (390°F). Can be recycled, blended to products, or upgraded by hydrocracking. Can be divided into light cycle oil (LCO) (boiling range ~200–340°C, 390–650°F) and heavy cycle oil (HCO) (boiling range ~340–455°C, 650–850°F).

Dealumination See Zeolite, Dealuminated.

Deasphalted Oil (DAO) Oil obtained by solvent extraction of vacuum bottoms. The deasphalting solvents used are low molecular weight hydrocarbons (e.g., liquid propane and butane), alcohols, and esters. The solvent dissolves the desirable oil, such as lubricating oil, and separates it from insoluble asphaltic and resinous material. The solvent is subsequently separated from the oil–solvent mixture in evaporators. When liquid propane is the solvent, the product is known as propane deasphalted oil (PDA oil). See also Solvent Extraction.

Debutanized Gasoline Hydrocracker gasoline from which butane and lighter hydrocarbons have been separated in a fractionating tower.

Dehydrocyclization Catalytic reaction in which a paraffin is dehydrogenated and converted to an aromatic compound. For example, heptane → toluene + hydrogen. Occurs during the catalytic reforming of paraffinic feedstocks.

Delayed Coking Process in which heavy feedstocks (fractionator bottoms, heavy catalytic cycle oil) are converted to coke by slow thermal cracking. The coking process takes place in cracking drums (soakers) and results in coke and overhead products (light gases, naphtha, gas oil). Coker gas oil is a potential hydrocracking feedstock. See also Fluid Coking.

Demet Process Process designed to remove metals (mainly vanadium and nickel) deposited on catalysts during catalytic process. The metals are converted to compounds, such as chlorides, that can easily be removed by washing. Applied primarily to catalysts used in the processing of high-metal feedstocks.

Demetallization Removing metals from a feedstock or catalyst. Demetallization of a feedstock is a catalytic process carried out under hydrogen in the presence of a catalyst usually containing cobalt and molybdenum or nickel and molybdenum on an alumina support. Metals usually removed are vanadium, nickel, iron, copper, and arsenic. Demetallization of a hydrocracking catalyst removes the heavy metals deposited on the catalyst during the hydrocracking process. See also Catalyst Reclamation.

Dense Bed Loading of Catalyst A technique to orient catalyst particles in a reactor by controlling falling rate and trajectory. Results in high bed density and avoids further settling of the catalyst during operation. However, may lead to a higher pressure drop than sockloading.

Desalting Process for removal of brine from crude oil prior to primary distillation. Consists in emulsifying the crude oil with water at ~120°C (250°F) under pressure, dissolving the salts in the water and separating the water and oil phases in a high-potential electric field.

Dewaxing The removal of wax (long-chain paraffins) from lubricating oils in order to lower their pour point. Can be accomplished by chilling and filtering (cold press filtration method), by a solvent extraction process (solvent dewaxing), or by selective catalytic cracking or hydrocracking of waxy paraffins. A common commercial solvent used for dewaxing is a mixture of methyl ethyl ketone and benzene (ketone dewaxing). ZSM-5 zeolite is often used as a dewaxing catalyst.

Diesel Fuels Light fuel oils, usually boiling in the range of ~175 to 370°C (350 to 700°F), with specifications on sulfur content, ignition quality, fluidity, volatility, and cetane number. Number 1 diesel fuel (cetane number >50, boiling range ~180–315°C, 360–600°F) is made from virgin stock and is used as fuel in trucks and buses. Number 2 diesel fuel (boiling range ~175–343°C, 350–650°F) is made by blending naphtha, kerosene, and light cracked oil. Railroad diesel fuels have higher boiling ranges (up to 370°C, 700°F) and lower cetane numbers (40–45).

Diesel Index (DI) Measures the ignition quality of a diesel fuel. Defined as:

$$DI = \frac{\text{aniline point (°F)} \times \text{gravity, °API}}{100}$$

Higher numbers are indicative of better burning quality.

Dimethyl Disulfide (DMDS) Organic sulfur compound used as feedstock additive for catalyst presulfiding [$(CH_3)_2S_2$].

Distillation Cut Point The temperature at which there is as much of the heavy product in the light cut as there is light product in the heavy cut while taking into account the yields of each of these cuts. In the case of perfect fractionation, this is the temperature at which the light cut ends and the heavy cut begins.

Distribution Tray Mechanical device in the reactor used to distribute the vapor and liquid reactants across the reactor cross-section.

Dry Gas Gas mixture consisting of methane, ethane, and hydrogen, often used as fuel gas. Obtained in different refining processes.

Dual Function Catalyst See Bifunctional Catalyst.

Ebullating Bed Expanded catalyst bed used in such processes as H-Oil and LC-Fining where the fresh feed, recycle oil, and hydrogen that are in an upward motion are put in contact with said bed of catalyst.

Efficiency In hydrocracking, efficiency of conversion to a specific fraction is defined as the ratio of yield (vol %) of that fraction and feedstock conversion (vol %) to products boiling below the endpoint of that fraction. For example,

$$\text{Diesel fuel efficiency (175–370°C fraction)} = \frac{\text{vol \% (175–370°C) fraction}}{\text{vol \% conversion to 370°C}-}$$

Endpoint (EP) Upper temperature limit of a distillation.

Endothermic Reaction A reaction in which heat must be added to maintain reactants and products at a constant temperature. Catalytic cracking of hydrocarbons is an endothermic reaction.

EOR End of run referring to unit operation at the end of a catalyst system's useful life prior to reloading with fresh catalyst.

Exothermic Reaction A reaction in which heat is evolved. Catalytic hydrocracking and catalytic hydrogenation of hydrocarbons are exothermic reactions.

Extraction The process of separating a material by means of a solvent into a fraction soluble in the solvent (extract) and an insoluble fraction. See also Solvent Extraction.

Extrudate In hydrocracking, one of the catalyst-forming products. Obtained by extruding a mixture of catalyst components. Can have different forms, depending on the die used (cylinder, trilobe, twisted trilobe, tetralobe).

FBP Final boiling point of a cut.

Feed/Recycle Ratio The relative amount of total feed (feed + recycle) compared to fresh feed that is charged to the reactor.

Feedstock (Feed) Crude oil or one of its fractions charged to any process equipment for further processing. Feedstocks used in hydrocracking units are straight run gas oil, vacuum gas oil, cycle stock, thermally cracked stock, coker gas oil, and solvent-deasphalted oil. Some of the heavier feedstocks are usually upgraded (by hydrotreating) prior to use in the hydrocracking unit.

Fixed Bed A static bed of catalyst. Used in catalytic hydrocracking processes.

Flash Point Temperature to which a product must be heated in a standard laboratory test at which the vapor above the sample will ignite if subjected to a flame. Is indicative of the fire or explosion potential of a product.

Flexicoking A fluid coker which also includes a gasifier. Provides effective sulfur control.

Flue Gas In general, a gas resulting from combustion, discarded to the flue or stack.

Fluid Catalytic Cracking (FCC) A widely used catalytic cracking process in which the oil is cracked in the presence of microspheroidal catalyst particles, which are maintained in a fluidized state in the reactor by the oil vapors. The cracking takes place at about 460–530°C (860–985°F) and catalyst regeneration at about 590–760°C (1100–1400°F). A variety of commercial processes and equipment designs are available for FCC of oil.

Fluid Coking Continuous process to convert residues to more valuable products, using a fluidization technique. The residue is coked by being sprayed into a fluidized bed of hot, fine coke particles. Coke and more valuable overhead products (light gases, naphtha, gas oil) are obtained. The latter ones are separated by fractionation. Coker gas oil is a potential hydrocracking feedstock. See also Delayed Coking.

Fluidized Bed A bed of catalyst aerated to or just above the point where all particles are in free but suspended motion in the aeration medium.

Freeze Point A low temperature property of aviation fuel. A low freeze point ensures reliable equipment operation both in flight and on the ground. Determined by a standard laboratory test.

Fresh Feed Rate Number of barrels per day of refinery feed charged to the hydrocracking unit. The term is used to differentiate between feed initially charged to the unit and those products that are recycled.

Fuel Oil Equivalent (FOE) Heating value of a standard barrel of fuel oil, equal to 6.05×10^6 Btu. On a yield chart, dry gas and refinery fuel gas are usually expressed in FOE barrels.

Fuel Oils Any of a large number of heavier products, usually boiling above ~220°C (430°F). Can be classified into domestic fuel oils, diesel fuel oils, and heavy fuel oils. Contain a distillate component and/or a residue component.

FVT Final vapor temperature of a cut.

Gas Oils The crude oil fractions boiling within the general range of 260–540°C (500–1000°F). Atmospheric gas oil (AGO) (260–345°C, 500–650°F), light vacuum gas oil (LVGO) (345–430°C, 650–800°F) and heavy vacuum gas oil (HVGO) (430–540°C, 800–1000°F) are obtained by atmospheric or vacuum distillation, respectively. These are hydrocracker and cat cracker feedstocks containing sulfur and nitrogen compounds that can be removed by hydrotreating.

Gasoline A refined petroleum product that is suitable for use as fuel in internal combustion engines. Often characterized by boiling range, octane rating, and volatility. In hydrocracking, product boiling in the range of C_5–200°C (C_5–390°F), corresponding to ~C_5–C_{11} composition. May also be defined as cut to other endpoints. Commonly sold as regular and premium gasoline, which differ primarily by their antiknock properties.

Gasoline Additives Such additives, used in small amounts, enhance certain desirable properties of gasoline or add new properties. For example, antiknock additives improve octane quality; deicing additives reduce ice formation in carburetors; anti-oxidants inhibit gum formation; anticorrosion additives inhibit rust formation; antiwear additives control wear of engine cylinders and pistons; carburetor detergents keep carburetor free of deposits; and so forth.

Gasoline Pool Gasolines of various qualities and origins blended in certain ratios to meet final product specifications. Gasoline pool components are cat cracker and hydrocracker gasolines, alkylate, reformate, polymer gasoline, and light straight-run gasoline. Oxygen-containing compounds, such as methanol, ethanol, and MTBE, are also added to the gasoline pool.

GC RON Research octane number as determined by a gas chromatographic method. The octane rating is calculated using hydrocarbon compositions obtained from chromatographic simulated distillation of liquid products. See also Octane Numbers.

Graded Catalyst Beds A technique to load different catalysts with varying porosity in a catalyst bed to mitigate potential pressure drop problems from feed contaminants.

Guard Bed Catalyst bed used to mitigate pressure drop problems in the reactor due to contaminants in the feedstock.

Gum Heavy polymers and resins formed in light petroleum products during heating or exposure to oxygen in storage.

H-Oil Commercial process developed by Texaco, using ebullating bed technique to hydrocrack residua.

Heavy Cycle Oil (HCO) See Cycle Oil.

Heavy Ends The highest boiling portion of a petroleum fraction.

Heavy Gas Oil (HGO) See Gas Oils.

Heavy Metals Present in trace amounts in crude oils, mostly as oil-soluble nitrogen complexes of the porphyrin type. Concentrated mainly in the residue fraction. Catalyst contaminants. The most important heavy metal contaminants are nickel, vanadium, copper, and iron, which are deposited on the catalyst during hydrotreating.

Heavy Naphtha A typical product stream boiling between light naphtha and kerosene. See also Naphthas.

Heavy Oil In general, an oil with an API gravity of less than 20° and with a sulfur content of over 2%.

Heavy Virgin Naphtha (HVN) Naphtha distilled directly from crude oil, usually fed to the catalytic reformer for octane improvement. The usual boiling range is ~90–210°C (200–410°F). Also referred to as heavy straight-run naphtha (HSRN).

High-Pressure Separator Vessel used to separate up to three phases (vapor, liquid hydrocarbons, water) at a pressure near reaction conditions.

HOC Process Heavy oil cracking process, developed by Phillips Petroleum and Kellogg Co. for processing heavy oils and residue feedstocks.

Hot Flash Separation Separation of vapor from liquid in the reactor effluent of a hydrocracker done at relatively high temperature. An effective HPNA management technique to eliminate plugging of the effluent cooler.

HPLC High-performance liquid chromatography. An analytical method which accomplishes the separation by partitioning the sample between a mobile liquid phase and an immobile solid support.

HPNAs Heavy polynuclear aromatics hydrocarbons. Usually aromatic molecules with more than 11 aromatic rings.

Hycon Residue conversion moving-bed process, developed by Shell, in which catalyst is added at the top of the reactor and withdrawn at the bottom, with catalyst regeneration occurring outside the reactor.

Hydrocracking High-pressure, high-temperature catalytic process for hydrocarbon cracking and hydrogenation of unsaturated hydrocarbons over bifunctional catalyst. Average reactor temperature: 290–400°C (550–750°F); average pressure: 85–140 bar (~1200–2000 psig); catalysts: Pd/zeolite, Ni-W/zeolite, Ni-Mo/zeolite in alumina matrix, or hydrogenation metals on amorphous SiO_2-Al_2O_3 support.

Hydrodenitrogenation Catalytic process for removal of nitrogen from a petroleum feedstock as ammonia in the presence of hydrogen. See Hydrotreating.

Hydrodesulfurization Catalytic process for removal of sulfur from a petroleum feedstock as hydrogen sulfide in the presence of hydrogen. See Hydrotreating.

Hydrogel Rigid, amorphous material containing a network of colloid particles and the imbibed liquid phase. Upon drying, is converted into xerogel.

Hydrogen Addition Process such as hydrocracking, LC-Fining, H-Oil, etc., which saturates polycyclic aromatics and converts them to lighter, usable products such as transportation fuels.

Hydrogen Consumption The amount of hydrogen consumed in the process of operating the unit. This includes chemical consumption, solution, and equipment losses.

Hydrogen Partial Pressure Pressure of hydrogen in a mixture of gases equal to the pressure hydrogen would exert if it occupied by itself the same volume as the mixture at the same temperature. Equal to the product of system pressure × mol % hydrogen.

Hydrogen Transfer Transfer of hydrogen from a hydrogen donor to a hydrogen acceptor. Hydrogen transfer often leads to coke formation.

Hydrogenolysis Catalytic reaction in which carbon–carbon bond cleavage and hydrogenation of the products takes place over monofunctional catalysts such as metals (Pt, Pd, Ni), metal oxides, or sulfides. Example: pentane + H_2 → propane + ethane.

Hydroprocessing Catalytic hydrogenation processes used for the conversion and upgrading of petroleum fractions and products.

Hydrotreating High-pressure, high-temperature catalytic process for sulfur, nitrogen, and heavy metals removal and hydrogenation of unsaturated hydrocarbons in petroleum feedstocks. Often used to upgrade hydrocracking feedstocks. Catalysts: molybdenum or tungsten sulfides, promoted by cobalt or nickel and supported on alumina or silica-alumina; average pressure: 70–140 bar (~1000–2000 psig); average reactor temperature: 370–400°C (700–750°F).

IBP Initial boiling point of a cut.

Inhibitor A substance that when present in small amounts in a petroleum product prevents or retards undesirable chemical changes in the product or in the condition of the equipment in which the product is used. Inhibitors usually prevent or retard oxidation or corrosion.

Inventory In hydrocracking, total weight of catalyst contained in the unit or specific reactor at any given time.

Isomerization Rearranging the structure of a molecule without altering its molecular formula. In petroleum refining, a catalytic process that selectively isomerizes C_5–C_6 low-octane paraffins to higher octane isomerate, mainly i-C_5 and 2,2- or 2,3-diisobutane. The isomerate is an important component of the gasoline pool.

IVT Initial vapor temperature of a cut.

Jet Fuels Light- to middle-distillate products with a boiling range of ~150–290°C (300–550°F) having specifications related to burning quality, freezing point, and gum formation. Usually blended from desulfurized kerosene, cracked and hydrocracked blending stock. Used in jet engines and aircraft turbine engines.

K Factor See Characterization Factor.

Kerosenes Middle distillate products usually boiling in the range of ~150–290°C (300–550°F) with specifications on smoke point, flash point, freezing point, and sulfur content.

Knocking Undesirable ignition of a portion of a motor fuel in the cylinder head of an internal combustion engine due to spontaneous oxidation reactions rather than to the spark. Is noisy and causes loss of power. Reduced or eliminated by antiknock additives.

LC-Fining Name of commercial process developed by Amoco using ebullating bed technique to hydrocrack residua.

Lead Susceptibility Variation of the octane number of a gasoline as a function of the tetraethyl or tetramethyl lead (TEL, TML) content. Paraffins and naphthenes have a greater octane response to alkyl lead than olefins and aromatics.

LHSV Liquid hourly space velocity. In a catalytic fixed bed reactor, the volume of liquid feed entering per hour per volume of catalyst.

Light Cycle Oil (LCO) A product from the cat cracker boiling in the range of ~200–340°C (390–650°F). LCO is usually blended with Number 2 domestic heating oil or diesel fuels. Also used as hydrocracking feedstock.

Light Ends Light gaseous products, usually C_4 and lighter hydrocarbons.

Light Virgin Naphtha (LVN) A light solvent-type naphtha distilled from crude, usually boiling in the range of C_5–100°C (212°F). May be used as a solvent or included in the gasoline pool following desulfurization. Has relatively good octane. Referred also as light straight-run naphtha (LSRN).

Liquefied Natural Gas (LNG) See Natural Gas.

Liquefied Petroleum Gas (LPG) Usually propane, butane, isobutane, or mixtures thereof. Held in liquid state by pressure to facilitate storage, transport, and handling. Generally used as heating fuel. Can also be converted to aromatics (e.g., Cyclar process).

Loss on Ignition (LOI) The loss in weight of a catalyst sample following heating for a specified time at a specified temperature.

Low-Pressure Separator Vessel used to separate up to three phases (vapor, liquid hydrocarbons, water) at a pressure significantly lower than reaction conditions.

Lube Oil Base Stock Any hydrocarbon fraction suitable for use in compounding lubricating oils.

Main Fractionator The primary distillation unit used to separate the condensable reaction products.

Make-up Hydrogen Hydrogen added to the unit to compensate for the hydrogen consumed and purged to maintain recycle gas purity and system pressure.

Maltenes Mixture of resins and oils obtained as filtrates from asphaltene precipitation with nonpolar solvents, such as *n*-pentane or *n*-hexane. Also soluble in CS_2 and CCl_4.

Mass Balance Relative mass of all product streams compared to all feed streams.

Matrix See Catalyst Matrix.

Mercaptans Organic, sulfur-containing compounds found in crude oils and in different petroleum fractions. Formed by thermal and catalytic decomposition of sulfur compounds during processing. Have the general formula R-SH, where R is an organic radical and SH is a thiol group. Removed from petroleum fraction by hydrotreatment.

Metal Passivation Process that reduces or eliminates the harmful effect of metals deposited on the catalyst during processing of metal-containing feedstocks. The passivating agent or passivator is often a compound that reacts with the contaminating metals forming a catalytically less active or inactive compound.

Metal Tolerance The ability of a catalyst to maintain its catalytic properties in the presence of metal contaminants.

Metals See Heavy Metals; Sodium.

Middle Distillates Petroleum fractions usually boiling in the range of ~150–370°C (300–700°F). The exact cut is determined by the specifications of the products. Include light fuel oils, jet and diesel fuels.

Mild Hydrocracking Low-conversion process used to hydrocrack and desulfurize vacuum gas oil to produce transportation fuels and feedstock for other cracking processes.

Molecular Sieve Natural or synthetic crystalline material having micropores of uniform size. Has a high internal surface area and is capable of separating molecules on the basis of their size or shape, by absorption and sieving. Can also have ion exchange and catalytic properties. See Zeolites.

MON+0 or MONC Motor octane number unleaded (clear).

Motor Fuel Pool See Gasoline Pool.

Motor Octane Number (MON) See Octane Numbers.

Moving Bed A process in which the catalyst is added and withdrawn from the reactor as needed and in which catalyst regeneration is performed outside the reactor.

MTBE Methyltertiarybutyl ether, $C_5H_{12}O$. High-octane-number, 110 (R+M)/2 gasoline additive produced from methyl alcohol and isobutylene.

Naphthas A variety of light petroleum distillates, usually in the boiling range of ~C_5–200°C (390°F). Virgin (or straight-run) naphthas are subdivided into light (~C_5–70°C, 160°F), intermediate (~70–138°C, 160–280°F), heavy (~138–165°C, 280–330°F), and very heavy (~165–200°C, 330–390°F) virgin naphthas. The boiling ranges of these fractions can vary, depending on specifications. Naphthas are generally gasoline boiling range cuts and need further processing (reforming, hydrotreating) to make suitable quality gasoline.

Naphthenes General term describing cyclic, saturated hydrocarbons found in oil. Also termed cycloparaffins and cycloalkanes. General formula: C_nH_{2n}.

Natural Gas Natural mixture of light gases, having methane as major component and ethane, propane, CO_2, and nitrogen as minor components. Sometimes contains H_2S (sour natural gas). Used as fuel and as raw material for the petrochemical industry. Often transported in liquefied form (LNG).

n-d-M Method A method to determine the carbon distribution (aromatic, naphthenic, paraffinic) and the average number of rings in a petroleum sample, by measuring the refractive index *n*, density *d*, and molecular weight *M* of the sample. The method is used to characterize refinery streams.

Nitrogen (Nitrogen Compounds) In petroleum refining used as a general term to describe organic, nitrogen-containing compounds found in crude oil or in different petroleum fractions (e.g., carbazoles, indoles, pyridines, quinolenes). Concentrated in heavy fractions and residues. Usually classified in basic and neutral nitrogen (compounds).

NO_x Mixture of nitrogen oxides (mostly NO and NO_2) formed during catalyst regeneration. Generated during combustion of nitrogen-containing coke on spent catalyst and from regeneration air. Potential air pollutant.

NPRA National Petroleum Refiners Association.

Octane Additives (1) Group of organic or organometallic compounds added in small amounts to gasoline to increase its octane rating. Examples: TEL (tetraethyl lead), TML (tetramethyl lead), MTBE (methyltertiarybutyl ether), ETBE (ethyltertiarybutyl ether), TBA (tertiarybutyl alcohol). (2) Inorganic compounds added in small amounts to cracking catalysts to increase the octane rating of FCC gasoline. Example: ZSM-5 zeolite.

Octane Barrel The yield in gasoline obtained in a hydrocarbon conversion process (expressed in barrels) multiplied by its octane rating (usually expressed as $(R + M)/2$).

Octane Number (Research and Motor) Measures the knocking characteristics of a gasoline as determined in a laboratory engine by a standard ASTM method. The knocking tendency of the fuel is compared with those of blends of *n*-heptane (0 octane) and iso-octane (100 octane) when run in the standardized engine under standard operating conditions. RON is measured at an engine speed of 600 rpm, whereas MON is measured at 900 rpm. The percentage of iso-octane in the blend having the same knocking characteristics as the fuel is taken as the octane number of that fuel.

Octane Pool See Gasoline Pool.

Octane Response The degree to which a gasoline octane number increases when an octane-boosting additive, such as tetraethyl lead, is added.

Octane Scale A series of arbitrary numbers, from 0 to 120.3, used for gasoline octane rating. The scale is defined by three reference materials: *n*-heptane, assigned the number 0; iso-octane (2,2,4-trimethylpentane), assigned the number 100; and iso-octane plus 6 ml tetraethyl lead, assigned the number 120.3.

Oil Drop Method Used in the manufacture of spherical hydrocracking catalysts. By pouring an aqueous sol of catalyst components into hot oil, catalyst spheres are formed. The spheres are aged, washed, dried, and calcined. Hydrogenation metals are then added by impregnation.

Oligomerization Catalytic reaction in which several (usually 2 to 4) light olefin molecules (e.g., ethylene, propylene, butadiene) react to form larger molecules, such as dimers, trimers, or tetramers. The reaction products are linear, branched or cyclic oligomers. Mixed oligomers result from the reaction between different unsaturated hydrocarbons.

Once-Through Unit operation utilizing no recycle liquid.

Operating Variables (Process Variables) In hydrocracking, the parameters that can be changed in order to optimize the operation of the unit. Examples: feed preheat temperature, feed liquid hourly space velocity, average reactor temperature, fraction cut points.

Organometallic Compound A compound in which a metallic atom or atoms are combined with an organic radical. Heavy metals in crude oil are often present as organometallic compounds.

Ovalenes Polynuclear aromatic compounds containing 10 aromatic rings. Unsubstituted ovalene: $C_{32}H_{14}$. Their derivatives are found in crude oil residues or are formed during heavy oil processing. Strong coke precursors with a deleterious effect on the hydrocracking process.

Oxygenated Compounds Present in crude oil and in different petroleum fractions. Can be of acidic type (carboxylic acids, phenols, cresols) or nonacidic type (esters, amides, ketones, benzofurans). Residue has a high concentration in oxygenated compounds.

PAHs. Polynuclear aromatic hydrocarbons, another name used commercially, describing compounds containing several condensed aromatic rings in the molecule. See also PNAs.

Paraffins General term describing open-chain, saturated hydrocarbons of the general formula C_nH_{2n+2}. Found in crude oil and in different oil fractions. Also termed alkanes. Can be straight chained (normal) or branched (iso). Major component of petroleum wax.

PIANO Analysis Analysis for paraffins, isoparaffins, aromatics, naphthenes, and olefins.

Plug Flow Liquid and gases moving concurrently at equal velocities.

PNAs. Polynuclear aromatic hydrocarbons. Aromatic molecules with multiple aromatic rings similar to PAHs.

Poisons See Contaminants.

Polymer Gasoline Product obtained by the acid-catalyzed polymerization of propylene and butylene. Phosphoric acid is commonly used as a catalyst in this process.

PONA Analysis Analysis for paraffins, olefins, naphthenes, and aromatics.

Pore Small opening in a material permitting admission, adsorption, or passage of a fluid. Often classified on the basis of size into micropores (diameters less than 20 Å), mesopores (diameters between 20 and 500 Å), and macropores (diameter larger than 500 Å).

Pore Size Distribution Distribution of pore volume as a function of pore radius. Determined usually from nitrogen adsorption isotherms, obtained for porous materials.

Pore Volume Void volume in an individual catalyst particle. Usually determined by nitrogen adsorption, mercury porosimetry or water saturation. Expressed as cm^3/g.

Pour Point Lowest temperature at which an oil will flow or pour when it is chilled, at a standard rate. A rough indicator of the oil's wax content.

Preheat Temperature Temperature to which the combined feed (fresh plus recycle feed) is heated in a feed preheater or heat exchanger prior to injection into the reactor.

Presulfiding See Catalyst Presulfiding.

Pyrolysis A chemical change brought about by heat. Applied to production of ethylene from hydrocarbon streams.

Pyrolysis Gasoline (Steam-Cracked Gasoline). Obtained by steam cracking of liquid hydrocarbon feedstocks. Formed as byproduct in the manufacture of ethylene and propylene by steam cracking naphtha. Has poor oxidation and gum stability due to its high concentration in diolefins. Also rich in aromatics.

Quench Gas A cold stream, usually recycle gas hydrogen, used to cool the reactants to control the exothermic hydroprocessing reactions.

Quench Zone A mechanical device in the reactor for introduction of cold quench gas and distribution of vapor and liquid hydrocarbons.

(*R* + *M*)/2 or (RON + MON)/2 Arithmetic average of research and motor octane numbers. Usually posted on gasoline pumps in service stations. Also called antiknock index.

Ramsbottom Carbon Residue Carbonaceous residue, expressed in wt %, obtained after evaporation and pyrolysis of an oil under standard conditions. See also Conradson Carbon.

Rare Earths Generic name for 14 elements of the lanthanide series with atomic numbers from 57 to 71, plus scandium and yttrium. Their chemical properties are very similar. Found in minerals bastnasite and monozite. The main elements are cerium, lanthanum, neodymium, and praseodymium. Rare earths–exchanged Y-type zeolites are components of some hydrocracking catalysts.

Raw Naphtha Untreated naphtha. Usually submitted to further processing.

RCC Process Reduced Crude Conversion Process. Developed by Ashland Oil Co. and UOP. A commercial, catalytic cracking process designed to convert reduced crude, including the 565°C+ (1050°F+) residue fraction, to gasoline.

RCD Unionfining Fixed-bed hydrotreating process developed by UOP to upgrade residua.

Reclamation (of spent catalyst). Processing of spent catalyst for partial or complete recovery of catalyst components. In hydrocracking, the end products from a catalyst reclamation process are usually metals (e.g., noble metals, electrolytic Ni or Co), metal oxides (e.g., MoO_3, V_2O_5, $Al_2O_3 \cdot nH_2O$), and metal salts (e.g., MoS_3, sodium molybdate, nickel chloride). The processes used for catalyst reclamation can be divided into hydrometallurgical, hydro/pyrometallurgical, and pyrometallurgical.

Recycle Gas Gas stream from the high-pressure separator that is compressed and reprocessed through the reactor circuit.

Recycle Oil Fractionated material reprocessed through the reactor circuit to achieve additional conversion.

Recycle Rate The barrels-per-day fractionator bottoms that are returned to the hydrocracking reactor for further conversion.

Reduced Crude A crude oil whose API gravity has been reduced by distillation of the lighter, lower boiling constituent. Sometimes defined as the 343°C+ (650°F+) fraction of a crude oil.

Refining Petroleum refining is the separation of crude oil into fractions and the subsequent upgrading of these fractions by different processes (e.g., cracking, hydrocracking, hydrotreating, reforming), in order to convert them to petroleum products.

Reformate Product from the catalytic reformer, usually a 90–105 octane aromatic naphtha boiling in the gasoline range (C_5^+ hydrocarbons).

Reforming Catalytic reforming is a refinery process in which a primarily paraffinic or naphthenic feedstock with a 95–200°C (200–400°F) boiling range is converted to a predominantly aromatic, high-octane gasoline blending stock (motor fuel reformate) or to pure aromatic products (BTX reformate). Hydrogen is also produced in the process. The catalyst is usually platinum, palladium, or platinum-rhenium on alumina, and operates under hydrogen.

Refractory Feed A feed that is difficult or impossible to convert catalytically. Recycle oil is often a refractory feed because it is more difficult to convert than the fresh feed.

Regeneration In hydrocracking technology, the process of coke burnoff from spent catalyst in order to restore catalyst activity. In the case of noble metal catalysts, regeneration is followed by rejuvenation for further activity enhancement.

Reid Vapor Pressure (RVP) The vapor pressure of a product determined in a volume of air four times the liquid volume at 38°C (100°F). Indicative of the starting characteristics of motor fuels, of the tendency to cause vapor lock in gasoline feed systems, as well as of explosion and evaporation hazards.

Rejuvenation In hydrocracking, redispersion of noble metals (Pt, Pd) on regenerated catalyst, in order to restore the hydrogenation function.

Relative Volumetric Activity (RVA) Efficiency of a catalyst to crack (or to convert hetero atoms) compared to a reference catalyst whose activity is arbitrarily set at 100%, at constant loaded volume. Defined by the expression:

$$\text{RVA} = \frac{k \text{ sample}}{k \text{ reference}} \times \text{RVA reference}$$

where k = reaction rate constant.

Relative Weight Activity (RWA) Same definition as relative volumetric activity except that the comparison with the reference catalyst is made on an equal weight basis and is usually done by calculation, based on compacted bulk densities.

Research Octane Number See Octane Numbers.

Resid Hydrocracking Processes designed to hydrocrack resid feedstocks, in order to convert them to higher value products. Examples of such processes: H-Oil process, HYCON process, LC-Fining, VCC process.

Residuum (Residue, Resid) The heavy, high-boiling material remaining after nondestructive crude oil distillation (atmospheric or vacuum bottoms). By one definition, residue is the fraction boiling above 525°C (~975°F). In a broader sense, the term residue covers a broad range of products, boiling above 345°C (650°F). Contain asphaltenes, resins, and oils. High in sulfur, nitrogen, oxygen, and metals. Have high viscosities and Conradson carbon, and low API gravities.

Resins Broadly defined, resins are the nonvolatile, *n*-pentane- or *n*-hexane-soluble fraction of a feedstock that is retained by chromatographic adsorbents such as clay. Insoluble in liquid propane. Polar substances found together with asphaltenes in residues from crude oil distillation, consisting of molecules similar to, but less aromatic than, asphaltenes. See also Asphaltenes.

Road Octane Number Measures the knocking characteristics of a gasoline as determined in specially equipped automobiles under actual road conditions. Is the percentage (by volume) of iso-octane which would be required in a blend of iso-octane and *n*-heptane to give incipient knock in an automobile engine operating under the same specified conditions as the gasoline being tested.

RON+0 or RONC Research octane number unleaded (clear).

ROSE Process. Residue Oil Supercritical Extraction is a process used for the upgrading of heavy oils to make them suitable for catalytic cracking and hydrocracking.

RVP See Reid Vapor Pressure.

SCF Standard cubic feet. Standard conditions in petroleum and natural gas usage refer to a pressure base of 14.696 psia and a temperature of 15.5°C (60°F).

Screening (1) Passing a catalyst through a series of selected wire mesh screens which sort catalyst particles, fines, and extraneous material by size. (2) Testing a series of catalysts in a catalytic testing unit in order to establish their ranking and suitability for a specific application.

Selectivity See Catalyst Selectivity.

Sensitivity Difference between research octane number and motor octane number of a gasoline or of its components (e.g., alkylate, reformate).

Severity Intensity of the operating conditions of a process unit. The severity of operation is increased by increasing reactor temperature, catalyst activity, or decreasing feedstock space velocity.

Single Stage Unit A hydrocracking unit where the conversion is carried out in a single reaction stage.

Sintering of metals Agglomeration of noble metals in catalyst during hydrocracking process, resulting in catalyst deactivation (see also Rejuvenation).

Skeletal Density Skeletal density of a catalyst particle is the density of the solid material, excluding pore volume.

Smoke Point Property related to combustion characteristics of aviation fuel. Determined by a standard laboratory test.

Sockloading of Catalyst One of several reactor loading techniques used for hydroprocessing catalysts. The catalyst is guided manually into the reactor through a sock.

Sodium Troublesome element found in crude oil, in some feedstocks, and in catalysts. Reduces catalyst activity by neutralizing its acid sites. Reduces catalyst stability by forming an eutectic with the zeolite, which fuses at a lower temperature.

Solvent Deasphalting See Deasphalted Oil.

Solvent Extraction Process in which certain liquids (hydrocarbons, phenol, liquid SO_2, ethylene glycol, sulfolane, etc.) are used to extract desirable or undesirable compounds or mixtures from petroleum streams. Used to upgrade some dewaxing feedstocks. See also Deasphalted Oils.

SOR Start-of-run, referring to initial unit operation with fresh catalyst.

Sour and Sweet Crudes General classification of crudes according to their sulfur content. Sour crudes usually contain >0.5–1% sulfur, whereas sweet crudes contain less sulfur.

SO_x Mixture of sulfur oxides (SO_2 + SO_3) formed when burning off sulfur-containing coke from spent catalyst. Formed whenever sulfur-containing feedstocks are processed. Potential air pollutant.

Space Velocity The weight (or volume) of gas or liquid passing through a given catalyst or reaction space per unit time, divided by the weight (or volume) of catalyst through which the fluid passes. Abbreviated WHSV (weight hourly space velocity) and VHSV (volume hourly space velocity). VHSV is used as a process variable in fixed-bed processing, such as hydrotreating, hydrocracking, or reforming. High space velocities correspond to short reaction times.

Specific Gravity A number equal to the ratio of the density of a substance to the density of a reference substance. In the case of solids and liquids, the reference substance is water. In the case of gases, the reference is air.

Spent Catalyst In hydroprocessing, catalyst deactivated by coke, metals, and hetero atoms (S, N), submitted to reactivation or reclamation.

Spiked Feedstock In hydrocracking, feedstock to which certain chemicals (spiking agents) have been added in order to enhance or suppress certain properties. For example, organic sulfur compounds may be added to a feedstock in order to enhance its sulfiding ability in the catalyst presulfiding process.

Stability (Catalyst Stability) Ability of a catalyst to withstand the changes in physical and chemical properties that occur during use and lead to catalyst deactivation. May be thermal, hydrothermal, or catalytic. Catalytic stability is characterized by the rate of change of activity and/or selectivity with time on-stream.

Stabilization (Light Ends Removal) Distillation process designed to remove the light ends from a liquid fraction in order to reduce the vapor pressure of that fraction. For example, it is used to separate light hydrocarbons from hydrocracker gasoline. See also Debutanized Gasoline.

Stacked Beds The combining of both hydrotreating and hydrocracking catalysts within the same reactor configuration.

Steam-Cracked Gasoline See Pyrolysis Gasoline.

Steam Cracking A thermal process that nonselectively converts various hydrocarbon streams into light olefins, primarily ethylene and propylene. See also Pyrolysis.

Steam Reforming High-temperature, catalytic process that converts low molecular weight hydrocarbons and steam primarily to hydrogen and carbon monoxide. Catalyst: nickel on refractory support.

Straight-Run Gasoline A nonprocessed gasoline fraction distilled from crude oil. A gasoline component with low octane rating.

Straight-Run Products Stocks or products separated from crude oil by distillation without further processing. Also called virgin stocks or products.

Stripping (1) A fractional distillation in the presence of steam, carried out on each side-stream product after it leaves the main distillation tower, in order to remove the more volatile components from those products. (2) Removal of entrained hydrocarbons from spent FCC catalyst in the stripping zone of a FCC unit, using steam-blown countercurrent to the flowing catalyst.

Sulfur (Sulfur Compounds) Undesirable compounds found in varying amounts in most crudes and hydrocracker feedstocks (e.g., thiophene derivatives, organic sulfides, disulfides, mercaptans, etc.). Contribute to catalyst deactivation. Some feedstock sulfur ends up in the coke of spent catalyst and is converted to SO_x upon catalyst regeneration.

Super Oil Cracking (SOC) A hydrothermal slurry phase, residue-cracking process, utilizing disposable catalyst loaded in tubular reactors.

Surface Area Total area of exposed surface of a catalyst or catalyst component, including the surface of its pores. Determined by nitrogen adsorption and expressed in m^2/g.

Sweetening Removal or conversion to innocuous substances of sulfur compounds from petroleum products by one of many processes.

Syn Gas (Synthesis Gas) Various mixtures of carbon monoxide and hydrogen. Can be produced, for example, by steam reforming of methane or by partial methane oxidation. Used in the synthesis of a variety of chemicals.

TAME Tertiaryamylmethyl ether. Octane-boosting gasoline additive. Produced from isoamylenes (2-methyl-1-butene and 2-methyl-2-butene) and methanol.

TBA Tertiarybutyl alcohol. Octane-boosting gasoline additive. Produced by catalytic hydration of isobutylene, or obtained as a byproduct from the manufacture of propylene oxide.

TBP See True Boiling Point Distillation.

Temperature Excursions High catalyst bed temperature beyond normal operation resulting from uncontrolled hydrocracking reactions.

Thermally Cracked Stocks Hydrocarbon streams derived from thermal conversion processes occurring in coking units, visbreaking units, or thermal cracking units.

Thiophene Heterocyclic, sulfur-containing compound (C_4H_4S), used as spiking agent in testing of hydrocracking catalysts.

Throughput Ratio See Combined Feed Ratio.

Topping Removal by distillation of the lighter constituents from crude oil, leaving only the heavier constituents. The product is called topped crude or reduced crude.

Transportation Fuels Hydrocarbon streams used in transportation such as automotive gasoline, diesel and aviation jet fuel.

Treat Gas Light gases, usually rich in hydrogen, used in refinery hydrotreating processes.

Trickle Bed See Fixed Bed.

True Boiling Point (TBP) Distillation A multiple-plate, high-reflux laboratory fractionation that measures the true boiling points of the components contained in an oil sample. Often requires a vacuum distillation with temperatures corrected to atmospheric pressure.

Turnover Number (Turnover Frequency) Number of molecules reacting per active site of catalyst in unit time.

Two Stage Unit Hydrocracking unit where conversion is carried out in two reaction stages.

Ultrastable Zeolite See Zeolite, Ultrastable.

Unit Cell Basic repeating unit in a regular crystal structure. Generally defined by six or fewer crystallographic parameters, showing dimensions and angles.

Unit Cell Size See Zeolite Unit Cell Size.

Vacuum Gas Oil See Gas Oils.

Vacuum Tower Bottoms (VTB) Residuum left at the bottom of the vacuum fractionator. Boiling range: over 565°C (1050°F). Can be processed by visbreaking, coking, or deasphalting to produce heavy oil or hydrocracking feedstock.

Vapor Lock Index Measure of the tendency of a gasoline to generate excessive vapors in the fuel line, causing interruption of normal engine operation.

VCC Veba Combicracking. See Combicracking.

Virgin Products (Virgin Stock) See Straight-Run Products.

Visbreaking Lowering or "breaking" the viscosity of a residue by mild thermal cracking (~480°C, 895°F), in order to improve its quality as a fuel oil component. During the visbreaking process, gas oil and gasoline boiling range products are also formed. The resulting gas oil can be used as hydrocracker feedstock.

Viscosity Measure of the internal friction or the resistivity to flow of a liquid. For petroleum products, it is usually expressed in the number of seconds required for a certain volume of oil sample to pass through a standard orifice under specified conditions.

Viscosity Index Empirical number for quality of lube oil which indicates how the viscosity of fluid changes with temperature.

Void Fraction The space not occupied by the solids in a fixed volume. This is usually applied to measure the volume not occupied by catalyst pills in a reactor bed.

VTB See Vacuum Tower Bottoms.

Wax High-boiling, straight chain or branched chain hydrocarbons, solid at room temperature, with melting points in the range 38–68°C (100–155°F).

Wet Gas Gas mixture obtained from different petroleum process units, containing C_1–C_4 hydrocarbons. C_3 and C_4 hydrocarbons are usually separated to be used as LPG.

WHSV See Space Velocity.

Xerogel Dried hydrogel that has lost the physically retained liquid phase.

Zeolite Natural or synthetic microporous, crystalline aluminosilicate with ion exchange, sorption, molecular sieving, and catalytic properties. Also referred to as molecular sieve. See Zeolite X and Y.

Zeolite, Dealuminated Zeolite in which some of the aluminum has been removed from the zeolite silica-alumina framework by hydrothermal and/or chemical treatment. Results in an increase in framework SiO_2/Al_2O_3 ratio and a decrease in zeolite unit cell size. Y-type dealuminated zeolites are commonly used in hydrocracking catalysts.

Zeolite, Modified Zeolite whose composition and/or structure has been modified as compared to the synthesized or natural form (e.g., by ionic exchange, dealumination, etc). Sometimes refers only to structural modification (changes in the zeolite framework).

Zeolite, Ultrastable Y Zeolite with framework SiO_2/Al_2O_3 ratio over 6.0, unit cell size 24.5 Å or less, and high thermal/hydrothermal stability. Prepared by high-temperature steaming of ammonium-exchanged Y zeolites. Used in formulation of hydrocracking catalysts.

Zeolite Unit Cell Size Important structural characteristic of zeolites used in hydrocracking catalysts. For Y- and X-type zeolites, the unit cell size is the length of one edge of the basic repeating cubic unit in the zeolite crystal. Normally determined by X-ray diffraction and is expressed in angstroms or nanometers. Closely related to the SiO_2/Al_2O_3 ratio in the zeolite framework and affects catalyst activity, selectivity, and stability.

Zeolites X and Y Have crystal structure similar to that of mineral faujasite. Consist of a negatively charged, rigid, three-dimensional aluminosilicate framework with charge-balancing cations located in the zeolite cages. The unit cell size (in the as-synthesized sodium form) is between 24.6 and 25.0 Å. It decreases with increasing framework silica-to-alumina ratio. The silica-to-alumina molar ratio in X zeolites can vary from 2 to 3, whereas in Y zeolites it can vary from 3 to about 6. Used as components of hydrocracking catalysts.

ZSM-5 Zeolite Member of the "pentasil" family of high-silica zeolites. Used as catalyst for dewaxing distillates and lube oil base. In FCC catalysis used as an octane-boosting additive. Also used as catalyst in a variety of commercial processes, such as conversion of methanol to gasoline, ethylbenzene manufacture, xylene isomerization, toluene disproportionation, etc.

Index

Catalyst fouling (*see also* Coking), 113